Nanoelectronics

Quantum Engineering of Low-Dimensional Nanoensembles

Nanoelectronics

Quantum Engineering of Low-Dimensional Nanoensembles

Vijay Kumar Arora

CRC Press
Taylor & Francis Group
Boca Raton London New York

CRC Press is an imprint of the
Taylor & Francis Group, an **informa** business

CRC Press
Taylor & Francis Group
6000 Broken Sound Parkway NW, Suite 300
Boca Raton, FL 33487-2742

Printed on acid-free paper
Version Date: 20141002

International Standard Book Number-13: 978-1-4987-0575-2 (Pack - Book and Ebook)

Library of Congress Cataloging-in-Publication Data

Arora, Vijay Kumar.
Nanoelectronics : quantum engineering of low-dimensional nanoensembles / author, Vijay Kumar Arora.
pages cm
Includes bibliographical references and index.
ISBN 978-1-4987-0575-2 (alk. paper)
1. Nanoelectronics. I. Title.

TK7874.84.A76 2014
621.3815--dc23 2014034837

Visit the Taylor & Francis Web site at
http://www.taylorandfrancis.com

and the CRC Press Web site at
http://www.crcpress.com

Let peace prevail on earth!

Om Shanti!

Contents

Preface

Nano in its various formats (or allotropes to borrow the term from carbon) has changed the landscape not only for electronics, but also for human enterprise in planning the future of a community, state, country, and the planet Earth. Nanoscience, nanotechnology, nanoengineering, and nanobusiness are all moving targets, creating new start-ups among unlikely partners. The complete nanoensemble, by its very nature, is a multi- and interdisciplinary venture. Perhaps there is a need for engagement not only for scientists and engineers, but also for thinkers, ethicists, lawyers, theologians, and the nation's politicians. Engineering, a process of synthesis, is an engine of innovations, inventions, and growth. To quote Theodore von Kármán, the California Institute of Technology's provost during its formative years: "Scientists discover the world that exists; engineers create the world that never was." In other words, science is about being driven by curiosity to understand the world. Engineering is about using science to transform the world. This book is an example of the unification of science and engineering covering a journey from quantum physics to nanoengineering, including a brief coverage of mind–body connections to capture the power of brain waves and communication skills through checks and balances in the information matrix in Appendix I. The book draws heavily from published papers by the author as listed in Appendix J. The author can be reached at vijay.arora@wilkes.edu for any comments and/or MATLAB® codes needed while reading the book. MATLAB is a vehicle that is heavily utilized for graphics and numerical work.

Nanoelectronics have captured the attention and imagination of the world's scientists and engineers to replace silicon with carbon-based electronics. It covers the domain below a characteristic length of 1 micrometer (μm) terminating with about 1 nanometer (nm). A nanometer is one-billionth of a meter, or three to five atoms in width. It would take approximately 40,000 nm lined up in a row to equal the width of a human hair. Below a nm lies the atomic world with a hydrogen atom as the basis for nomenclature of quantized (digitized) energy levels. Quantum engineering at nm scale is closely related to nanotechnology, an umbrella term that encompasses all fields of science that operate on the nanoscale. As is well known, carbon gives biology, and silicon gave geology and semiconductor technology. However, nano gives different structures or allotropes of carbon and other materials appearing for the application. The transition to carbon is bringing biology into the mainstream of engineering as a process of synthesis, integrating ideas from several disciplines. Carbon nanotubes (CNTs), a few nm in diameter, are rolled-up versions of graphene sheets that can extend in length to a few μm. That is why engineering is emerging to be a converging science or it is sometimes labeled as an applied art of liberal arts. For those who are not well versed with liberal arts, the classical liberal arts comprise quadrivium (astronomy, arithmetic, geometry, and music) and trivium (grammar, logic, and rhetoric). These are all needed in the study of nanoelectronics as we practice the text of liberal arts through quadrivium and the context in which liberal arts are practiced through trivium. In fact, these were marks of an educated person in medieval times. And, now, these attributes are embedded in the desired outcomes by Washington Accord countries (washingtonaccord.org).

The developed countries are trying to cash in on the intellectual property generated as devices are reduced down to nanometer scale and their potential applications are explored. Since the time de Broglie proposed the dual nature of matter in 1927, physicists have been unraveling the nature of these waves. It was not until around the turn of the 21st century that engineers got seriously involved in unraveling the mystery of quantum waves. Traditionally, engineers are trained through the channel of physics, chemistry, and mathematics. It is not difficult to conjecture that biology is an essential basic science for engineers to include in their future design applications. The development of manpower resources through education and commercialization is becoming an important component of the educational delivery process. A brief introduction to these traits is included in Chapter 1.

Chapter 2 sets the foundation by drawing heavily from the literature and bringing the band structure of carbon-based devices into the limelight. Quantum effects are well known in physics books. However, Hamiltonian formulation leading to density function theory remains in the realm of a few experts. The superoperator theory developed by the author and its application to density matrix for quantum transport is briefly discussed in Chapter 9. This superoperator theory (*Physical Review B*, Volume 21, 1980, p. 876; *Physica B*, Volume 106, 1981, p. 311) has a close connection with the nonequilibrium Green's function (NEGF) that is heavily discussed in the literature as are Monte Carlo procedures. These methodologies were driven by the availability of supercomputers in the 1980s and funding opportunities that were provided by the U.S. government and its agencies in the ensuing years; so, supercomputers remain engaged to derive the benefits from taxpayers' money. NEGF and Monte Carlo simulations are embedded in the minds of scholars who graduated from institutions so funded. Equally good is the name Las Vegas simulations as it is indeed a gamble to get good results based on which input parameters are tweaked. NEGF and density function theory can take one from an atomic level to determine the band structure. However, its usefulness ends as soon as the band structure is identified. For example, identification of *K*–*K*′ Dirac points in graphene and the related Dirac cone is the starting point for any nonequilibrium study. In fact, this band structure based on 6-*K* points has been experimentally verified. Nonequilibrium Arora's distribution function (NEADF) is the starting point from band theory to equilibrium to extreme nonequilibrium carrier statistics, discussed in Chapters 3 and 4. Chapter 5 converts the electric-field response of drift velocity into current–voltage relationships that are driven by the presence of critical voltage and saturation current arising from the unidirectional drift of carriers. Chapter 6 applies the effect of these scaled-down dimensions to nano-MOSFETs (metal–oxide–semiconductor field-effect transistor) that continue to be a means of transportation of ideas in transforming the mindset for the future design of ultra large-scale integration (ULSI) in its various formats. MOSFET has also brought some fundamental physics to the forefront, for example, the quantum Hall effect. Chapters 7 through 10 consider specialized applications that can be tried through a number of suggested projects that are all feasible with MATLAB codes; so, no complicated computing engine is required. In fact, Ohm's law is transformed into a tanh law that is applicable for all dimensionalities, including the allotropes of carbon. The resistance $R = V/I$ takes a different meaning in the ballistic domain as well as in the high-field domain, exceeding the critical value of the electric field that is thermal voltage divided by the long-channel mean-free path (mfp), easily obtainable from the measured ohmic mobility. Quantum resistance is identified only in a one-dimensional (1D) nanostructure with Buttiker transmission defined in terms of ballistic injection from the contacts.

The book covers amazing feats of nanoelectronics and where the field might be headed. It also covers the less appreciated by-product of the nanoelectronic revolution, namely, the deeper understanding of current flow, power consumption, and device operation in the ballistic domain, following the bottom-up approach, starting from Chapter 2 to the more advanced in Chapters 7 through 10. The question often posed is: "What would happen if a resistor is shorter than an mfp so that an electron travels ballistically like a bullet hitting the target without an intermediate medium scattering a carrier in its ballistic path?" This question is very well answered starting from three-dimensional (3D) to two-dimensional (2D) to 1D to carbon allotropes giving directions for further exploration. The sections on ballistic transport shatter the ill-conceived belief that once scattering is eliminated, mobility must enhance and resistance should ideally go to zero. In fact, mobility is experimentally found to degrade and resistance enhanced, which is well explained by nonstationary or transient transport. Many properties of long conductors can be understood in simple terms by viewing them as a series of elastic resistors, each equal to the length of an mfp, thereby enhancing the importance of the energy gained in an mfp, a basis of NEADF. Considering these facts, the book presents a comprehensive review of nanoelectronic transport starting from quantum waves, to ohmic and ballistic conduction, and to saturation-limited extreme nonequilibrium conditions, including the phonon and photon emission.

Concerted efforts have been planned and made to facilitate the conversion of research and development into products. Many U.S. entrepreneurs are looking toward Asia to develop intellectual property in order to profit from the technology. This Asia-America (East meeting West) amalgamation is encouraging government–industry–business–academia interactions around the globe with a view to help each other to reap the benefits of the emerging technology, in concert with the title of this module. Some of the application areas include aerospace, agriculture and food, homeland security and defense, energy, environment, quantum computing, information processing, medicine and health, drug delivery, and understanding the nature of deoxyribonucleic acid (DNA) to eradicate diseases, even estimating the year of life for a particular disease to appear. DNA atlas, brain atlas, body energy waves, and so on are catching momentum as the domain of nanoengineering expands. Noteworthy are nanoelectronic chips likely to find commercial applications by implanting these chips in human beings. The chip, in addition to information-processing elements, will contain sensors to monitor the presence of chemicals, giving indications of enhanced sugar or blood pressure level, for example. These nanochips will also contain molecular drug chambers to release the drug as needed. In addition, signals from the nanochip can be transmitted to a computer or the Internet and can be deposited in a place where it can send an emergency signal that can trigger an action plan.

One application estimate of the National Nanotechnology Initiative (NNI) is the development of smart materials. Smart materials are dynamic in the sense that a material can change its most basic properties in response to the outside stimulus, akin to Newton's Law of Action and Reaction. There are a number of sensors that are being investigated to monitor temperature, humidity, light, sound, electricity, and the presence of certain chemicals, bacteria, explosives, or DNAs. The most important fact is that these sensors are being integrated onto a chip so that a variety of nanoelectromechanical systems (NEMS) can be designed. Other applications include self-assembled biostructures with growing biomedical and pharmaceutical applications. There is also a shift in the award of Nobel Prizes on technologies that had the most impact on humans. Integration of the artificial (man-made) and the natural (divine) in a variety of applications will test human ingenuity as well as stretch its operational capabilities to a level never anticipated before.

Nanoelectronics, the major emphasis of this module, will see a dramatic transformation. Molecular, polymer, and DNA electronics are alternative paradigms for nanoelectronics. Major advances have included inexpensive organic transistor chips to act as tags for the identification of products, parcels, and postage. One outcome is the recent introduction of radio-frequency identification (RFID) that, coupled with geographical positioning system (GPS), can track the mobility of living entities and materials. The following list identifies a number of problems that can be considered in designing nanolayers and their applications:

1. To design fabrication conditions such that the growth takes place in closed chambers with minimal exposure to the atmosphere at an affordable cost
2. To develop lithographic techniques to cut dimensions in the order of a nanometer without distortions in the geometry of nanostructures

3. To develop etching processes so that the undercutting under the mask pattern can be reduced to increase the device density as well as to have better control over the device and circuit topology
4. To develop a comprehensive transport theory for quantum processes at the nanometer scale
5. To develop a high-field distribution function that could be utilized for multivalley structures of materials used in the nanofabrication process
6. To design and optimize the shape and also the size of the quantum well for high-frequency applications extending beyond 200 GHz
7. To carry out numerical calculations based on the quantum transport framework to understand the role of quasi-ballistic electrons determining the *I–V* characteristics and high-frequency limitations of ultra-submicron gate lengths
8. To investigate the low-frequency and high-frequency noise behavior and identify the noise sources in nanometer-scale quantum wells
9. To study the electric field and velocity distributions in the active and parasitic regions of the device
10. To explore the possibility of designing a multistate logic device by using quantum states in a given quantum well
11. To manage the nanomachines, interconnects, and information flow from one stage to the other stage without noise added due to electromagnetic inductions both from external sources and from chip components, creating high signal-to-noise ratio.

Although not exhaustive, the above list forms a vantage point for the global outlook of the arena of future nanoelectronics devices and ULSI. The emphasis in the module is on theoretical ideas needed to characterize the nanolayers and to understand their operational characteristics. As electrons in these nanostructures lose their classical character in one or more of the three Cartesian directions, the classical theories utilizing the 3D semiclassical nature of the electrons become invalid. Circuit theory textbooks rely heavily on the applicability of Ohm's law, which collapses as electronic components reach micro- and nanoscale dimensions. Circuit analysis based on Ohm's law becomes invalid and critical voltage takes an increasing importance to determine the domain of the validity of Ohm's law. The default value of critical voltage is infinity in the ohmic regime, but is as low as a fraction of a volt when linear current–voltage characteristics become sublinear and the resistance surges due to current saturation effects. For two resistors of the same ohmic values but of differing lengths, the shorter resistor is more susceptible to this effect. In addition, the power consumed in this regime is a linear function of voltage when compared to quadratic behavior in the ohmic regime.

The material presented in the book can be used in a variety of courses and in the application of physics to electrical engineering. The author has tested the book in a variety of courses at Wilkes University and Universiti Teknologi Malaysia (UTM). Students ranging from senior-level undergraduate to PhD level are interactively involved in these courses and in research and development projects, using the archetype set forth in this book. While this material is peripherally discussed in a number of undergraduate and graduate textbooks, based on empirical evidence, this paradigm is derived from statistical and theoretical considerations that are in accordance with the empirical evidence and hence can be easily extended to interpret data from a variety of sources. Parameters extracted from these interpretations can be input into computer-aided design/engineering (CAD/CAE) simulation programs to give an understanding of fundamental processes at the micro-/nanoscale. A number of projects are added to each chapter and solved problems are included in the first six chapters with follow-up problems for students to attempt to get ready for a research-based education.

I leave these introductory remarks with a note of gratitude to all those who put in countless hours of volunteer work to make sure that we succeed in our mission and vision in producing this book. Michael L. P. Tan, Faraz Najam, Elnaz Akbari and Arkaprava Bhattacharyya from the UTM, Ismail Saad from the Universiti Malaysia Sabah, and Desmond Chek (now in Intel Corporation, Penang) have provided magnificent support to keep the author afloat as he was trying to complete this book. At Wilkes University, the students taking the EE271: Semiconductor Devices course have been particularly helpful in pointing out errors and inconsistencies. Without students' input and assessable impact, it is difficult to know whether ideas are communicated to them to grasp the newly emerging concepts and train their minds to shape the future of nanoengineering. After all, it is the attitudes residing in the mind (brain's software) that train the brain's hardware in programming thinking processes of the future. The book has also been tested as notes given for a short course to a number of forums under the IEEE-EDS Distinguished Lectures Program of the Electron Devices Society (EDS). EDS support, although a token, in fact is highly valued as it allowed the author to make connections throughout the globe, making him indeed a global citizen. The book encourages seasoned and budding scholars to stay connected as a community in service to humankind, as noted in Appendix K.

A special word of gratitude is due to Dr. Gagandeep Singh, senior acquisitions editor (engineering/environmental sciences), CRC Press/Taylor & Francis Group for his concerted effort to make the book publication ready. To all the readers of this book who will become my global family members, my obligation is to make sure the book remains a part of the incredible growth by ensuring fair and constructive dialog. Dr. Singh initiated that constructive dialog though an impartial peer-review system. I am proud that this book is one where new as well as established members of the scientific community will find new knowledge, not present elsewhere in the published books by protuberant authors. For young scholars, I have tried where possible to allow for constructive revisions so that the book could be more accessible and appreciated. Young scholars are the purveyors of pure science (see Appendix K for impure science) and are the future of global scientific advances and a critical component of their continued future success. To established readers, I have strived to maintain and enhance the scientific quality and relevance of the book so that the work published is not work lost. This is the spirit, essence, and soul of future nanoelectronics. The constructive dialog with Dr. Singh was worth the time spent toward that effort.

Last but not the least, I am indeed very grateful to the UTM higher administration in hosting my productive stay in turbulent environments of the UTM, which changed its status to a research university during my stay. I am also proud to be connected to them through the spirit of the Washington Accord as Malaysia became a full signatory member of the Washington Accord (washingtonaccord.org) in 2009. As I traveled from UTM to Sri Lanka and India to propagate the ideals of the Washington Accord, 2014 saw both of them being admitted to the Accord as full signatory members. It was Dato' Seri Professor Dr. Ir. Zaini Ujang, vice chancellor at the UTM (now secretary general to the Malaysian Ministry of Education) who found time to travel with us to India to strike the Memorandum of Understanding (MoU) between India's Indian Institute of Technologies (IITs) as well as sponsor our travel (with Professor Halim) to the National Institute of Technical Teachers' Training and Research (NITTTR) in Bhopal, where the spirit of the Accord was propagated from real experiences of Malaysia being admitted into the Accord as narrated by Professor Abdul Halim Yatim, then the dean of the Faculty of Electrical Engineering. Dato' Seri Zaini, as we call him, also brought the Harvard business case studies to the realm of UTM that are also an essential component of liberal values of problem/project-based learning (PBL). Datin Paduka Professor Dr. Siti Hamisah, then the deputy vice chancellor of academics and internationalization (now the deputy director general at the Malaysian Ministry of Education) was a superb host for our splendid stay at UTM and a key figure in bringing Malaysia to the Accord. Her successor, the present deputy vice chancellor of academics and internationalization, Professor Dr. Rose Alinda made sure that we do not feel the vacuum created by the departure of Datin Paduka Siti. It was her efforts that brought us back to the UTM on an extended contract for which I am deeply grateful to her.

All readers are welcome to contribute problems and projects for the book that will be implemented in the next edition of the book or on the website of the book with an acknowledgment as "The project/problem contributed by XXXXX." It is also the spirit of the Washington Accord that education should be PBL. Suggestions for improvement are always welcome by sending an e-mail to the author at vijay.arora@wilkes.edu.

MATLAB® is registered trademarks of The MathWorks, Inc. For product information, please contact:

The MathWorks, Inc.
3 Apple Hill Drive
Natick, MA 01760-2098 USA
Tel: 508 647 7000
Fax: 508-647-7001
E-mail: info@mathworks.com
Web: www.mathworks.com

1

Nanoengineering Overview

The next big thing in emerging technologies is going to be extremely small. The typical scale of this smallness is a few nanometers (a nanometer, abbreviated as nm, is one-billionth of a meter). To appreciate the smallness of a nm, let us assume that the diameter of the human hair is 100,000 nm. These ultrasmall nanoelectronic devices are the subject of intensive research directed toward system design for practical applications. Nanoengineering is little known to the general public, but in the science and public policy community its promise is exhilarating. Coming from the Stone Age of 5000 BC, we have made tremendous progress toward improving lifestyle [1,2]. For about 200 years after American independence, we followed Adam Smith's model of *An Enquiry into Nature and Causes of the Wealth of Nations* (1776), which implied that the wealth is created by laissez-faire economy and free trade. Over the next 50 years or so, the economy was dictated by John Maynard Keynes as outlined in *The General Theory of Employment, Interest and Money*. Keynes concluded that the wealth is created by careful government planning and stimulation of the economy. In the 1990s, according to Paul Romer, the wealth is created by innovations and inventions, such as computer chips. The semiconductor industry is the heart of this evolution and revolution. The field is growing at a tremendous rate. Carbon-based materials, for example, graphene, organic polymers, and other smart meta materials, are appearing on the world stage for custom-made nanoelectronic circuits in the future. Nanoengineering can be understood only if we delve into the wave character of an electron.

1.1 Quantum Waves

Nanoengineering originated from the work of Louise de Broglie in 1926 who postulated that an electron with its classical momentum $p = m^*v$, where m^* is the effective mass and v is the velocity, is connected to its wave character through the wavelength λ_D, termed the de Broglie wavelength, given by [3]

$$\lambda_D = \frac{h}{p} \tag{1.1}$$

The momentum is related to the kinetic energy E by

$$E = \frac{1}{2}m^*v^2 = \frac{p^2}{2m^*} \Rightarrow p = \sqrt{2m^*E} \tag{1.2}$$

With the application of Equation 1.2, the thermal wavelength λ_{Dth} for thermal energy $E = k_BT$ is given by

$$\lambda_{Dth} = \frac{h}{\sqrt{2m^*k_BT}} \tag{1.3}$$

Here, k_B is the Boltzmann constant and T is the ambient temperature. Equation 1.3 is suitably adapted to the numerical calculations as follows:

$$\lambda_{Dth} = \frac{7.63\,\text{nm}}{\sqrt{(m^*/m_o)\times(T/300\,\text{K})}} \tag{1.4}$$

λ_{Dth} for a free electron ($m^* = m_o$) is 7.63 nm at room temperature ($T = 300$ K). However, in a semiconductor, for example, in GaAs ($m^* = 0.067m_o$), its value is 29.5 nm. The wave character becomes more prominent as the temperature is further reduced. The de Broglie wavelength for Si ($m^* = 0.26m_o$), GaAs ($m^* = 0.067m_o$), and GaN ($m^* = 0.19m_o$) as a function of temperature is depicted in Figure 1.1. Quantum effects are more pronounced at low temperatures and for materials with low effective mass.

Figure 1.2 shows the spectrum going from macro- to nanoengineering scale. The nanometer size of the de Broglie wavelength demonstrates why it is important to consider the wave character of an electron. As body cells and DNA are of nanometer size, nanoengineering is of special significance for applications in biosystems and hence the anticipated DNA or bioelectronics.

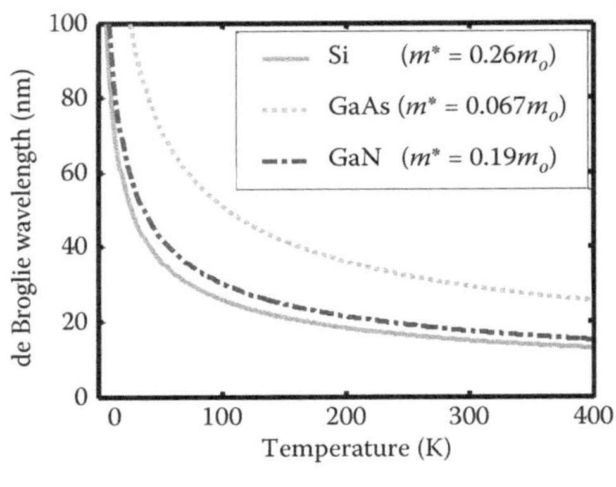

FIGURE 1.1
The de Broglie wavelength λ_D as a function of temperature for Si, GaAs, and GaN.

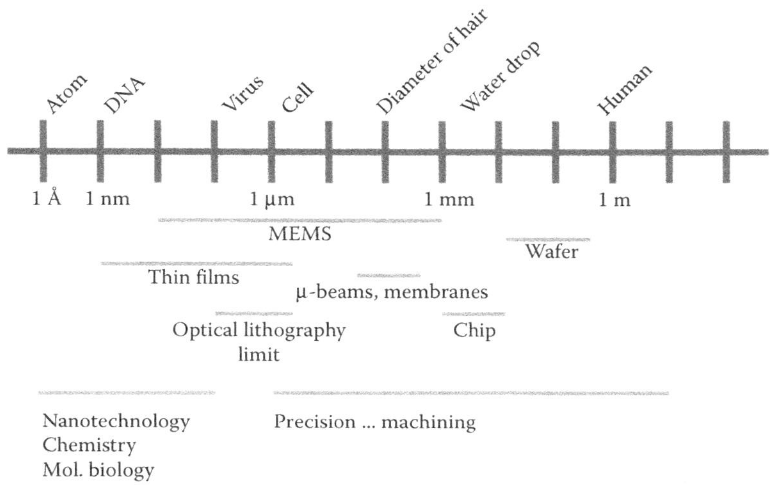

FIGURE 1.2
Downsizing of the scale.

Modern crystal growth techniques involving a variety of materials and fine line lithographic systems are allowing us to integrate electronic, optical, mechanical, and even biofunctions on a single chip. Design teams are trying to squeeze the whole system on a chip comprising application-specific integrated circuits (ASICs). Computer-aided design/engineering (CAD/CAE) is extensively utilized to understand the complexities of interconnects, heat dissipation, and packaging. Silicon has been the device vehicle for the semiconductor industry for a long time and still continues to enjoy its dominance in the semiconductor market. However, several new materials are being searched and researched for nanoengineering applications.

1.2 Nanoengineering Circuits

The metal–oxide–semiconductor field-effect-transistor (MOSFET), as shown in Figure 1.3, is the building block of the semiconductor industry. It is a three-terminal device, which acts like a conductor, modulated by the gate while current flows from the source to the drain or vice versa. The fourth body terminal (not shown in Figure 1.3) is used to adjust the threshold voltage of the device. Threshold voltage is the minimum voltage applied to the gate so that a detectable, although negligible, amount of current can flow in the channel. Although the physical structure of the transistor remains the same, the dimensions are decreasing with a new technology coming out

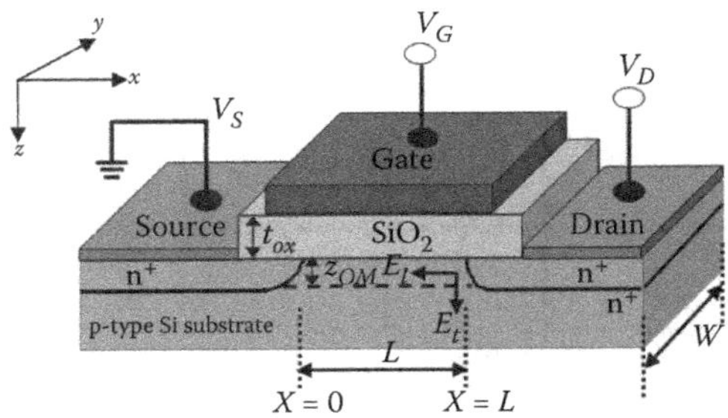

FIGURE 1.3
Schematics of an MOS transistor. The current flows from the drain to the source when the gate voltage is above the threshold voltage. The body terminal is used to adjust the threshold voltage.

every 3 years (one generation). In two generations, the size of the transistor is reduced approximately by a factor of 2. Figure 1.4 shows the projected decrease in size with the increase in the number of instructions computed per second. The size of the channel has now reached sub-0.1 μm (less than 100 nm) domain, while the number of bits processed per second is more than

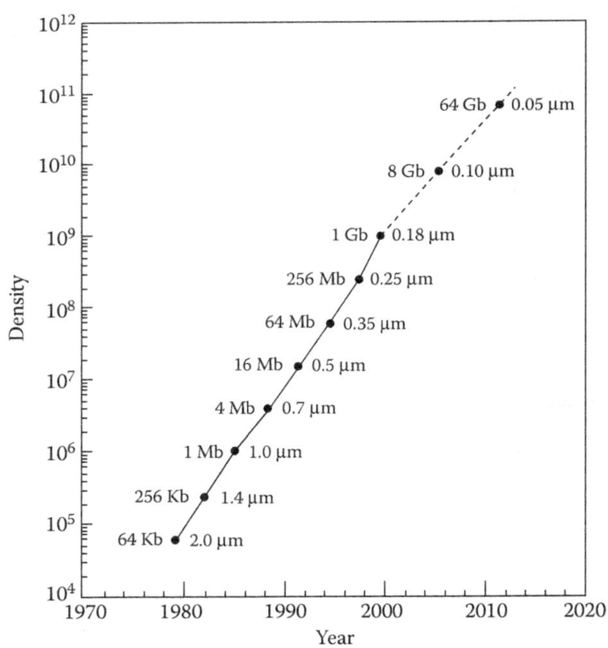

FIGURE 1.4
Exponential increase of dynamic random access memory density versus year based on the Semiconductor Industry Association (SIA) roadmap.

several billions. Beyond 0.1 μm, the road is bumpy and many envision the end of *Moore's law* for exponential reduction in size and exponential growth in terms of functionality and number of transistors packed on a typical 1 cm × 1 cm square chip. Some label exponential growth in terms of the density of bytes on a given chip. *More than Moore* is the slogan the semiconductor industry is chanting as the landscape brings new materials and new players to compete with silicon.

With the advent of nanofabrication technology and the scaled-down dimensions of modern microcircuits, progressively larger electric fields are being encountered in a channel as well as in the parasitic regions forming the contacts and those isolating one device from the other on a 1 cm × 1 cm square chip. These devices are tightly packed and their functions integrated in ultra-large-scale integration (ULSI) as more and more components are being squeezed. As chip real estate is becoming expensive from a processing point of view, there is a move to go vertical and 3D chips are appearing on horizon. Parasitic regions are necessary to isolate one device from another and also to make metallic contacts for circuit applications. These parasites, as one may call them, are a necessary evil as it may seem. Without these parasites, we cannot keep functions isolated. However, we need to design parasitic regions in such a way, so these do not rob the action of the conducting channels for current to flow. These undesired parasites offer unwanted resistance, capacitance, and inductance degrading the signal propagation in a device or through a system of devices. Under the influence of high electric fields, the velocity response to the electric field is highly nonlinear (in fact, sublinear) and velocity eventually saturates (more on it in later chapters). This nonlinearity has a substantial effect on device operation, such as in the switching delay of an inverter or increase in parasitic resistance of the source and drain region of a field effect transistor (FET). The normal long-channel circuit model makes the extensive use of Ohm's law. However, continued use of Ohm's law in short channels scaled below 100 nm gives wrong predictions for signal delay and hence operating frequency of a circuit. It is in these scale-down dimensions, device fabrication and geometries are being challenged to evade or at least minimize unwanted hot electron degradation and other deleterious effects. Nothing comes free, bringing to focus the engineering process of optimization. For example, lightly doped regions are added to the source and drain end to avoid hot-electron degradation, which considerably increase the series resistance 5–10-fold under ohmic conditions, which may increase by another factor when velocity saturation effects are also phased in.

Complementary metal–oxide–semiconductor (CMOS) technology controls roughly 80% of the integrated-circuit (IC) market [3,4], as indicated in Figure 1.5. In CMOS transistors, the mobility of electrons is roughly twice that of the holes. The lower mobility of holes is compensated by increasing the width of the p-type metal–oxide–semiconductor (PMOS) transistor to twice that of an n-type metal–oxide–semiconductor (NMOS) transistor. GaAs lacks

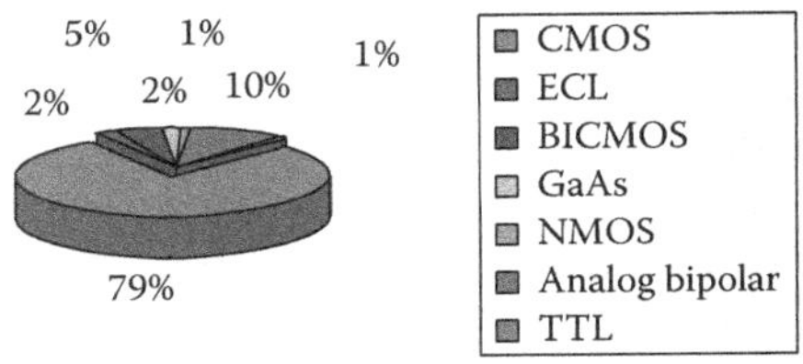

FIGURE 1.5
Market share of various integrated circuit (IC) technologies.

that advantage (in GaAs, the mobility of electrons is approximately 15 times that of holes) and is also more expensive. An enhancement-depletion type of complementary design is being explored for GaAs, which can also be used for CMOS-type applications. However, Si offers a considerable cost advantage over all materials. A serious disadvantage of Si is its poor optoelectronic properties where other so-called direct-band-gap semiconductors offer distinct advantages. The knowledge base built up from silicon technology is still advantageous as custom-made circuits of different materials and different geometric configurations are explored.

Physical structure of CMOS, as shown in Figure 1.6, can be implemented in a variety of ways during the fabrication process. One possible mode is to start with an n-type substrate and make a p-well for NMOS by polarity inversion. Or, one may start with a p-type substrate and make an n-well for PMOS.

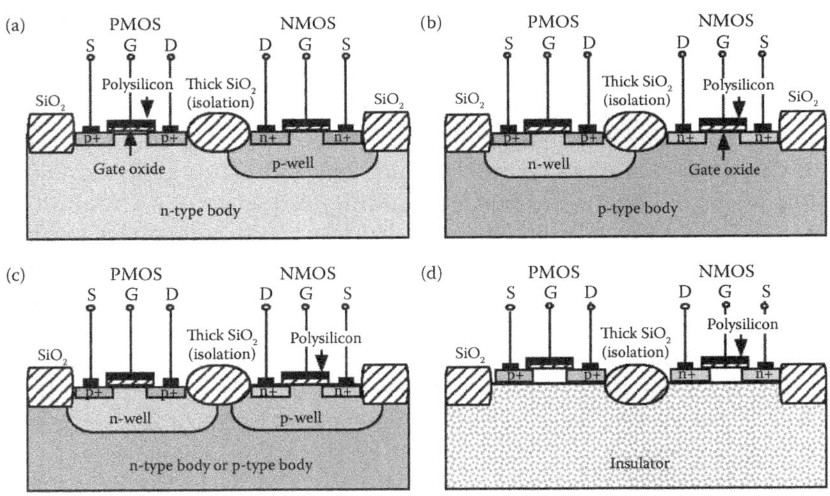

FIGURE 1.6
Cross-section of CMOS IC: (a) the NMOS transistor is formed in a separate p-type region, known as p-well, (b) the PMOS transistor is formed in a separate n-type region, known as n-well, (c) the NMOS and PMOS transistors formed in either n-type or p-type region, known as twin well, and (d) SOI uses a substrate that is an insulator.

Another variation is to start with a twin-well structure on either n- or p-type substrate which offers independent control of NMOS and PMOS. But, the most promising silicon-on-insulator (SOI) technology emerging these days involves a pair of complementary transistors on an insulator, making it possible to tightly pack CMOS logical elements. The major disadvantage of SOI is that the MOSFET body floats, so its threshold voltage cannot be adjusted by applying body voltage. Insulating the substrate eliminates a large number of parasites and makes it radiation-resistant. This is attractive for devices used in space flights. Many material problems remain to be solved in making it the dominant technology of the future.

The advantage of using CMOS, in contrast to using NMOS or PMOS technologies, is its low power consumption in standby mode. The power is consumed only during the switching operation when both transistors turn on momentarily (Figure 1.7). Figure 1.7 also indicates the circuit diagram of

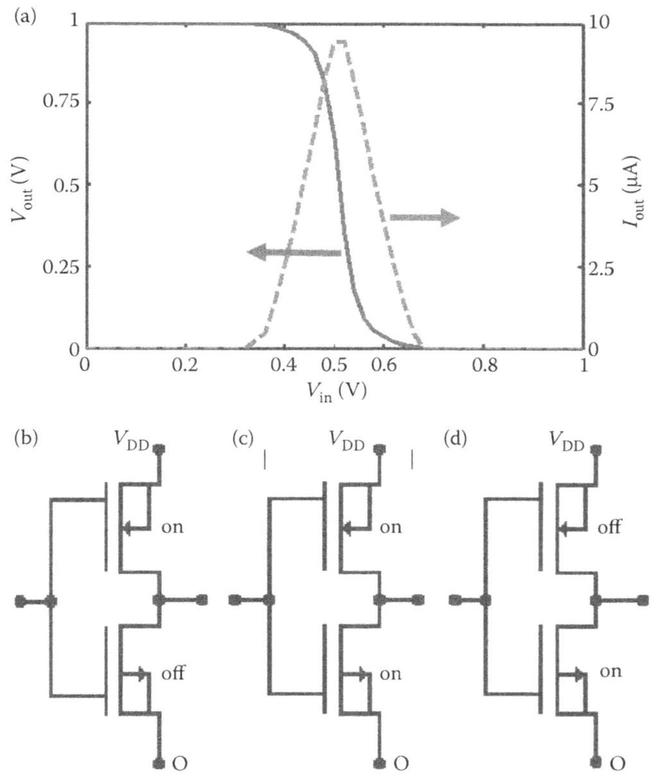

FIGURE 1.7
Operation of a CMOS inverter: (a) voltage transfer characteristics (VTC) of the three stages, (b) low voltage applied at the input, (c) switching from low voltage to high voltage creating an infinite resistance enhancing the RC time delay of the circuit, and (d) high voltage applied at the input giving low output.

a CMOS inverter. When low voltage is applied to the input, the lower NMOS (pull-down) transistor is off because of the required positive-threshold voltage. The upper PMOS transistor with negative-threshold voltage is "on," thereby connecting the output to V_{DD} which is normally 5 V in multimicron structures with scaled-down values as low as 1.0 V. The threshold voltage of a transistor is an important design parameter that can be adjusted through material properties and applications of the voltage to the substrate. The current does not flow if the applied gate voltage is less than the threshold voltage. When high voltage is applied to the input, the upper transistor turns off and the lower transistor turns on, connecting the output to the ground. Turn-on resistances of both transistors must be reduced below specified limits (ideally zero). Another factor that inhibits the delay is the RC time delay. The load capacitance of the following gates is driven by the CMOS inverter in a ULSI environment. The speed of signal propagation through the inverter depends on how many transistors (called fan-out) in the next stage are driven by the output of the previous stage. The high-frequency operation of the propagating digital signal is limited by the RC time constants. If parasitic capacitance is reduced, substantial improvement in the speed of ULSI circuit is possible.

Although the literature on nanoengineering may refer to nano-electromechanical systems (NEMS), including self-replicating machines built at the atomic level, it is admitted that an assembler breakthrough will be required for widespread use of NEMS. The smallest mechanical machines readily available in a wide variety of forms are really on the millimeter scale, as in conventional wristwatches. Figure 1.8 shows how etching a silicon area and bonding and thinning can create cavities that can be further designed by the photolithographic process to get a small diaphragm for a blood pressure sensor, or a free-standing cantilever to sense the acceleration (or deceleration) as in an automobile so it will open an airbag, or a cavity resonator to get desired frequencies. These sensors capitalize on two major physical properties: the dependence on the change in the capacitance of the fabricated sensor (as shown in Figure 1.8) and also on the piezoelectric properties attributed to piezoelectric effect, where a signal appears on exerting pressure on a crystal. In each case, an external stimulus is transformed into an electrical signal that triggers the output to take certain action, for example, open a drug cell to release blood-pressure-lowering medicine as it senses higher blood pressure.

The idea of limiting the size scale of a miniaturized technology is fundamentally interesting for several reasons. As sizes approach the atomic scale, the relevant physical laws change from the classical to the quantum-mechanical laws. The changes in behavior from classical, to mesoscopic, to atomic scale are broadly understood in contemporary physics, but the details in specific cases are complex requiring further exploration through transformed physical laws. While the changes from classical physics to nanophysics may mean that some existing devices will fail, the same changes open up possibilities for new devices, encouraging innovations by enhancing the ingenuity of a human mind.

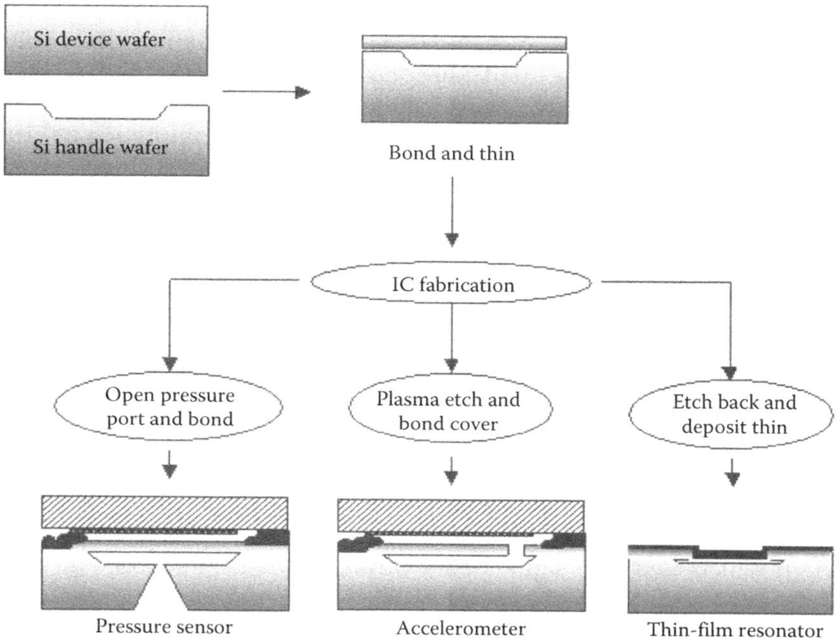

FIGURE 1.8
Fabrication process of nano-electro-mechanical systems (NEMS) to produce pressure sensors, sensors for cars to monitor change in speed or direction, and resonator for generating frequency.

1.3 Bioapplications

A primary interest in nanotechnology comes from its connections with biology. The smallest forms of life—bacteria, cells, and the active components of living cells—have sizes in the nanometer range. In fact, it may turn out that the only possibility for a viable complex nanotechnology is that represented by biology. Certainly, the present understanding of molecular biology has been seen as an existing proof for "nanotechnology" by its pioneers and enthusiasts. In molecular biology, the self-replicating machines at the atomic level are guided by deoxyribose nucleic acid (DNA), replicated by ribonucleic acid (RNA), and specific molecules are assembled by enzymes and cells. DNA is a molecule that encodes the genetic instructions used in the development and functioning of all known living organisms and many viruses. DNA is a nucleic acid; alongside proteins and carbohydrates, nucleic acids compose the three major macromolecules essential for all known forms of life. Most DNA molecules are double-stranded helices, consisting of two long biopolymers made of simpler units called nucleotides—each nucleotide is composed of a nucleobase (guanine, adenine, thymine, and cytosine), recorded using the letters G, A, T, and C, as well as a backbone made of alternating sugars (deoxyribose) and phosphate groups (related to phosphoric acid), with the nucleobases (G, A, T, C) attached to the sugars. RNA is a ubiquitous family of large biological molecules that perform multiple vital roles in the coding, decoding, regulation, and expression of genes. Together with DNA, RNA comprises the nucleic acids, which, along with proteins, constitute the three major macromolecules essential for all known forms of life. Like DNA, RNA also is assembled as a chain of nucleotides, but is usually single-stranded. Cellular organisms use messenger RNA (mRNA) to convey genetic information. Nature is replete with molecular-scale motor-in-action entities with life and soul. Ion channels that allow (or block) specific ions (e.g., potassium or calcium) to enter a cell through its lipid wall appear to be exquisitely engineered molecular-scale devices where distinct conformations of protein molecules define an open channel versus a closed channel. Naturally, nanowires and DNA/RNA chains resemble and may form a topic for comparative studies as DNA/RNA are much simpler to grow than nanowires. Arteries and blood vessels can be designed as tiny capacitors through which electrical signals pass through. Quantum capacitance may hold the key to this phenomena.

The Human Genome Project (HGP) is an international scientific research project with a primary goal of determining the sequence of chemical base pairs that make up human DNA and of identifying and mapping the total genes of the human genome from both a physical and functional standpoint [1]. It remains the largest collaborative biological project [2]. A parallel project was conducted outside of government by the Celera Corporation, or Celera Genomics, which was formally launched in 1998. Most of the government-sponsored sequencing was performed in universities and research centers from the United States, the United Kingdom, Japan, France, Germany, Spain, and China [6]. Researchers continue to identify protein-coding genes and their functions; the objective is to find disease-causing genes and possibly use the information to develop more specific treatments. It may also be possible to locate patterns in gene expression, which could help physicians glean insights into the body's emergent properties. The "genome" of any given individual is unique, and mapping *the human genome* involves sequencing multiple variations of each gene. There are approximately 20,500 genes in human beings, the same range as in mice. Understanding how these genes express themselves will provide clues to how diseases are caused. The human genome has significantly more segmental duplications (nearly identical, repeated sections of DNA) than other mammalian genomes. The research shows that fewer than 7% of protein families appeared to be vertebrate-specific. In real life, we focus more on 7% differences and ignore the 93% that is similar and hence the conflict in the human domain. Nanotechnology is expected to open pathways to understand these minute phenomena that make us think differently, triggering the creative usage of untriggered brain power.

Biological sensors such as the rods and cones of the retina and the nanoscale magnets found in magneto-tactic bacteria appear to operate at the quantum limit of sensitivity. Understanding the operation of these sensors require application of nanophysics. One might say that Darwinian evolution, a matter of odds of survival, has mastered the laws of quantum nanophysics, which are famously probabilistic. Stochastic processes and quantum nanophysics displaying probabilistic interpretations entail molecular building blocks that may trigger the design of man-made sensors, motors, and much more, with expected advances in experimental and engineering techniques for nanotechnology.

Figure 1.9 shows various areas of human brain for information processing. The brain rewires itself as one area remains unutilized or underutilized. Future advances in nanotechnology may help us in understanding the mystery of natural intelligence and how artificial intelligence may supplement the information processing in the brain. There are uncharted worlds inside a human head, but nanoengineering is drawing up a map to capture the best of the human thinking processes where 100 billion jabbering neurons create the knowledge—or illusion—that we are here. Not only can the brain learn new tricks, but it can also change its structure and function—even in old age. When combined with psychophysics, it may allow us to understand the frequencies that create a wanted signal (e.g., musical frequencies) or a noise (irritant signal). The phenomena of total integration as various fields merge under the umbrella of nanotechnology are yet to be seen. The mind–body connection of the diseases will bring many mystics into the realm of nanoengineering [2]. The mind (the collection of attitudes) is the software of the brain hardware. It is conjectured that disease or malfunction starts at the mind level and propagates to the body level. In the *Time* article containing the left of Figure 1.9, the evidence of a mind–body connection is presented: Train your mind, create your brain. A healthy gross body (*sthul srira* in Sanskrit literature), with our physical features noted and visible to others, contains a subtle or microscopic body (*sukhsam srira*) where our emotions

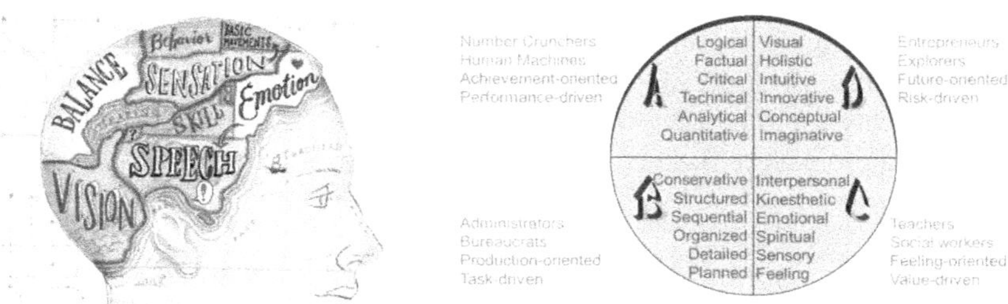

FIGURE 1.9
The mystery of the human brain as a signal processor of wide variety. Also shown is the four-quadrant model espoused by Ned Hermann showing dominance in each quadrant and suitability of a profession to operate from each quadrant. (From *Time* Magazine, January 18, 2007.)

and values reside. Beyond the sukhsam srira lies causal body (*karan srira*) that is the cause of both our emotions and beliefs. It is the karan srira from which the belief systems and religions of the world originate and complement the whole field of nanoengineering.

As molecular-scale machines are now achievable, the most optimistic strategic thinkers note that these invisible machines could be engineered to match the size scale of the molecules of biology. Medicinal nanomachines might then be possible, which could be directed to correct defects in cells, to kill dangerous cells, such as cancer cells, or even, most fancifully, to repair cell damage present after thawing of biological tissue, frozen as a means of preservation. It may even be possible to kill the disease at the level of *sukhsam srira,* so residual effects, as known in medical circles, can be completely eliminated. That should trigger the formation of a healthy and peaceful planet Earth. Too good to be true, but HGP is drawing the map. Nanoengineering, perhaps, will bring our focus to how similar we are in the mind–body complex.

Nanoengineering cannot be complete unless this triology of body–mind–spirituality electronics is explored where quantum waves may provide some answers. A spiritual person will see a religion as a means to understand the Godly state of *karan srira* (causal body) rather than see an end. He/she will embrace each and every religion as his own because he or she understands the spiritual power behind all that exists. Einstein was an extremely spiritual person whose vision of space–matter integration clearly indicates that a Godly creation embedded in a human body cannot be distinct from its creator and originator. With nanoengineering, a body–mind–spirituality connection will perhaps expand to understand the meaning of death as a renewal of the subtle body to discard the old in search of a new subtlety. The right half of Figure 1.9 shows the brain dominance in each of the four quadrants as researched by Ned Hermann. Most of the right handers think from the left side of the brain which is more verbal in terms of equations, words, and so on. However, the right half is holistic and is important in making connections with other thinking minds. That part is to be developed through scholarly discourses. Nanoengineering will encourage mapping the right side of the brain, while the left side is suppressed, so the right has a chance to show its creativity and innovation. Meditative powers as are common these days will be encouraged in order to see the changes, as the right half expands to where the future of nanoelectronics resides.

1.4 Growth and Decay

Economists talk about explosion of wealth, educators talk about explosion of knowledge, and social scientists talk about explosion of population and depletion of resources. In nanoelectronics, Moore's law talks about the upsizing in a variety of forms (memory, functions, etc.) and downsizing in terms of the size of a device. Most of these growth and decay problems hold their quantitative origin in an exponential function as given below:

$$y = y_o e^{\pm \alpha x} \tag{1.5}$$

where y is the dependent variable, N is the number of transistors on a chip, B the bits processed per second, L the size of a transistor, and so on. The independent variable x could be the time t or space (x) variable. For example, in physics $N = N_o e^{-\alpha t}$ represents the number of atoms remaining after disintegration in a radioactive process with a decay rate α, starting from initial number N_o.

The most common application of an exponential function is in the growth of capital through compounding and depreciation or depletion of resources through its usage over a period of time. If principal P_o is kept in the bank that earns an interest rate α (e.g., $\alpha = 0.05$ or 5%), then at a simple (linear rate of growth), the future value $P(t)$ will be

$$P(t) = P_o(1 + \alpha t) \tag{1.6}$$

where $P_o \alpha t$ is the interest earned on P_o in t years. However, if the rate α is compounded in m periods, the growth in the first period is $P_1 = P_o(1 + \alpha/m)$, and the future value at the end of first period becomes the initial value for the second period giving $P_2 = P_1(1 + \alpha/m) = P_o(1 + \alpha/m)^2$. As the process is continued, the amount will be $P_m = P_o(1 + \alpha/m)^m$ at the end of the year (m periods). For t number of years, the amount accumulated is

$$P(t) = P_o\left(1 + \frac{\alpha}{m}\right)^{mt} \tag{1.7}$$

If the size of the period is miniscule so that $m \to \infty$, the growth is continuous. In that limit, $\underset{m\to\infty}{\text{limit}}\left(1 + (\alpha/m)\right)^m = e^{\alpha}$. For a continuous growth, Equation 1.7 results in

$$P(t) = P_o e^{\alpha t} \tag{1.8}$$

where $(e^{\alpha} - 1) = \alpha_{eff}$ is sometimes referred to as the effective rate of interest as yearly ($t = 1$) interest is $(P_o e^{\alpha} - P_o) = P_o(e^{\alpha} - 1) = P_o \alpha_{eff}$. The banks normally advertise the annual percentage rate (APR). So, if the APR is 5% ($\alpha = 0.05$), the effective rate will be $e^{0.05} - 1 = 0.0513$ or 5.13%. Think of the credit card companies: when 25% (0.25) interest is quoted, the effective interest is in fact 28.4%. The same law with negative exponent $e^{-\alpha t}$ applies to depletion or decay, with α being the decay rate.

Two features of this model are worth examining: the doubling time $t = T_2$ for the growth $P(T_2) = 2P_o = P_o e^{\alpha T_2}$ and half-time $t = T_{1/2}$ $P(T_{1/2}) = (1/2)P_o = P_o e^{-\alpha T_{1/2}}$ for decay. The natural logarithm of $e^{\alpha T_2} = 2$ or $e^{-\alpha T_{1/2}} = 1/2$ yields

$$T_2 \text{ or } T_{1/2} = \frac{0.693}{\alpha} \approx \frac{0.7}{\alpha} = \frac{70}{\alpha}(\%) \tag{1.9}$$

This is sometimes called the law of 70. If an offered interest rate of 5% is compounded daily ($m \to \infty$ assumption applies here as well), it will take 14 years to double the principal. If it is 10%, it will take only 7 years. Credit card companies charging a 25% rate will double the credit balance in less than 3 years. Any other explosion or depletion can be similarly modeled by the exponential model or its complement $1 - e^{-\alpha t}$, which has a steady-state value of 1 as $t \to \infty$. $1 - e^{-\alpha t}$ is known to give limited growth with a ceiling of 1.

In an RC circuit with a capacitor C is connected in series with a voltage source V and a resistance R (Figure 1.10), the charge on the capacitor $Q = CV$ does not accumulate instantly. The charge accumulation is slow and is given by the complementary exponential function [5]

$$q(t) = CV(1 - e^{-t/\tau_{RC}}) \tag{1.10}$$

where $\alpha = 1/\tau_{RC}$ with the time constant $\tau_{RC} = RC$ (known as RC time constant). τ_{RC} gives the rate of the charge accumulation. In one time constant $t = \tau_{RC}$, the accumulated charge $q(\tau_{RC}) = 0.63CV$ is 63% of the final value CV. As the charge accumulates on the capacitor, the current $i = dq/dt$ approaches zero in long-time limit. The voltage $v_c(t)$ across the capacitor and the current $i(t)$ is given by

$$v_c(t) = V(1 - e^{-t/\tau_{RC}}) \tag{1.11}$$

$$i(t) = \frac{dq}{dt} = \frac{V}{R} e^{-t/\tau_{RC}} \tag{1.12}$$

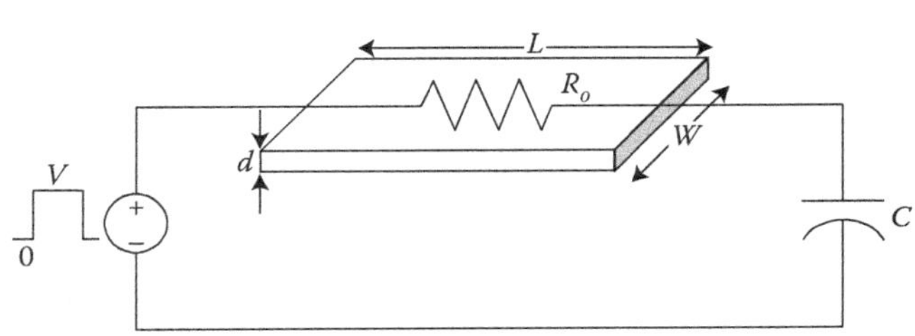

FIGURE 1.10
A prototype RC circuits with a resistor and a capacitor.

The transient phenomena as described by exponential and other functions in response to external stimulus are noted on circuit elements. These transients have special significance in information processing on and off the chip.

1.5 Scope

It may seem attractive to cover all these interesting effects, but to make it a tractable volume, scope must be limited. Physics is the backbone of modern nanoengineering and also of traditional liberal arts. In medieval times, the mark of an educated person was the dominance of seven liberal arts, divided into quadrivium (big four) and trivium (little three). Quadrivium comprises arithmetic, geometry, astronomy, and music, whereas trivium comprises grammar, logic, and rhetoric [6]. Many consider nanoengineering as the new liberal art that synthesizes traditional liberal arts with science and engineering. The application of physics to nanoengineering necessarily requires the knowledge of arithmetic (including mathematics of all kind), geometry (including graphics of all nature), astronomy (understanding the balancing forces of the nature to keep the universe in equilibrium), and the music of nature (how various frequencies can be synthesized to create a signal with a high signal-to-noise ratio). With nanoengineering so developed through the application of quadrivium, it is necessary for nanoengineers to communicate their findings to a diverse audience, which is where trivium comes into play. The correct grammar (or syntax) is needed in transmitting a message (signal) without ambiguity. The correct logic is required to organize ideas in a logical sequential format, so ideas flow fluently from one sentence to the other, one paragraph to the other, from one section to the other, and so on. Rhetoric forms the backbone of communication, both in written as well as spoken forms. The art of communication is now an integral part of the engineering culture, in the true sense of synthesis. Scientists study the world as it exists. However, engineers create a new world either on or off the chip using nanodevices. To create that world on a chip requires good communication channels among devices on a chip and among team players designing the chip.

The classical theories ignore quantum waves as a wavelength is minuscule compared to the size of a device. However, modern devices are smaller than the size of a quantum wavelength and hence, conducting channels act like waveguides for quantum waves. Most circuit designs are based on Ohm's law. Ohm's law is the basis of the linear drift velocity response to the driving electric field in a device or circuit. Theoretical considerations, now confirmed by a variety of experiments, show that the velocity follows a sublinear behavior in the presence of a high electric field and eventually saturates. Most nanoscale designs must take this velocity saturation into account. The applications of quantum physics are required to study nanoelectronic devices, systems, and components [7,8]. Quantum waves naturally come into play when device dimensions are reduced to the nanometer-scale. The electric field on these scaled-down devices is also high, which makes the familiar Ohm's law invalid. This is the major focus as we proceed from atoms to bands and on to performance evaluation of low-dimensional nanosystems. The appearance of quantum waves on nanoscale and the failure of Ohm's law are integrated as we explore current response to an applied voltage. Let the journey begin!

EXAMPLES

E1.1 Calculate the thermal de Broglie wavelength of Si and GaAs at room temperature (300 K), 77 K and 4.2 K.

$$\text{Si at } T = 300\text{ K: } \lambda_{Dth} = \frac{7.63\text{ nm}}{\sqrt{(0.26) \times (300\text{ K}/300\text{ K})}} = 15.0\text{ nm}$$

$$\text{GaAs at } T = 300\text{ K: } \lambda_{Dth} = \frac{7.63\text{ nm}}{\sqrt{(0.067) \times (300\text{ K}/300\text{ K})}} = 29.5\text{ nm}$$

At 77 K, the scaling factor $\sqrt{300/77} = 1.97$ gives for Si $\lambda_{Dth} = 29.6$ nm and for GaAs $\lambda_{Dth} = 58.2$ nm.

At 4.2 K, the scaling factor $\sqrt{300/4.2} = 8.45$ gives for Si $\lambda_{Dth} = 126.8$ nm and for GaAs $\lambda_{Dth} = 249.3$ nm.

At low temperature, the thermal velocity due to energy $E = k_BT$ goes over to the Fermi velocity due to the Fermi energy E_F. The above equation thus may not be valid at low temperatures. It is particularly true for metals. The examples of metals and their de Broglie wavelength (Fermi wavelength) are given in Appendix E.

E1.2 An electron's trajectory in a magnetic field is helical (as described in physics books). The stable orbits are those that contain an integer number of de Broglie waves. Calculate the possible digital energies of an electron when put in an orbit following the laws of physics.

An electron traveling at an angle to an applied magnetic field makes a spiral motion with the axis of the spiral in the direction of the magnetic field, as shown in the following figure. (a) The velocity vector in a magnetic field applied in the x-direction being decomposed into components parallel and perpendicular to the magnetic field. (b) Perpendicular motion is affected by the magnetic field resulting in a circular orbit, the circle being carried away by parallel component resulting in a spiral. Only the velocity component

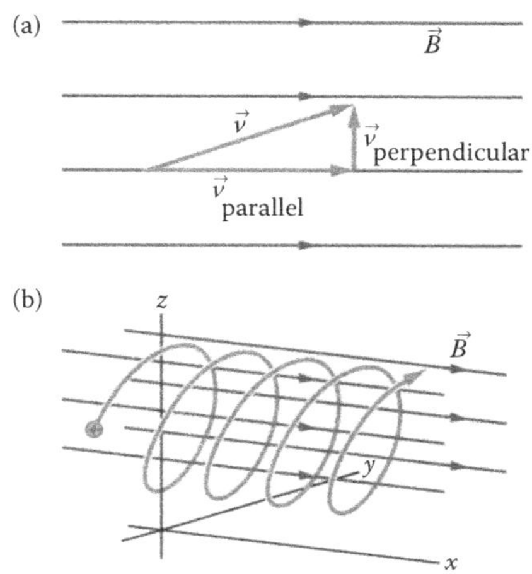

perpendicular to the magnetic field is affected by the magnetic field $\vec{B}$. The component parallel to the magnetic field is unaffected. The velocity $v_x = \hbar k_x / m_n^*$ is therefore of analog type. However, the perpendicular component follows the periodic circular motion that is quantum in character when the radius of the orbit is comparable to electron's de Broglie wavelength.

The magnetic force $q\vec{v} \times \vec{B}$ provides the centripetal force in the circular orbit. The magnetic force exists due to the perpendicular-to-the-magnetic-field component $v_\perp$ of the velocity vector $\vec{v}$ and is given by

$$qv_\perp B = \frac{m_n^* v_\perp^2}{R} \Rightarrow v_\perp = \frac{qRB}{m_n^*} \tag{E1.1}$$

Each orbit contains only the integer number of de Broglie waves for sustained and stable orbits. This resonant condition for constructive superposition of quantum waves in an orbit yields

$$2\pi R_n = n\frac{h}{m_n^* v_\perp} \Rightarrow v_\perp = \frac{h}{m_n^* R_n} \tag{E1.2}$$

where n is an integer ($n = 1, 2, 3, \ldots$). Elimination of $v_\perp$ from Equations E1.1 and E1.2 yields the quantized orbit the radius:

$$R_n = \left(\frac{n\hbar}{qB}\right)^{1/2} = n^{1/2} R_1, \quad R_1 = \left(\frac{\hbar}{qB}\right)^{1/2} \tag{E1.3}$$

The quantized kinetic energy $E_{\perp n}$ is given by

$$E_{\perp n} = \frac{1}{2} m_n^* v_{\perp n}^2 = n\hbar\omega_c, \quad \omega_c = \frac{qB}{m_n^*} \tag{E1.4}$$

In addition, there is a potential energy as confinement to magnetic field presents an analogous situation to that of a harmonic oscillator whose energy is $(1/2)h\omega_c$. The total energy is then

$$E = E_{\perp n} + E_{\parallel} = \left(n + \frac{1}{2}\right)\hbar\omega_c + \frac{\hbar^2 k_x^2}{2m_n^*}, \quad n = 0, 1, 2, 3, \ldots \tag{E1.5}$$

where $E_{\parallel} = \frac{1}{2} m_n^* v_{\parallel}^2 = \frac{\hbar^2 k_x^2}{2m_n^*}$ with $v_{\parallel} = \frac{\hbar k_x}{m_n^*} \cdot p_x = \hbar k_x$ is the momentum in wave terminology.

E1.3 A certain silicon-gate NMOS transistor occupies an area of 25 λ^2, where λ is the scaling constant (feature size) in VLSI technology. How many NMOS transistors can fit in a 5 mm × 5 mm chip if (a) $\lambda = 1$ μm? (b) 0.25 μm? (c) 0.10 μm?

1 mm = 1000 μm. Hence, 5 mm = 5000 μm

a. $N = 5000^2/25(1^2) = 1$ million transistors
b. $N = 5000^2/25(0.25^2) = 16$ million transistors
c. $N = 5000^2/25(0.1^2) = 100$ million transistors

E1.4 The cost of processing a wafer in a particular process is \$1000. Assume a yield of only 35% of the chips that are good. Find the cost per good chip for a (a) 75-mm wafer and (b) 150-mm wafer.

a. The number $N = \pi R^2/L^2$ after making corrections by discarding approximately $2\pi R/\sqrt{2}L$ for the partial chips at the edges. The discarding assumption for the partial chips is that roughly the diagonal of the partial chip lines up on the periphery. For a 75-mm wafer, it gives approximately 130 chips. The cost per good chip is \$400/(0.35 × 130) = \$8.79.
b. The 150-mm wafer has a total of 600 chips yielding a cost of \$400/(0.35 × 600) = \$1.90 per good chip.

E1.5 Cosmic ray bombardment of the atmosphere produces neutrons, which in turn react with nitrogen to produce radioactive carbon-14 (${}_6C^{14}$). Radioactive carbon enters all living tissues through carbon dioxide, which is first absorbed by plants. As long as a plant or animal is alive, ${}_6C^{14}$ is maintained in the living organism at a constant level. Once the organism dies, however, ${}_6C^{14}$ decays according to the equation

$$A = A_o e^{-0.000124t}$$

where A is the amount present after t years and A_o is the initial amount at $t = 0$ (when tissue dies). If 500 mg of ${}_6C^{14}$ are present in the sample from a skull at the time of death, how many milligrams (mg) will be found in the sample after (a) 15,000 years and (b) 45,000 years. (c) How many milligrams of ${}_6C^{14}$ would have to be present at the beginning in order to have 25 mg present after 18,000 years.

a. $A = 500\text{ mg } e^{-0.000124\times 15000} = 77.84\text{ mg}$
b. $A = 500\text{ mg } e^{-0.000124\times 45000} = 1.89\text{ mg}$
c. $A = A_o\ e^{-0.000124\times 18000} = 25\text{ mg} \Rightarrow A_o = 25\text{ mg} \times e^{0.000124\times 18000}$ $= 233\text{ mg}$

PROBLEMS

P1.1 When Equations 1.1 and 1.2 are combined, the de Broglie wavelength as a function of energy is written as $\lambda_D = (\lambda_o(\text{nm})/\sqrt{(m^*/m_o)E(\text{eV})})$. Give the numerical value of $\lambda_o(\text{nm})$. Hint: $\lambda_o(\text{nm})$ is the wavelength for $m^*/m_o = 1$, $E(\text{eV}) = 1$. Calculate the de Broglie wavelength in nm for a free electron of energy 1 eV. Does your result agree with that given by Equation 1.4 at room temperature ($T = 300$ K)? Room-temperature thermal energy is $E = k_BT = 0.0259$ eV.

P1.2 What is the de Broglie wavelength (in nm) of an electron at 100 eV? What is the wavelength for electrons at 10 keV? In an electron microscope, the electron is accelerated through a potential difference of 100 kV. What advantage would such a high voltage yield to the resolution of the electron microscope? Comment.

P1.3 In aluminum, the Fermi energy is $E_F = 11.63$ eV. (a) Assuming the effective mass in a metal is that of a free electron, calculate its Fermi velocity v_F (in m/s) corresponding to the kinetic energy equal to the Fermi energy $E_F = 11.63$ eV. (b) Calculate the de Broglie (Fermi) wavelength $\lambda_F = \hbar/m_o v_F$ (in nm) corresponding to the Fermi velocity.

P1.4 A typical chip size is 1 cm × 1 cm. The available silicon wafers have grown from 1″ (25 mm) diameter to now 12″ (300 mm) in diameter. Assuming that the silicon wafers are round, estimate the number of chips that can be fabricated on a 1″ wafer? 12″ wafer? Account for any correction needed to your results as a chip on the edge of the round wafer is not a perfect square.

P1.5 A CMOS circuit with contacts is of the order of 1 μm × 1 μm. (a) How many CMOS circuits can be accommodated on a 1 cm × 1 cm? How many circuits are on a 1 cm × 1 cm square chip if the CMOS circuit size is reduced to 0.25 μm on each side? How many 0.25-μm circuits are on if the chip size is increased to 2 cm × 2 cm?

P1.6 A chip with a typical size of 1 cm × 1 cm area contains 10 million CMOS circuits in the present state of technology. Estimate the size (in nm and in μm) of a single CMOS circuit. State your assumptions about the shape of a CMOS circuit to determine the size.

P1.7 Search the International Technology Roadmap for Semiconductors (ITRS) on the Internet and jot down a few points regarding the present state of size of the device, chip, and a CMOS circuit. Are the predictions of Figure 4.4 for year 2010 consistent with the roadmap?

P1.8 Computer chips are pervasive in the sense that we do not even notice their presence. Make a list of at least 10 items where you identify nanoscale electronics is playing a dominant role in its operation. Write a few sentences to justify their size and operation based on nanoelectronics.

P1.9 The number of memory bits per chip are described by the exponential function $B = 20.0\,e^{0.455(y-1960)}$. Estimate the memory bits per chip in year 2020.

P1.10 The number of transistors in a microprocessor follows an exponential growth given by $N = 1027\,e^{0.347(y-1960)}$. What will be the number of transistors on a microprocessor in year 2020?

P1.11 How many years does it take for memory bits per chip to double based on $B = 20.0\,e^{0.455(y-1960)}$? How many years for a factor of 10?

P1.12 A rise time t_r of a square pulse is defined for charge on the capacitor to change from 10 to 90%. Show that $t_r = 2.2RC$.

CAD/CAE PROJECTS

C1.1 *Limited growth.* People assigned to assemble circuit boards for a computer manufacturing company undergo on-the-job training. From the past it was found that the learning curve for the average employee is $N = 40(1 - e^{-0.12t})$, where N is the number of boards assembled per day after t days of training.

a. Plot this function using Excel or MATLAB® for $0 \le t \le 30$.

b. What is the maximum number of boards an average employee can be expected to produce in a day?

C1.2 *Growth and decay.* As circuits become complex, CAE is being used extensively to identify relations between dependent and independent variables. Figure 1.4 gives memory growth ($M(t)$) and downsizing ($L(t)$) over time t (in years) on a chip that can be modeled by an exponential function: $M = M_o e^{\lambda_M t}$ and $L = L_o e^{-\lambda_L t}$, where M_o is the memory and L_o the length of a transistor at $t = 0$ (year 1979 in Figure 1.4). λ_M is the growth parameter and λ_L is the downsizing parameter (both in y^{-1}). Tabulate the $M(t)$ and $L(t)$ data labeled in Figure 1.4 as a function of calendar time and time t ($t = 0$ is the calendar year 1979). Fit this data with an exponential graph and determine the parameters M_o, L_o, λ_M, and λ_L. What is the doubling time t_2 for the memory ($M(t_2) = 2M_o$) and half-time $t_{1/2}$ for the length (size) ($L(t_{1/2}) = (1/2)L_o$)? Using the model, predict the memory M and transistor size L for the year 2012. Discuss your results.

C1.3 *Exploration project.* Determine the present state of technology as far as the feature size is concerned. What lithographic techniques and tools are being employed for downsizing a chip? Evaluate and assess possibilities of replacing silicon as a desired material as carbon-based and organic materials are becoming competitive both from the cost-effective point of view as well as ease in implementing lithographic technologies.

C1.4 *Liberal arts project.* As nanotechnology expands, there are many advocates, and some opponents of combining natural (nature given) with artificial (man-made). Supplementing the brain by tapping the uncharted territories is one such area to combine natural intelligence with artificial intelligence. Other possibilities are to implant chips in your body that will sense the deficiency of an ingredient and release as necessary the required amount from a computer chip. Explore and summarize arguments used by those against and for nanotechnology development. If you have problem starting, read the paper entitled "Engineering the soul of management in the nano era," *Chinese Management Studies*, vol. 3, pp. 213–234, 2009 by V. K. Arora.

References

1. M. Shur, *Introduction to Electronic Devices*. New York: John Wiley, 1996.
2. V. K. Arora, Engineering the soul of management in the nano era, *Chinese Management Studies*, 3, 213–234, 2009.
3. J. Walker, *Physics*, 4th ed. Upper Saddle River, NJ: Pearson/Prentice-Hall, 2010.
4. B. L. Anderson and R. L. Anderson, *Fundamentals of Semiconductor Devices*. New York, NY: McGraw-Hill, 2005.
5. M. Tan et al., Enhancement of nano-RC switching delay due to the resistance blow-up in InGaAs, *NANO*, 2, 233, 2007.
6. V. K. Arora and L. Faraone, 21st century engineer-entrepreneur, *IEEE Antennas and Propagation Magazine*, 45, 106–114, 2003.
7. A. T. M. Fairus and V. K. Arora, Quantum engineering of nanoelectronic devices: The role of quantum confinement on mobility degradation, *Microelectronics Journal*, 32, 679–686, 2001.
8. V. K. Arora, Quantum engineering of nanoelectronic devices: The role of quantum emission in limiting drift velocity and diffusion coefficient, *Microelectronics Journal*, 31, 853–859, 2000.

2

Atoms, Bands, and Quantum Wells

Charge carriers (electrons, holes, or ions) in multimicron mesoscopic systems behave like particles. Their energies have analog (or classical) character meaning that their energy variables are continuous. However, when one or more of the three Cartesian directions are constrained to a nanometer scale, the carriers exhibit a wave (or quantum) character. In this setup, the energies are digitally (or quantum) separated by forbidden regions. Analog characteristics can be built from digital energies by expanding a nanoscopic system to macro- or microscopic scales in which case the quantum jumps are miniscule. This transformation leads to the concept of density of states (DOS), which is the number of quantum states per unit volume (or area or length) per unit energy.

2.1 Birth of a Quantum Era

The origin of the quantum era is traced to the work of Louis de Broglie in the early part of the twentieth century. He postulated that just like a photon is a packet (quantum) guided by electromagnetic waves, electrons are guided through the existence of quantum waves. This wave–particle duality is represented by the relation

$$\lambda_D = \frac{h}{p} \tag{2.1}$$

where wave property, represented by the de Broglie wavelength λ_D, is connected to the particle property represented by crystal momentum of a carrier $p = m^*v$. Here, m^* is the effective mass and v is the carrier velocity. m^* includes the effect of the force field of the crystal lattice on a carrier and hence differs from the free electron mass m_o. $m^* = 0.067m_o$ for gallium arsenide (GaAs). m^* is actually a tensor as it varies from one direction to the other in the crystal lattice. However, for most engineering approximations, it is taken to be a scalar that is averaged depending on the application.

Equation 2.1 has a close resemblance to that of a photon (quantum of light) with energy $E = hf = mc^2$ and momentum $p = mc$, where f is the frequency of a photon and m is its relativistic mass. The Newtonian laws do not apply to photons for which the theory of relativity defines the energy–mass connection ($E = hf = mc^2$). The mass as known in Newtonian mechanics has an entirely different meaning when electron speed is near that of light. The mass increases with the speed of the particle. As speed approaches the speed of light $c = \lambda/T = f\lambda$, the mass tends to rise to infinity unless the rest mass of the photon is negligible. This zero mass or zero momentum of a photon enters into the description as selection rule for electrons making a transition must necessarily have zero change in momentum for photon to be emitted or absorbed. Otherwise, the probability of emission or absorption of a photon is greatly reduced. The wavelength of a photon is similarly related to that of the massless (rest mass $m_o = 0$, not moving mass m) photon as

$$\lambda = \frac{c}{f} = \frac{hc}{hf} = \frac{hc}{mc^2} = \frac{h}{p} \tag{2.2}$$

In a crystal with an ensemble of carriers (electrons or holes), the average energy is thermal energy k_BT. In fact, it depends on the degrees of freedom. For one-dimensional (1D), two-dimensional (2D), and three-dimensional (3D) motion, it is $(1/2)k_BT$, k_BT, and $(3/2)k_BT$, respectively, or $(d/2)k_BT$, where $d = 3$ (bulk), 2 (layer), or 1 (line). The thermal de Broglie wavelength λ_{Dth} corresponding to the thermal momentum $p_{th} = m^* v_{th} = \sqrt{2m^*E} = \sqrt{2m^*k_BT}$ for an energy $E = k_BT$ is given by

$$\lambda_{Dth} = \frac{h}{m^* v_{th}} = \frac{h}{\sqrt{2m^* k_BT}} \tag{2.3}$$

with thermal velocity equal to

$$v_{th} = \sqrt{\frac{2k_BT}{m^*}} \tag{2.4}$$

Equation 2.2 for photons and Equation 2.3 for electrons are easily adaptable for numerical computation if $E = hf$ for a photon or $E = p^2/2m^*$ for an electron is expressed in eV:

$$\lambda = \frac{hc}{E(\text{eV})} = \frac{1240\,\text{eV}\cdot\text{nm}}{E(\text{eV})} \quad \text{(photon)} \tag{2.5}$$

$$\lambda_D = \frac{1.23\,\text{nm}\cdot(\text{eV})^{1/2}}{\sqrt{(m^*/m_o)E(\text{eV})}} \quad \text{(electron)} \tag{2.6}$$

Equation 2.5 gives, for a photon of energy $E = 2.0$ eV, the wavelength $\lambda = 620$ nm. Equation 2.6, for GaAs ($m^* = 0.067\, m_o$) at room temperature ($E = k_B T = 0.0259$ eV), gives for electron a de Broglie wavelength of $\lambda_D = 29.5$ nm. Equations 2.5 and 2.6 are distinct in the sense that Equation 2.5 is applicable to a relativistic particle (an Einsteinian boson following Bose–Einstein statistics). Equation 2.6 is applicable to a classical (Newtonian) particle, called Fermion obeying the Fermi–Dirac statistics. Hence, caution should be engaged not to interchange one with the other. Appendix E lists the de Broglie wavelength of several metals where the top energy known as the Fermi energy is several eV. For example, copper (Cu) has a Fermi energy of 7.0 eV. The corresponding Fermi velocity is $v_F = 1.57 \times 10^6$ m/s and the Fermi wavelength is $\lambda_F = 0.464$ nm. The carrier statistics, as will come later, is distinctly different for electrons, photons, or phonons. A phonon is a quantum of energy associated with lattice vibrations that has nonzero momentum.

2.2 Hydrogen-Like Atom

The hydrogen (H) atom is the smallest atom with one proton in the nucleus and one electron outside the nucleus revolving in a circular orbit. The hydrogen atom model works well with reasonable success for an atom with Z (atomic number) protons in the nucleus surrounded by an equal number of electrons in various orbits surrounding the nucleus. It is named hydrogen-like for the many-electrons atom, as the placement scheme of the hydrogen atom can be used to place many electrons in various orbits. The electrons in the outermost orbit define the electronic and chemical properties of a given atom.

According to the primitive thinking of the atom, the electrons are buried in the atom like seeds in a watermelon (positive charge cloud spread over the sphere) as shown in Figure 2.1. This watermelon model was discarded in favor of the planetary model shown in Figure 2.2 that treats an electron revolving around the positive nuclear charge Zq. Z is the atomic number of the hydrogen-like atom and $q = 1.6022 \times 10^{-19}$ C is the fundamental quantum of charge that exists on a proton ($+q$) with positive charge and on an electron ($-q$) with negative charge. In the normal state, the atom is neutral and hence Z is the number of protons as well as the number of electrons. An electron is held in the orbit by the Coulomb force of attraction between the positively charged nucleus and a negatively charged electron:

$$F = \frac{1}{4\pi\varepsilon_o}\frac{Q_1 Q_2}{r^2} = \frac{1}{4\pi\varepsilon_o}\frac{(Zq)(-q)}{r^2} = -\frac{1}{4\pi\varepsilon_o}\frac{Zq^2}{r^2} \tag{2.7}$$

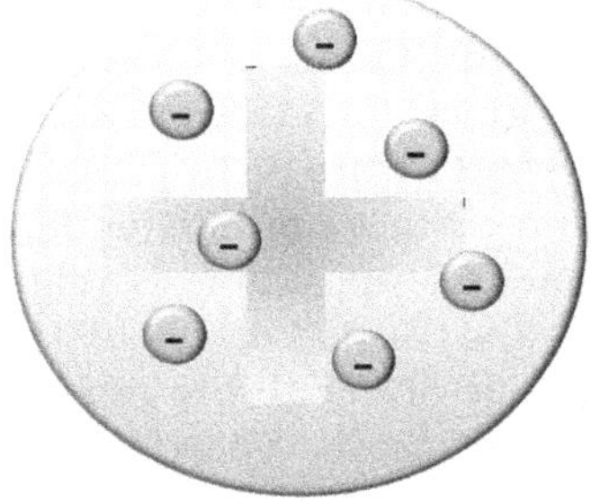

FIGURE 2.1
Watermelon model of the atom with electrons buried in the positive charge cloud. (Adapted from J. Walker, *Physics*, 4th ed. Upper Saddle River, NJ: Pearson/Prentice-Hall, 2010; B. L. Anderson and R. L. Anderson, *Fundamentals of Semiconductor Devices*. New York, NY: McGraw-Hill, 2005.)

where $\varepsilon_o = 8.85 \times 10^{-12}$ $C^2/N\ m^2$ is the permittivity of the free space. $Q_1 = Zq$ is the nuclear charge and $Q_2 = -q$ is the charge on an electron. Therefore, the force shown in Equation 2.7 is what is felt by a single electron in the nuclear charge field. The unit $C^2/N\ m^2$ of ε_o is equivalent to F/m.

In a plane, the circular orbit looks as shown in Figure 2.3. As the force is the negative gradient of the potential energy (PE) U ($F = -dU/dr$), the potential energy is the energy stored at $r' = r$ as work is done in bringing an electron from $r' = \infty$ to $r' = r$. This work is given by

$$U = \int_{\infty}^{r} F(-dr') = \int_{\infty}^{r} -\frac{1}{4\pi\varepsilon_o}\frac{Zq^2}{r'^2}(-dr') = -\frac{1}{4\pi\varepsilon_o}Zq^2\left.\frac{1}{r'}\right|_{\infty}^{r} = -\frac{Zq^2}{4\pi\varepsilon_o r} \quad (2.8)$$

The negative sign in Equation 2.8 is indicative of the fact that the electron–nucleus system is a bound system and energy (named the ionization energy)

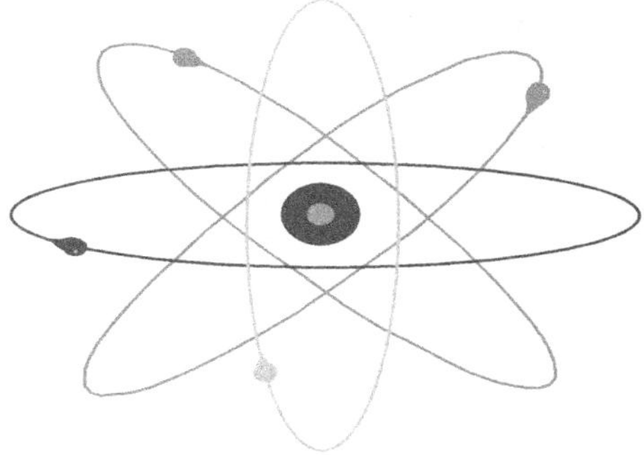

FIGURE 2.2
The planetary model of the atom with positive charge at its center (nucleus) around which the electrons revolve.

FIGURE 2.3
The flat orbit in a plane showing the work done as it is brought from infinity and kept in an orbit of radius r and hence has the potential energy (PE) due to its position at $r' = r$.

must be supplied to break it into separate entities (electron detached from positively charged nucleus) or the electron is taken to infinite distance from its normal location (ground state).

The kinetic energy (KE) of the electron in the orbit is evaluated by equating the Coulomb force of Equation 2.7 to the centripetal force $F_c = m_o v^2/r$ with the result

$$\frac{1}{4\pi\varepsilon_o}\frac{Zq^2}{r^2} = \frac{m_o v^2}{r} \Rightarrow \text{KE} = \frac{1}{2}m_o v^2 = \frac{1}{8\pi\varepsilon_o}\frac{Zq^2}{r} \tag{2.9}$$

where $m_o = 9.11 \times 10^{-31}$ kg is the free electron mass in the vacuum. In a crystal environment, an electron encounters the forces of crystal atoms and its mass is conveniently defined as an effective mass that can be measured. Its measured values are listed in tables (e.g., see Appendices B through D).

The total energy $E = \text{KE} + \text{PE}$ is then given by

$$E = \text{KE} + \text{PE} = \frac{1}{8\pi\varepsilon_o}\frac{Zq^2}{r} - \frac{1}{4\pi\varepsilon_o}\frac{Zq^2}{r} = -\frac{1}{8\pi\varepsilon_o}\frac{Zq^2}{r} \tag{2.10}$$

The waves after traversing the orbit must meet in phase if the de Broglie hypothesis of electron waves is accepted. Therefore, the orbit is expected to contain integer number of de Broglie waves:

$$2\pi r = n\lambda_D = n\frac{h}{p} \Rightarrow rp = rm_o v = n\frac{h}{2\pi} = n\hbar, \quad n = 1, 2, 3, \ldots \tag{2.11}$$

Equation 2.11 states that the standing quantum waves are only possible when electrons are in orbit where propagating waves meet in phase at the same point in the orbit. $\vec{L} = \vec{r} \times \vec{p}$ is the angular momentum whose magnitude is $rp = rm_o v$ as $\vec{r}$ and $\vec{p}$ are orthogonal to each other. This was the condition given by Neil Bohr stating that the angular momentum of the electron in an orbit is quantized as being equal to integer times $\hbar$. de Broglie's hypothesis gives it a distinct meaning that only those orbits survive or are in resonant state for which quantum waves meet in phase after traversing the complete orbit. With $p = m_o v$, the quantized velocity v_n from Equation 2.11 can be written as

$$v_n = \frac{n\hbar}{m_o r_n} \tag{2.12}$$

The quantized radii of the orbits are found by utilizing Equation 2.12 into Equation 2.9 to eliminate v_n,

$$\frac{1}{2} m_o \left(\frac{n\hbar}{m_o r_n} \right)^2 = \frac{1}{8\pi\varepsilon_o} \frac{Zq^2}{r_n} \Rightarrow \frac{1}{r_n} = \frac{1}{4\pi\varepsilon_o} \frac{Zm_o q^2}{n^2\hbar^2} \tag{2.13}$$

Equation 2.10 for the total energy E then quantizes (digitizes) and transforms to give

$$\begin{aligned} E_n &= -\frac{1}{8\pi\varepsilon_o} \frac{Zq^2}{r_n} = -\frac{Zq^2}{8\pi\varepsilon_o} \frac{1}{4\pi\varepsilon_o} \frac{Zm_o q^2}{n^2\hbar^2} = -\frac{1}{2} \frac{1}{(4\pi\varepsilon_o)^2} \frac{m_o Z^2 q^4}{n^2\hbar^2} \\ &= -2.18 \times 10^{-18}\,\mathrm{J} \frac{Z^2}{n^2} = -13.6\,\mathrm{eV} \frac{Z^2}{n^2} \end{aligned} \tag{2.14}$$

The quantized radius r_n of Equation 2.13 is given by

$$r_n = n^2 \frac{4\pi\varepsilon_o\hbar^2}{Zmq^2} = r_1 \frac{n^2}{Z}, \quad n = 1, 2, 3, \ldots \tag{2.15}$$

with

$$r_1 = \frac{4\pi\varepsilon_o\hbar^2}{m_o q^2} = 0.053\ \mathrm{nm} \tag{2.16}$$

Similarly, quantized velocity is given by

$$v_n = \frac{n\hbar}{m_o r_n} = \frac{n\hbar}{m_o} \frac{1}{4\pi\varepsilon_o} \frac{Zm_o q^2}{n^2\hbar^2} = v_1 \frac{Z}{n} \tag{2.17}$$

with

$$v_1 = \frac{q^2}{4\pi\varepsilon_o\hbar} = 2.2 \times 10^6\ \mathrm{m/s} \tag{2.18}$$

2.3 Photon Emission and Absorption

Equation 2.14 is central to understanding emission and absorption of a photon. The emitted energy $\Delta E_{n_u \to n_\ell}$ when an electron falls from a higher level with the quantum number n_u to the level with the quantum n_ℓ is released in the form of a quantum of light (photon) of energy.

Similarly, a photon can be absorbed to take an electron from a lower level to a higher level provided the energy $hf = \Delta E_{n_u \to n_\ell}$ of the photon is exactly equal to the energy spacing of the two levels. Using Equation 2.14 along with Equation 2.5, the emission wavelength for hydrogen ($Z = 1$) is given by

$$\lambda = \frac{1240 \text{ eV}\cdot\text{nm}}{13.6\,(\text{eV})\left((1/n_\ell^2) - (1/n_u^2)\right)} = \frac{91.18 \text{ nm}}{\left((1/n_\ell^2) - (1/n_u^2)\right)} \tag{2.19}$$

If white light (with all wavelengths or frequencies present) is incident on hydrogen gas, the photons with digitized wavelengths given by Equation 2.19 will be absorbed by the hydrogen atoms and other wavelengths (or frequencies) will pass through unhindered. One may say that hydrogen gas when excited (e.g., through an electric discharge in a hydrogen lamp) is able to emit photons of precisely those wavelengths. This is the origin of emission spectrum. Hydrogen atoms are able to absorb photons from the white light making those wavelengths missing as white light passes through the chamber containing hydrogen gas. This forms absorption spectrum. The visible part of the electromagnetic spectrum spans wavelengths from 400 (violet) to 800 nm (red) as shown in Figure 2.4. The visible part comprises most prominent rainbow colors (VIBGYOR): violet, indigo, blue, green, yellow, orange, and red. For optoelectronic applications, transition between the energy

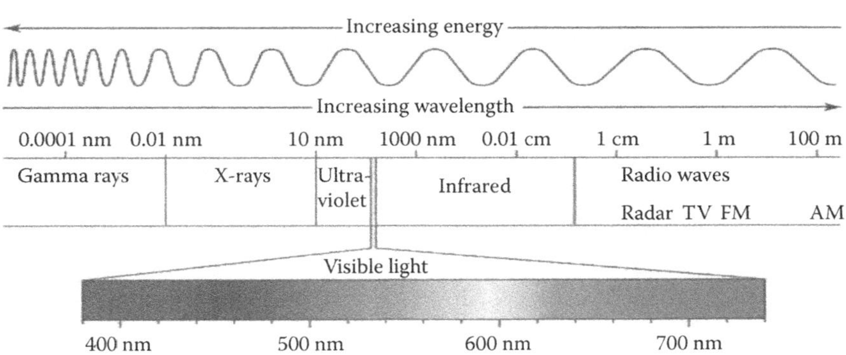

FIGURE 2.4
Electromagnetic spectrum.

level with or without the change in electron momentum is responsible for the emission of light (photon) (no change in momentum) or a mechanical quantum of energy phonon (with the possible change in momentum). The quantum character of energy has special significance in optoelectronic applications. Most of the wavelengths in the visible part of the electromagnetic spectrum arise from the Balmer series when an electron falls from any of the higher levels to $n_\ell = 2$ level. Other series that can be similarly defined are Lyman ($n_\ell = 1$), Paschen ($n_\ell = 3$) series, and so on.

For the Balmer series ($n_\ell = 2$), the first four wavelengths emitted are 653, 486, 434, and 410 nm corresponding to $n_u = 3, 4, 5, 6$, respectively. The emission and absorption spectrum of hydrogen is indicated in Figure 2.5. The emission and absorption spectra complement each other. In an emission spectrum, the atomic electrons in a hydrogen lamp containing gaseous plasma are excited to the higher level as energy is transferred from the passing electric current to the electrons. The excited electron has a limited life span in the excited state and drops to one of the lower states until it reaches the stable ground state and hence, various frequencies (or wavelengths) are emitted that conform to the pair of quantized energy levels. In the absorption process, a white light (containing all frequencies) is incident on the gaseous hydrogen. Only those photons are absorbed (snatched from the white light) by the hydrogen gas, which have energy exactly equal to that of two quantized levels. Others will pass through unabsorbed. One can also say that the hydrogen gas is transparent to all colors (frequencies) except those that are absorbed. That is the reason why in an absorption spectrum the background colors that cannot be absorbed are visible. In the emission spectrum, the background is dark except for those colors that are emitted.

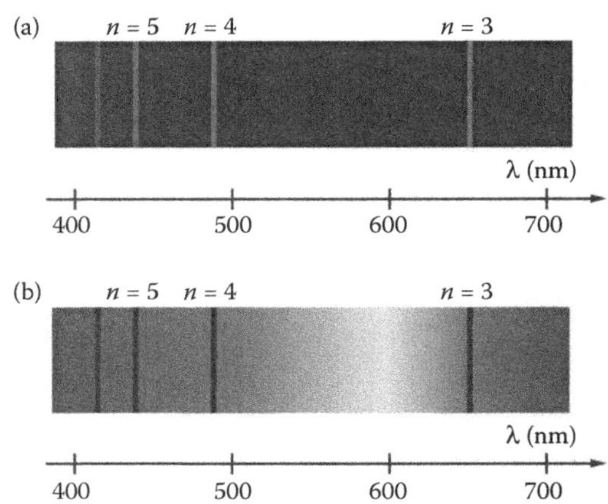

FIGURE 2.5
(a) Emission and (b) absorption lines of the Balmer series in the hydrogen atom.

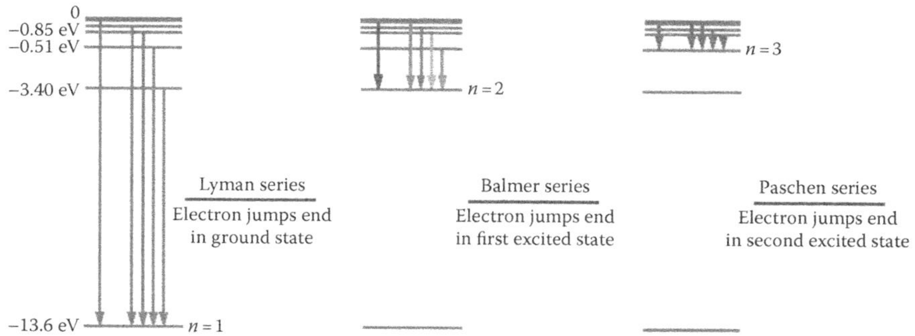

FIGURE 2.6
Transition energy levels for the first three series of the hydrogen atom (n_ℓ = 1 (Lyman), 2 (Balmer), and 3 (Paschen)) sequence). (Adapted from J. Walker, *Physics*, 4th ed. Upper Saddle River, NJ: Pearson/Prentice-Hall, 2010.)

The transitions are well described by what is known as the energy level diagram, shown in Figure 2.6.

2.4 Spherical Hydrogen-Like Atom

As shown in Figure 2.2, the orbits are not constrained to a plane. The sub-orbits reside on a sphere that can be characterized by the principal quantum number n as stated above. The orientation of the orbits on the sphere is given by another quantum number ℓ (azimuthal angular momentum) whose value changes from 0 to $(n - 1)$: 0 when the orbit is in the reference plane (say plane of this paper with angular momentum component equal to zero) and $\ell = n - 1$ when the orbit is perpendicular to the reference plane. The magnitude of the angular momentum is given by $L = \sqrt{\ell(\ell + 1)}\ \hbar$. The magnitude of the angular momentum may not be so important for bean counting as the number of available quantum states for occupation by electrons is identified. The orientation is important for counting the possible orbits. The component of L in a reference direction (say z-axis) is given by another quantum number $m_\ell = -\ell, \ldots 0 \ldots. +\ell$ (called magnetic quantum number). For every positive $m_\ell = +\ell$ quantum state, there is a corresponding $m_\ell = -\ell$ state, making 2ℓ possible values of m_ℓ. There are $(2\ell + 1)$ possible orientations when $\ell = 0$ is also included. In addition, electrons can spin on their own axis clockwise or counterclockwise like a spinning top or ball for which the spinning angular momentum is $S = m_s\hbar$ with $m_s = \pm 1/2$ (clockwise or counterclockwise). Again, it is sufficient to know that there are two states (sometimes these are referred to up and down to denote the direction of the spinning angular momentum). In summary, that is how the quantum numbers are characterized [3]:

$$\begin{gathered} \text{K, L, M,} \ldots \\ n = 1,\ 2,\ 3, \ldots, \infty \end{gathered} \tag{2.20}$$

$$\begin{gathered} \text{s, p, d, f} \\ \ell = 0, 1,\ 2, 3 \ldots, (n-1) \end{gathered} \tag{2.21}$$

$$m_\ell = -\ell,\ -\ell+1, \ldots\ 0, \ldots, +\ell \quad (2\ell+1) \text{ states} \tag{2.22}$$

$$m_s = \pm\frac{1}{2} \quad \text{(2 spin states)} \tag{2.23}$$

$n = 1, 2, 3, \ldots$ are also known as K, L, M, … shells and $\ell = 0, 1, 2, \ldots$ are named as s, p, d, … orbitals.

Each quantum state is described by the unique set ($n\ell m_\ell m_s$). Further, there is a restriction imposed by the Pauli Exclusion Principle that each quantum state is to be filled by only one electron. The total number of quantum states for a shell with a principal quantum number n is given by

$$2\sum_{\ell=0}^{\ell=n-1}(2\ell+1) = 4\sum_{\ell=0}^{\ell=n-1}\ell + 2\sum_{\ell=0}^{\ell=n-1}1 = 4\frac{(n-1)n}{2} + 2n = 2n^2 \tag{2.24}$$

Each subshell of quantum number ℓ contains $2(2\ell+1)$ orbitals. The factor 2 accounts for two spin states. $n = 1$ state has a maximum $\ell = 0$(s). Therefore, only two electrons can be accommodated in $n = 1$ shell. The next shell ($n = 2$) has subshells $\ell = 0$(s) (two electrons) and $\ell = 1$(p) ($2[2 \times 1 + 1] = 6$ electrons), with a total of eight electrons ($2n^2 = 2 \times 2^2 = 8$). Table 2.1 gives the filling number of first nine elements. The outermost shell holds the key for the chemical properties of an atom.

With helium (Z = 2), the $n = 1$ shell is full and cannot accommodate more electrons. Hence, helium is chemically inert gas. Boron is from Group III of the periodic table as it has three electrons in the outermost shell. As it is observed later, this can be useful for doping p-type silicon, which comes from Group IV (Z = 14). Chemically, silicon is represented as $_{14}Si^{28}$ which means that there are 14 electrons in silicon, while the total number of nucleons (14 protons and 14 neutrons) is present in the central nucleus. The electronic placement (configuration) of 14 electrons in Si results in $1s^22s^22p^63s^23p^2$. 3p orbital actually can accommodate six electrons, but all 14 silicon electrons have been already placed. So, four quantum states are unoccupied in the Si atom. In a Si crystal, these quantum states will be shared by four nearest neighbors by forming covalent bonds for each Si

TABLE 2.1

Electronic Configuration from Hydrogen to Fluroine

Atomic Number (Z)	Element (Symbol $_{Z}X^{A}$)	Electronic Configuration
1	Hydrogen ($_{1}H^{1}$)	$1s^1$
2	Helium ($_{2}He^{4}$)	$1s^2$
3	Lithium ($_{3}Li^{6}$)	$1s^22s^1$
4	Beryllium ($_{4}Be^{8}$)	$1s^22s^2$
5	Boron ($_{5}B^{10}$)	$1s^22s^22p^1$
6	Carbon ($_{6}C^{12}$)	$1s^22s^22p^2$
7	Nitrogen ($_{7}N^{14}$)	$1s^22s^22p^3$
8	Oxygen ($_{8}O^{16}$)	$1s^22s^22p^4$
9	Fluorine ($_{9}F^{18}$)	$1s^22s^22p^5$

atom; the quantum states appear completely filled. In a silicon crystal, the highest filled level will expand to a band of energy levels equal to the number of electrons forming a completely filled valence band. Multielectron atoms are complex and hence the energy level is split into sublevels that sometimes cross over to the next shell, as shown in Figure 2.7. In many cases, level-crossing occurs to the next shell.

For a qualitative description of the periodic table and in grouping the atoms in the periodic table, it is sufficient to follow the electronic configuration for a multielectron atom. That is why one-electron hydrogen atom model works equally well for multielectron atoms that can be described as hydrogen-like.

The electronic configuration of silicon is given by $1s^22s^22p^63s^23p^2$. The filled core ($1s^22s^22p^6$). is inert. The active chemical properties arise from partially filled $n = 3$ shell ($3s^23p^2$). Even though a 3p orbital can accommodate six electrons, supply of electrons ($Z = 14$) is exhausted. These four vacancies will be shared by another silicon atom in the crystal environment that will create tetravalent bonds. Si shares the same configuration ($1s^22s^22p^2$) in the periodic table as C (core + $2s^22p^2$), Ge (core + $4s^24p^2$), and Sn (core + $5s^25p^2$). These are known as Group IV elements. Table 2.2 shows some Group IV elements along with Group III and Group V elements.

2.5 Atoms to Crystals—Bands and Bonds

A hydrogen-like model of an atom is useful in understanding the collection of atoms to form a crystal. The energy levels of each atom remain distinct for two noninteracting atoms as shown in Figure 2.8. The potential energy well represented by $U = -Zq^2/4\pi\varepsilon_o r$ drops to $-\infty$ at $r = 0$ (the center of the atom) and goes to zero (or E_{vac}) at $r = +\infty$. Until now, the vacuum level was taken to

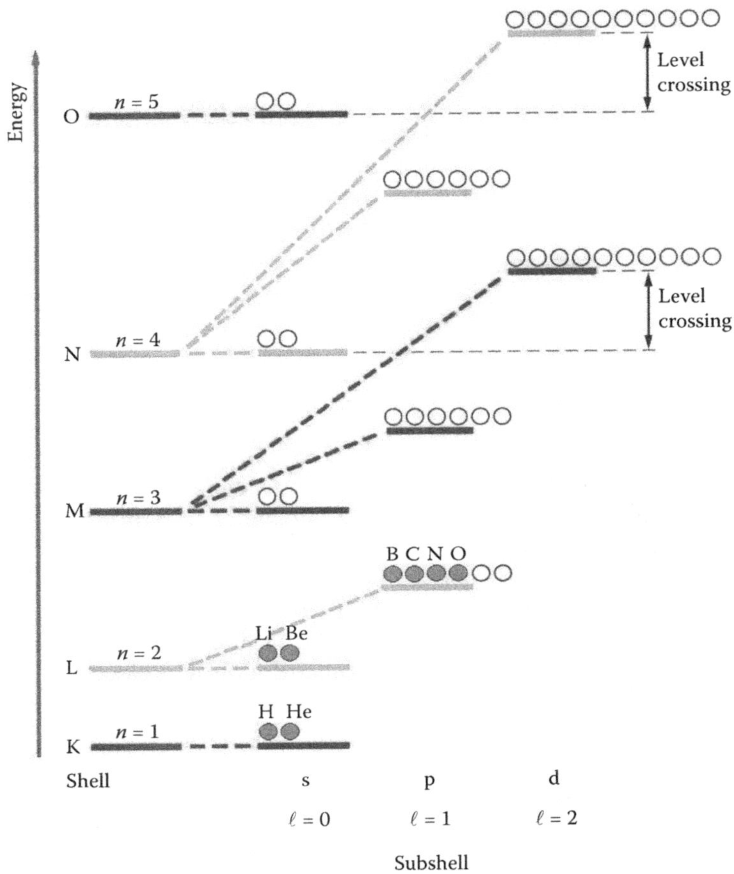

FIGURE 2.7
Filling of the first eight atoms of the periodic table. (Adapted from J. Walker, *Physics*, 4th ed. Upper Saddle River, NJ: Pearson/Prentice-Hall, 2010.)

TABLE 2.2
Partial Periodic Table from Group III (Left), Group IV (Center), and Group V (Right)

B	C	N
Al	Si	P
Ga	Ge	As
In	Sn	Sb

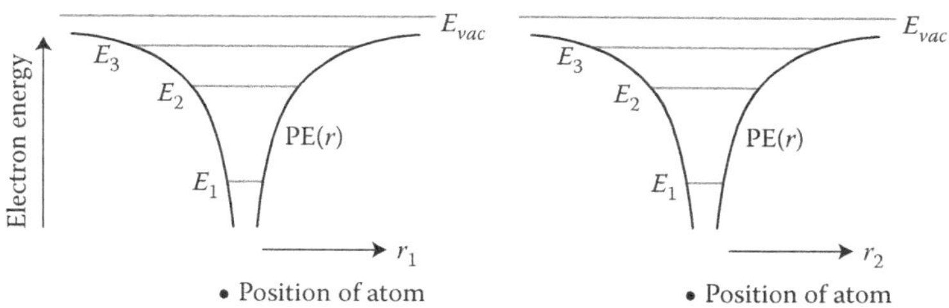

FIGURE 2.8
Two noninteracting hydrogen-like atoms, each keeping its own energy levels. (Adapted from B. L. Anderson and R. L. Anderson, *Fundamentals of Semiconductor Devices*. New York, NY: McGraw-Hill, 2005.)

be zero ($E_{vac} = 0$). As we move to interacting atoms, this vacuum level is of special significance. With this transformation, Equation 2.14 transforms to

$$E_n = E_{vac} - 13.6\ \text{eV}\frac{Z^2}{n^2} \tag{2.25}$$

By reducing the distance between two atoms, the atoms interact. This interaction leads to sharing of the energy levels that try to group together, as shown in Figure 2.9.

As interatomic distance decreases further, there is broadening of the atomic levels and energy levels cluster into bands, as shown in Figure 2.10. These bands present an analog-type character of the energy levels with spacing between digitized levels so small that they appear to be continuous in energy. In this scenario, the DOS concept becomes important as energy levels are filled according to the Pauli Exclusion Principle. The highest filled cluster is known as the valence band where all quantum states are occupied at 0 K. The next higher unfilled cluster with unoccupied quantum states is the conduction band. Each Si atom in the Si crystal feels as if it has eight electrons, thereby filling the $3s^23p^2$ outermost orbits (four from $3s^23p^2$ atomic configuration and

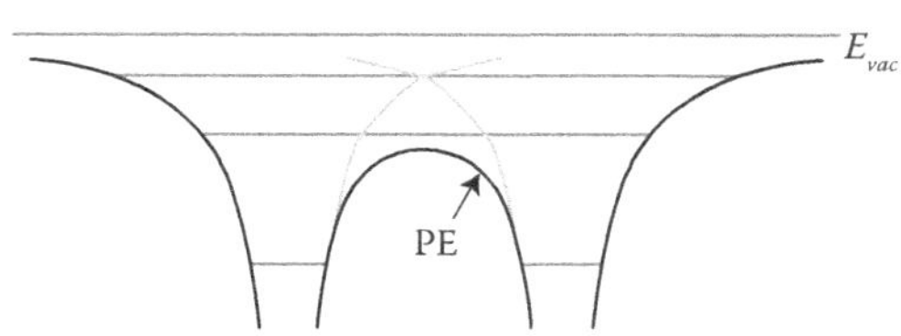

FIGURE 2.9
Two interacting hydrogen-like atoms sharing the energy levels those tend to cluster in bands. (Adapted from B. L. Anderson and R. L. Anderson, *Fundamentals of Semiconductor Devices*. New York, NY: McGraw-Hill, 2005.)

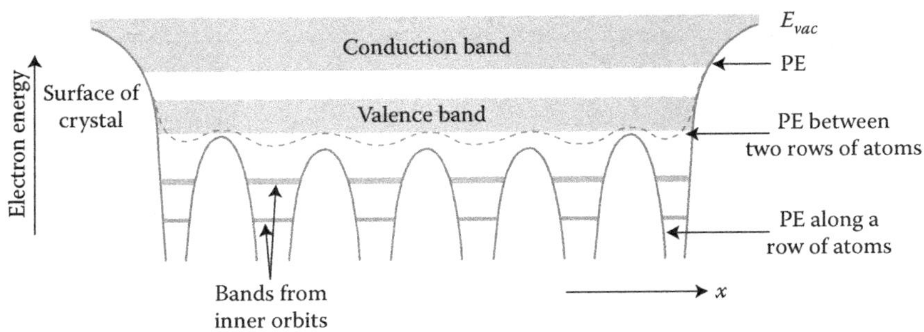

FIGURE 2.10
Interacting hydrogen-like atoms clustering into bands. (Adapted from B. L. Anderson and R. L. Anderson, *Fundamentals of Semiconductor Devices*. New York, NY: McGraw-Hill, 2005.)

four from sharing the resources with neighbors), thereby filling the valence band at 0 K. As the temperature rises, some of these electrons may have energy equal to that of the bandgap and hop over to the conduction band creating a hole in the valence band that can be characterized as a positively charged particle. In intrinsic (pure) silicon, the number of electrons and holes are equal. These are termed as intrinsic carriers. Approximately, 10^{10} cm^{-3} of electron–hole (e–h) pairs are generated at room temperature in silicon ($T = 300$ K); many electrons acquire the energy equal to the bandgap $E_g = 1.12$ eV compared to an average energy $(3/2)k_BT = 0.039$ eV per electron. So a very small number of valence band electrons have the opportunity to do so.

The energy distance between the lowest level in the empty cluster (conduction band) and the highest level in the filled cluster (valence band) is the energy bandgap $E_g = E_c - E_v$ as shown in Figure 2.11, which also shows other features. First, the conduction band edge E_c is now the potential energy level of the electrons and kinetic energy can be in any of the closely clustered analog-type energy levels. $\gamma = E_{vac} - E_v$ is the ionization energy and $\chi = E_{vac} - E_c$

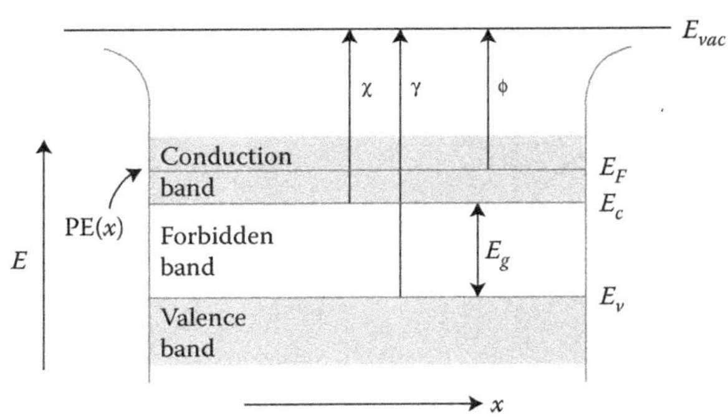

FIGURE 2.11
Definitions of vacuum energy E_{vac}, electron affinity χ, ionization energy γ, and the energy gap E_g.

is the electron affinity. A work function ϕ is defined as the energy that can take an electron from the Fermi level (the highest filled level at 0 K) and the vacuum level (when electron is detached from the crystal or ionized).

For a tetravalent atom, for example, silicon, the four electrons of partially filled $3s^23p^2$ share their electrons with their four nearest neighbors. This sharing of the resources makes each atom feel as if it has a full orbit of electrons. This is the origin of covalent bonds in the language of chemistry and hence a crystal is formed out of silicon atoms. Although strictly not correct as four bonds form an angle in 3D space, a 2D lattice structure shown in Figure 2.12 captures the essence of what is known as a covalent bonding scheme. This happy situation arises only at $T = 0$ K. As the temperature rises, broken bonds will become apparent at a finite T. Each broken bond yields a wandering (itinerant or conducting) electron in the crystal leaving a hole (void) in an otherwise perfect bonding structure. The positively charged hole can be filled by a localized electron, hopping from another site. In this description, it appears as the hole moving from one site to the other site, even if it is a localized electron that is hopping from one site to the other. As these localized electrons (or holes) hop slowly, the holes are known to have a lower mobility.

Two modifications in the energy levels of an electron are necessary. The first is that the electron is now not free. Its motion is affected by the potentials of the periodic atomic structure, facing peaks and troughs as it moves through the crystal. The free electron mass is replaced by the effective mass m^* ($m_o \to m^*$). The other transformation of the medium through which electron must travel replaces permittivity of the free space ε_o by $\varepsilon_o\varepsilon_r$ ($\varepsilon_o \to \varepsilon_o\varepsilon_r$), where ε_r is the relative permittivity (also known as the dielectric constant

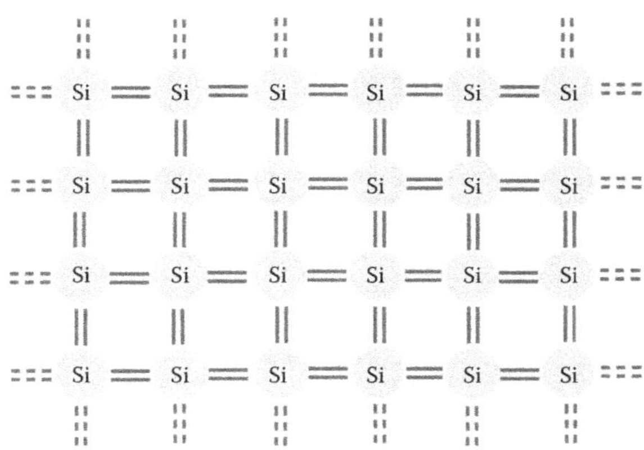

FIGURE 2.12
A 2D model of tetravalent atom forming a crystal with bonding of four of its electrons with the four surrounding nearest neighbors. Note the dangling bonds at the surface that are quenched by an annealing process. (Adapted from B. G. Streetman and S. K. Banerjee, *Solid State Electronic Devices*, 6th ed. Upper Saddle River, NJ: Prentice-Hall, 2006.)

that is normally represented by κ). With this transformation, Equation 2.25 becomes

$$E_n = E_{vac} - \frac{13.6\,\text{eV}}{\varepsilon_r^2}\frac{m^*}{m_o}\frac{Z^2}{n^2} \tag{2.26}$$

This dilution of binding energy in the crystal environment is responsible for freeing electron from the bound states because of the thermal agitation. Here, thermal energy k_BT plays a prominent role whose value at room temperature ($T = 300$ K) is 0.0259 V.

The creation of electrons and holes as a carrier of information is discussed next.

2.6 Thermal Band/Bond Tempering

The otherwise perfect crystal environments of Figure 2.12 and the related band structure of Figure 2.11 exist only at absolute zero ($T = 0$ K). In the intrinsic silicon represented by Figure 2.12, the thermal processes may knock one electron out of perfection, creating an empty state, as shown in the related band diagram of Figure 2.13. This may make an electron itinerant, as an empty state (a hole) that can be filled by one of the localized electron, giving the appearance as if the hole is moving from one atom to the other atom. Because of the sluggish nature of the localized electrons to move, the holes are less mobile than electrons. The creation of e–h pair and subsequent movement of

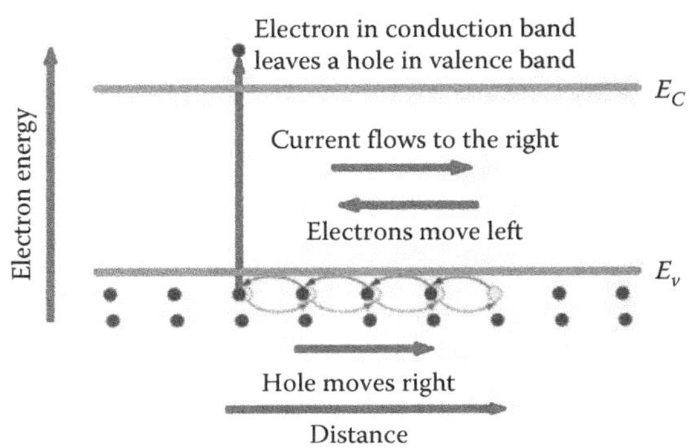

FIGURE 2.13
Creation of an e–h pair as an electron acquires energy equal to that of the bandgap. Movement of many localized electrons in the valence band is equivalent to movement of a single hole. (Adapted from B. G. Streetman and S. K. Banerjee, *Solid State Electronic Devices*, 6th ed. Upper Saddle River, NJ: Prentice-Hall, 2006.)

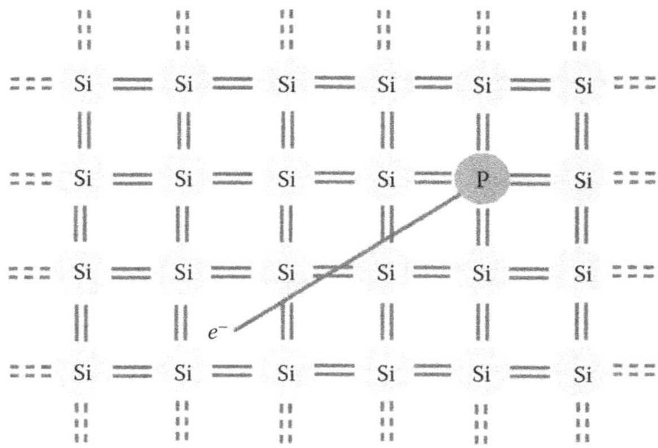

FIGURE 2.14
Creation of an electron as an element from Group V replaces a silicon atom (substitutional impurity).

holes from one localized electron place to the next is depicted in Figure 2.13. The energy of an electron is statistically distributed according to the carrier statistics discussed Chapter 3. Only a small percentage of electrons are able to break away from the valence band in a manner indicated in the diagram of Figure 2.13. In the bonding scheme of Figure 2.13, a broken band is equivalent to an electron being released from the atom as its binding energy weakens in the crystal environment. As an example in a silicon ($m^* = 0.26m_o$, $\varepsilon_r = 11.8$), the binding energy of the electron in $n = 3$ orbit is given by

$$E_3 = E_{vac} - 13.6\text{ eV}\frac{0.26}{(11.8)^2}\frac{1}{3^2} = E_{vac} - 2.82 \times 10^{-3}\text{ eV} = E_{vac} - 2.82\text{ meV} \quad (2.27)$$

E_{vac} may be the conduction band edge E_c or E_v. This description will become clear as one proceeds with tempering of the perfect semiconductor culture further by bringing alien (foreign) impurities into otherwise untempered intrinsic bands, creating impurity levels below the conduction band or above the valence band in the otherwise forbidden bandgap (see Figures 2.14 and 2.15).

2.7 Impurity Band/Bond Tempering

Figures 2.14 and 2.15 are indicative of the creation of an itinerant electron when an element from Group V of the periodic table replaces a silicon atom (substitutional impurity). Because of the low binding energy as evaluated from Equation 2.27 compared to the thermal energy $k_BT = 25.9$ meV, virtually

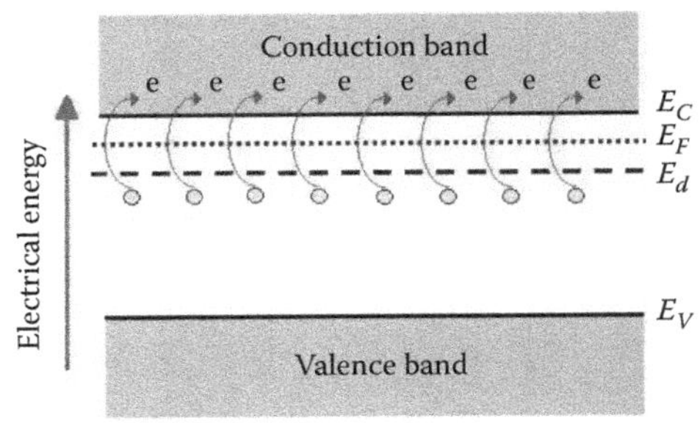

FIGURE 2.15
The donor energy level of the fifth electron in the forbidden bandgap and its thermal excitation to the conduction band.

all electrons from similar substitutional impurities are excited to the conduction band filling the empty states. Electrons in their empty states are hopping from one quantum state to the other. A phosphorus (P) atom replacing a Si forms the substitutional impurity that releases a free electron at room temperature. Phosphorus is a donor atom. It donates a wandering (itinerant) electron to the conduction band as its binding energy and the parent site is miniscule when compared to the thermal energy k_BT.

Figures 2.16 and 2.17 are indicative of the creation of an itinerant hole when an element from Group III of the periodic table replaces a silicon atom (substitutional impurity). The missing electron level can be filled by thermal excitation of the electron from valence band edge to the so-called acceptor level that resides in the forbidden bandgap. As an example, boron (B) is the acceptor atom that accepts electrons from the valence band which become localized, leaving behind the itinerant hole state that is characterized by a hole-effective mass similar to an electron-effective mass.

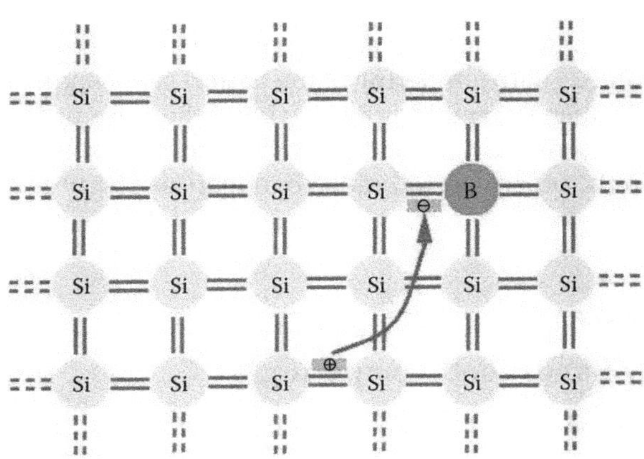

FIGURE 2.16
Creation of a hole as an element from Group III replaces a silicon atom (substitutional impurity).

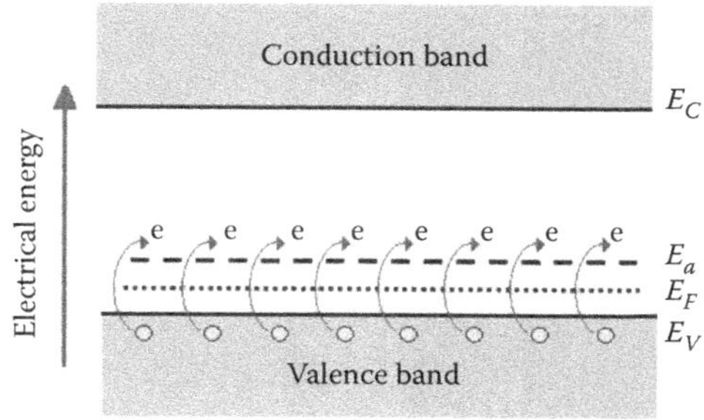

FIGURE 2.17
The acceptor energy level of the missing electron (hole) and its thermal excitation from valence band edge to the acceptor state in the bandgap.

2.8 Compound Semiconductors

Silicon (Si) and germanium (Ge) are elemental semiconductors. Silicon is the king of electronics because of its tremendous cost advantage over other semiconductors. However, it is not a suitable element for optoelectronic applications because of the momentum change as an electron falls from the conduction band to the valence band or vice versa. The relativistic theory of the photon indicates it to be virtually of zero rest mass and almost zero momentum. This makes interband transitions impossible in Si and Ge because momentum is not conserved. Si and Ge are in the class of the so-called indirect bandgap semiconductors where the top of the valence band is at zero momentum in k-space (phase space), but a minimum of the conduction band is at a nonzero value of k. A III–V compound semiconductor, for example, GaAs with Ga from Group III and As from Group V, is a suitable choice for optoelectronic applications. It is a direct bandgap material with no momentum change as an electron transfers from conduction band to valence band, with a high transition probability of photon emission as an electron makes a transition from the conduction to valence band. The bandgap of GaAs is 1.42 eV, so it emits a photon of wavelength $\lambda = 1240\ \text{nm}/1.42 = 873\ \text{nm}$, which is closer to the red end of the electromagnetic spectrum. To obtain photons in the visible range, a number of wide bandgap semiconductors are being engineered for possible applications. GaN and SiC are wide bandgap semiconductors whose bandgap can be varied by alloying. GaAs is the most experimented by alloying it with AlAs giving $Al_xGa_{1-x}As$ with bandgap changing and with the fraction x of aluminum going from $x = 0$ (GaAs) to $x = 1$ (AlAs). This bandgap engineering is very popular among material engineers as they discover new combinations for a variety of applications. A number of compound semiconductors for possible engineering of light-emitting diodes (LEDs) and lasers are shown in Figure 2.18.

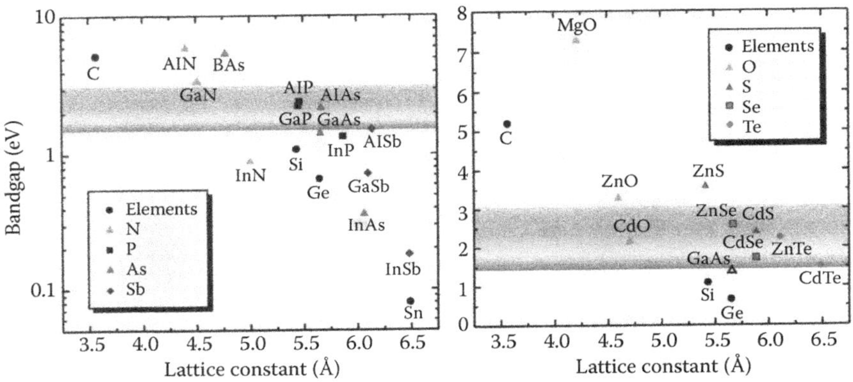

FIGURE 2.18
III–V and II–VI compound semiconductors for applications in optoelectronic applications.

The LED, or laser diodes, used in fiber-optic communication systems is one possible area of bandgap engineering. Figure 2.19 is an example where a fiber can run thousands of kilometers with service huts every 100–200 km. Normally, the decaying signal loss is replenished by splicing the main fiber with erbium-doped fiber inserted by cutting the fiber. This optical amplifier brings the signal to the original state and signal propagates with repeated replenishments.

Figure 2.19 shows an application of engineered optoelectronic devices for signal propagation with a minimum loss. The analog signal in the form of voice, text, or graphics is first encoded into digital format and then to optical pulses through the LED or laser. As an optical pulse travels through the fiber of Figure 2.19 there is naturally a loss in the amplitude of the signal that has

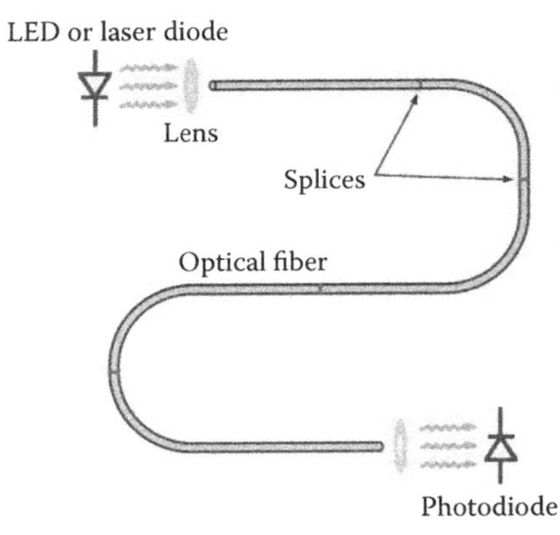

FIGURE 2.19
Typical setup for an optical fiber. (Adapted from B. L. Anderson and R. L. Anderson, *Fundamentals of Semiconductor Devices*. New York, NY: McGraw-Hill, 2005.)

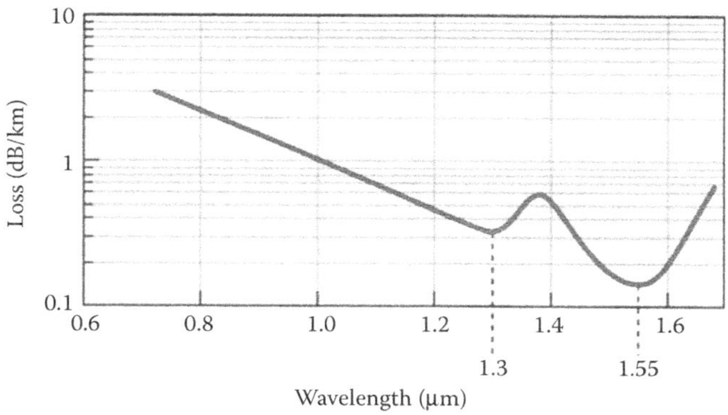

FIGURE 2.20
Low-loss wavelengths for optical fiber transmission. (Adapted from B. L. Anderson and R. L. Anderson, *Fundamentals of Semiconductor Devices*. New York, NY: McGraw-Hill, 2005.)

to be revitalized. Two wavelength windows where optical loss is minimal are identified to be 1.3 μm (1300 nm) and 1.55 μm (1550 nm) in Figure 2.20. The optical design for the bandgap to cater to these windows minimizes the loss in the fiber. Therefore, the bandgap suited for these windows is desired to be 0.80 eV for 1.55 μm wavelength and a slightly larger bandgap for 1.3 μm source.

Figure 2.21 shows the example of how bandgap plays a prominent role in determining the properties of a semiconductor material; whether or not a

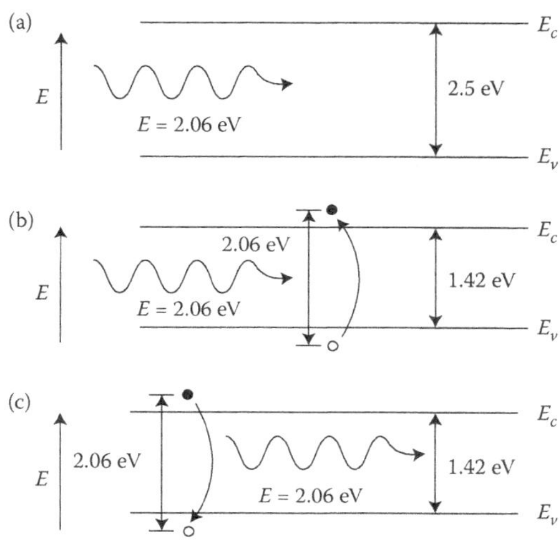

FIGURE 2.21
Absorption and emission of a photon in a direct bandgap semiconductor. (a) Photon of energy 2.06 cannot be absorbed in a bandgap material of 2.5 eV and hence material is transparent for this wavelength. (b) The photon is absorbed; hence bandgap material of 1.42 eV is opaque to photon. (c) Electron recombines with a hole releasing a photon of energy 2.06 eV.

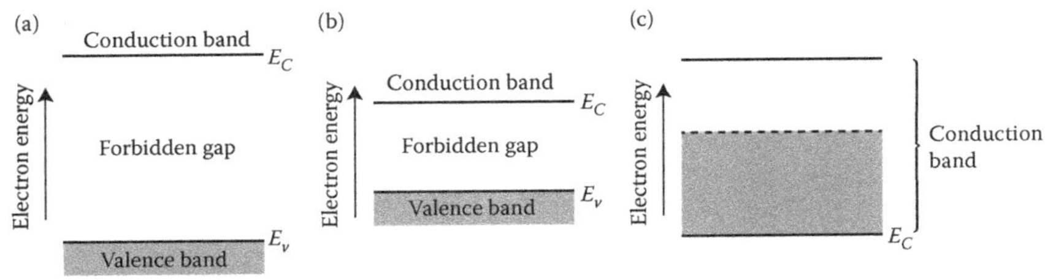

FIGURE 2.22
Energy band diagrams for (a) an insulator, (b) a semiconductor, and (c) a metal. The energies in the shaded regions are in general occupied.

material is transparent and opaque for photons depends on its energy relative to the bandgap. A photon of energy 2.06 eV ($\lambda = 602$ nm) incident on a semiconductor of bandgap $E_g = 2.5$ eV cannot be absorbed and hence the material will be transparent to that photon. However, the same photon can be easily absorbed by a semiconductor of bandgap $E_g = 1.42$ eV by producing an e–h pair with excess going into kinetic energy. Similarly, when an e–h pair combines with an energy difference of 2.06 eV, a photon of 602 nm is emitted.

The bandgap is also used to characterize materials according to their conducting properties. The large gap material (Figure 2.22a) with the bandgap in the range of several eV is an insulator, as no electrons or holes can be easily generated in these materials. SiO_2, an oxide of silicon, is one prominent insulator extensively in use. It is an amorphous glass-like insulator with non-crystalline characteristics. Most semiconductors with a characteristic bandgap shown in Figure 2.22b are of the order of 1–2 eV. For a class of materials known as semimetals (not shown), the bandgap can become negative, so conduction and valence bands overlap. In a metal, the highest filled band is only half-full, so electrons can hop from one quantum state to another. The metals are known to have a very high density of occupied quantum states and therefore offer a very low resistance. Also, since photons cannot be easily absorbed, most metals reflect photons.

2.9 Bands to Quantum Wells

As we move from bands to quantum wells or mesoscopic systems, further engineering of semiconducting thin films is desired. The most tried approach in this case is to build atomic layers on a given substrate, as shown in Figure 2.23 in an arrangement known as molecular beam epitaxy (MBE). Figure 2.24 shows layers of GaAs and AlGaAs alternately grown to create a synthetic crystal known as a superlattice or multiple quantum wells depending on the thickness of the layers.

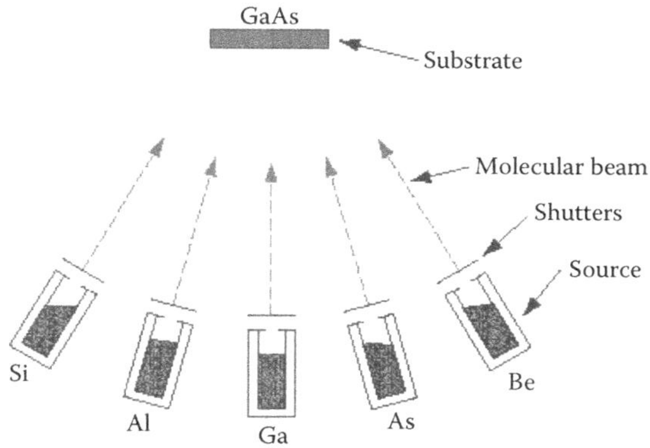

FIGURE 2.23
The computer-controlled cells of an MBE system. (Adapted from B. G. Streetman and S. K. Banerjee, *Solid State Electronic Devices*, 6th ed. Upper Saddle River, NJ: Prentice-Hall, 2006. Courtesy of Bell Laboratories.)

In an MBE, the solid materials from computer-controlled cells are evaporated in the right proportion to grow a layer of an alloy on a substrate. For example, a GaAs layer is alternated with a AlGaAs layer to make a synthetic crystal. An oft-quoted quantum well consists of a GaAs layer sandwiched inbetween two layers of AlGaAs, as shown in Figure 2.25. AlAs has a larger bandgap than that of GaAs. By alloying AlAs with GaAs, it is possible to vary the bandgap depending on the percentage x of AlAs to make a layer of $Al_xGa_{1-x}As$. The fabrication of this alloy is attractive due to an almost perfect match of the lattice constants of the AlAs and GaAs.

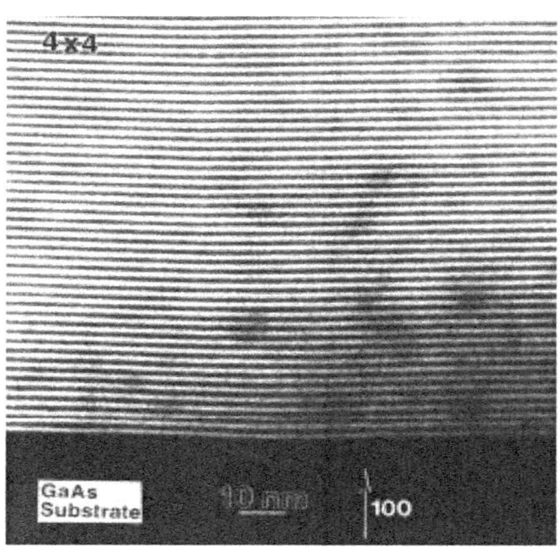

FIGURE 2.24
The example of growth of layers of alternating materials. (Adapted from B. G. Streetman and S. K. Banerjee, *Solid State Electronic Devices*, 6th ed. Upper Saddle River, NJ: Prentice-Hall, 2006. Courtesy of Bell Laboratories.)

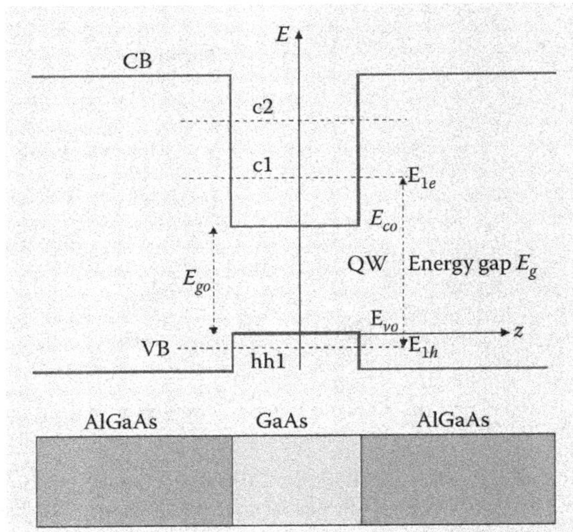

FIGURE 2.25
A prototype $Al_{0.3}Ga_{0.7}As/GaAs/Al_{0.3}Ga_{0.7}As$ quantum well where GaAs (bandgap $E_g = 1.43$ eV) is sandwiched between two wide bandgap layers of $Al_{0.3}Ga_{0.7}As$. Two quantized levels in conduction band and one (heavy hole) is shown in the valence band. (Adapted from B. G. Streetman and S. K. Banerjee, *Solid State Electronic Devices*, 6th ed. Upper Saddle River, NJ: Prentice-Hall, 2006.)

A prototype GaAs quantum well of 5 nm sandwiched between two layers of $Al_{0.3}Ga_{0.7}$ As is shown in Figure 2.25. Two-thirds of the bandgap difference 1.85 eV − 1.43 eV = 0.42 eV goes to the conduction band edge and one-third to the valence band edge. Therefore, quantum confinement of an electron in this quantum well is much stronger than that of a hole (note $E_{1e} - E_{co}$ is much larger than $E_{v0} - E_{1h}$). The effective bandgap E_g is now larger than that in a bulk (3D) semiconductor ($E_g = E_{go} + E_{1e} + E_{1h}$). 2D or 1D materials distinct from 3D will be considered as we proceed further.

The bandgap, electron, and hole effective mass of $Al_xGa_{1-x}As$ can be interpolated from the corresponding parameters of AlAs and GaAs by

$$E_g(Al_xGa_{1-x}As) = 1.426 \text{ eV} + 1.247 \text{ eV}\, x \tag{2.28}$$

$$m_n^*(Al_xGa_{1-x}As) = 0.067m_o + 0.083m_o x \tag{2.29}$$

$$m_{hh}^*(Al_xGa_{1-x}As) = 0.62m_o + 0.14m_o x \tag{2.30}$$

The relative permittivity (dielectric constant) is $\varepsilon_r = 13.18$.

Other MBE systems are dedicated to other alloys, for example, Si and Ge to form strained layer superlattices as Si and Ge lattice constants do not match.

Good superlattices have been produced using the mismatched lattice constants, provided the thickness of a layer is below a critical value. These are known as pseudomorphic layers where strain is accommodated by stretching or compressing the lattice at the interface. These are strained pseudomorphic layers that offer a distinct advantage as straining alters the filling of the bands.

The momentum $p = h/\lambda = \hbar/\lambdabar$, whether it is an electron or a photon, can be expressed in the form of a wavevector $k = 2\pi/\lambda = 1/\lambdabar$ as given by

$$p = \frac{h}{\lambda} = \frac{\hbar}{\lambdabar} = \hbar k \tag{2.31}$$

The momentum of a photon is $k \approx 0$. A photon can be emitted by any transition among energy levels without any change in momentum. This is a crucial point for the materials where conduction band minimum and valence band maximum involve a change in momentum ($k \neq 0$). Such materials with what is called an indirect bandgap must necessarily involve a phonon (a quantum of lattice vibrations).

The photon is a quantum of electromagnetic waves that for a monochromatic source is described by oscillating electric and magnetic fields perpendicular to each other and perpendicular to the direction of propagation. An oscillating electric field, $\mathcal{E}(x, t)$ is expressed as a sinusoidal wave

$$\mathcal{E}(x,t) = A\sin(kx - \omega t) \quad \text{or} \quad A\,e^{j(kx-\omega t)} \tag{2.32}$$

with $\omega = 2\pi f = E/\hbar$ being the angular frequency. A is the amplitude of the wave.

The same pattern is used for the de Broglie waves $\Psi\,(x, t)$ (known as quantum or matter waves)

$$\Psi(x,t) = A\sin(kx - \omega t) \quad \text{or} \quad A\,e^{j(kx-\omega t)} \tag{2.33}$$

The nature of the quantum waves is probabilistic as $|\Psi|^2\,dx = \Psi^*\Psi\,dx$ is the probability density of finding the electron between x and $x + dx$ along the x-axis at time t. Ψ^* is the complex conjugate of Ψ by replacement of j with $-j$. The total probability of finding an electron anywhere between $-\infty$ and $+\infty$ is 1:

$$\int_{-\infty}^{+\infty} |\Psi|^2\,dx = 1 \tag{2.34}$$

This is known as the normalization condition that allows one to calculate A in Equation 2.33.

2.10 A Prototype Quantum Well

The prototype quantum well (Figure 2.25) is often modeled as the one with infinite boundaries because the bandgap discontinuity ΔE_c (the fraction of the bandgap difference in the conduction edges) is much larger than the thermal energy $k_B T$. In fact, a part of the wavefunction does penetrate the regions of AlGaAs. In that case, the effective width L is larger than the physical width a. L, in terms of a, is given by [4]

$$L = \left(1 + \frac{1}{P}\right)a, \quad P = \frac{a}{2}\sqrt{\frac{2m^*\Delta E_c}{\hbar^2}} \tag{2.35}$$

where P is the strength of the quantum well. P defines the extent to which the wavefunction penetrates into the classically forbidden region, $P \to \infty$ being the ideal.

The compound wavefunction $\Psi(x,t) = \psi(x)T(t) = \psi(x)e^{-jEt/\hbar}$ is separable into $\psi(x)$, the function of position only, and $T(t) = e^{-jEt/\hbar}$, the function of time only. The quantum well given in Figure 2.25 is represented as one with infinite boundaries and in Figure 2.26 with boundary condition

$$\psi(0) = \psi(L) = 0 \tag{2.36}$$

This boundary condition clearly indicates that the wavefunction vanishing at the boundary means a particle is confined in the quantum well and its probability of existing outside the walls of the quantum well is zero. The vanishing wavefunction at the boundaries indicates the nonexistence (probability zero) of electrons in the forbidden region ($x < 0$ and $x > L$).

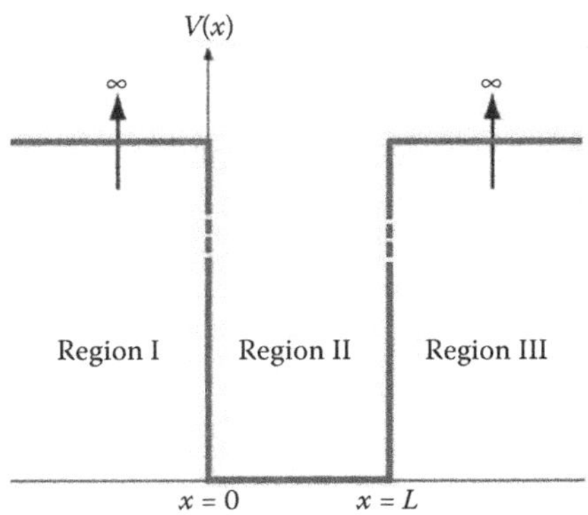

FIGURE 2.26
The prototype quantum well with the infinite boundaries.

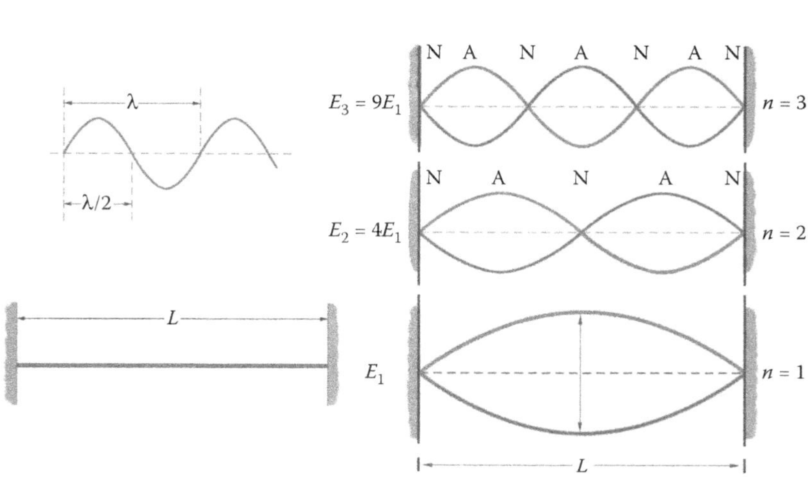

FIGURE 2.27
The wave function vanishing at the quantum well creating standing wave pattern as waves are reflected back from the edge of the quantum well.

As depicted in Figure 2.27, the vanishing quantum waves at the edge of the quantum well requires that only an integer number of half waves can be accommodated in the well of length L giving

$$L = n\frac{\lambda_D}{2} \Rightarrow \lambda_D = \frac{2L}{n}, \quad n = 1, 2, 3, \ldots \tag{2.37}$$

When combined with $\lambda_D = h/p$, the quantized values of the momentum are obtained as

$$p_n = \frac{h}{\lambda_D} = \frac{nh}{2L} \Rightarrow m^* v_n = \frac{nh}{2L} \tag{2.38}$$

The kinetic energy of the electron is then quantized as

$$E_n = \frac{1}{2}m^* v_n^2 = \frac{p_n^2}{2m^*} = \frac{1}{2m^*}\left(\frac{nh}{2L}\right)^2 = n^2\frac{h^2}{8mL^2} = n^2\frac{\pi^2\hbar^2}{2mL^2} = n^2 E_1 \tag{2.39}$$

with

$$E_1 = \frac{\pi^2\hbar^2}{2mL^2} \tag{2.40}$$

E_1 is the ground-state energy that lifts the conduction band edge of the bulk semiconductor so $E_c = E_{co}(\text{bulk}) + E_1$.

To appreciate the wave character of an electron, it is necessary to solve the time-independent Schrodinger equation for potential energy U given by

$$\frac{d^2\psi}{dz^2} + \frac{2m^*}{\hbar^2}(E - U)\psi = 0 \tag{2.41}$$

Assuming $U = 0$ in Region II and $U = \infty$ in Regions I and III of Figure 2.26, Equation 2.41 simplifies to

$$\frac{d^2\psi}{dz^2} + \frac{2m^*E}{\hbar^2}\psi = 0 \Rightarrow \frac{d^2\psi}{dz^2} + k^2\psi = 0, \quad E = \frac{\hbar^2k^2}{2m^*} \tag{2.42}$$

with the boundary condition $\psi(0) = \psi(L) = 0$. The nontrivial solution of Equation 2.42 is simple:

$$\psi(z) = A\sin(kz), \quad k = \frac{n\pi}{L} \tag{2.43}$$

The quantum condition in Equation 2.43 when substituted in the energy $E = \hbar^2k^2/2m^*$ expression of Equation 2.42 gives the quantized energy as expressed in Equation 2.39. The amplitude A of Equation 2.39 is obtained from the normalization condition noting that the wave function vanishes in Regions I and III:

$$\int_0^L \psi^*\psi \, dz = 1 \Rightarrow A = \sqrt{\frac{2}{L}} \tag{2.44}$$

The normalized wavefunction (also known as eigenfunction) and quantized energy values (also known as eigenvalues) are then given by

$$\psi(z) = \sqrt{\frac{2}{L}}\sin\left(\frac{n\pi z}{L}\right), \quad E_n = n^2\frac{\pi^2\hbar^2}{2m^*L^2} = n^2E_1 \tag{2.45}$$

Figure 2.28 gives the description of the energy levels in (a), the quantized wavefunction in (b), and probability density of the electrons in (c). The energy spacing increases as energy moves higher. On the other hand, the levels tend to come closer together in a hydrogen atom. The spectrum indicated is thus an inverted spectrum of that of the hydrogen level. This digitized nature of the energies as compared to analog (continuous) type gives the distinct flavor and makes devices low-dimensional as these are scaled down to the nanometer regime.

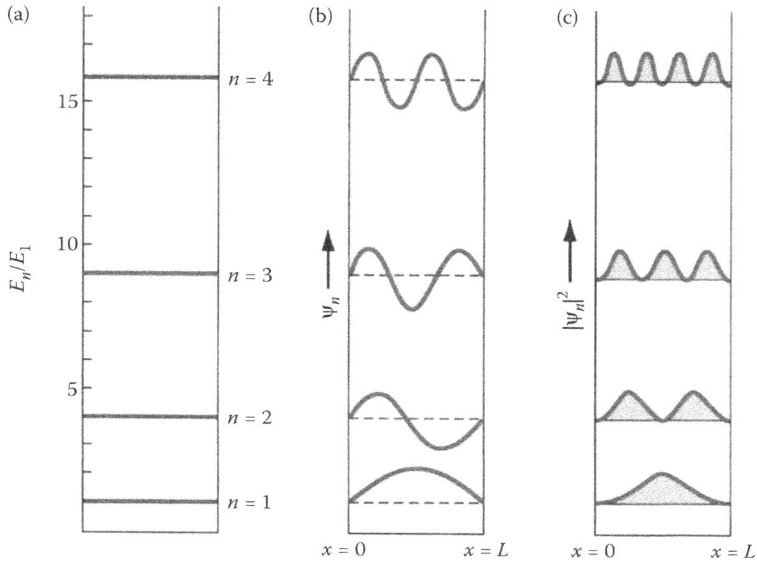

FIGURE 2.28
(a) The energy levels in the quantum well of Figure 2.26, (b) the wave function corresponding to the energy levels of (a), and (c) the probability wave distribution with area under each curve equal to 1.

2.11 3D (Bulk) Density of States

The world in which we reside is a 3D one as described by three Cartesian *x*-, *y*-, and *z*-directions orthogonal to each other. The digitized energy levels will be closely spaced and can be approximated as continuous analog states if the sample size exceeds the de Brogile wavelength in all directions. In this scenario, with E_{co} the potential energy of the quantized bands, the energy is analog in all three directions:

$$E_k^{c(v)} = E_{c(v)o} \pm \frac{\hbar^2(k_x^2 + k_y^2 + k_z^2)}{2m_{n(p)}^*}, \quad -\infty < k_{x,y,z} < +\infty \tag{2.46}$$

where E_{co} or E_{vo} is the conduction (valence) band edge unaffected by quantization in 3D bulk material. There is no restriction on *x*-, *y*-, or *z*-component $k_{x,y,z}$. These analog variables can have any value with no restrictions on their magnitudes being integer. Since the energy levels are closely spaced, the concept of the energy DOS comes in handy here. As shown in Equation 2.43, $k_{x,y,z} = n_{x,y,z}\pi/L_{x,y,z}$ are the digitized momenta in all directions when the confinement in all three directions is considered. The energy is the same for $\pm k_{x,y,z}$ as it depends on the square of $k_{x,y,z}$. The quantum condition of Equation 2.43 considers only positive $+k_{x,y,z}$ because of the quantum box from $x = 0$ to L (positive half-space). The change in momentum is $\Delta k_{x,y,z} = 2\Delta n_{x,y,z}\pi/L_{x,y,z}$ with a factor of 2 added when both positive and negative $k_{x,y,z}$ are considered. In other words, the number of quantum states contained in a continuous spectrum of $\Delta k_x \Delta k_y \Delta k_z$ are

$$\Delta n_x \Delta n_y \Delta n_z = \frac{L_x L_y L_z}{(2\pi)^3} \Delta k_x \Delta k_y \Delta k_z \tag{2.47}$$

Since energy $E - E_{co} = \hbar^2 k^2 / 2m_n^*$ depends only on the magnitude of k with $k^2 = k_x^2 + k_y^2 + k_z^2$, the use of polar coordinates is appropriate. In a polar coordinate system, $\Delta k_x \Delta k_y \Delta k_z = k^2 dk \sin\theta d\theta d\phi$. Integrating over $\theta = 0 \rightarrow \pi$ and $\phi = 0 \rightarrow 2\pi$ covers the complete energy sphere with radius k. In k-space, this angular integration yields

$$\frac{\Delta n_x \Delta n_y \Delta n_z}{L_x L_y L_z} = 2\frac{1}{(2\pi)^3} k^2\, dk \int_0^{\pi} \sin\theta\, d\theta \int_0^{2\pi} d\phi = 2\frac{4\pi}{(2\pi)^3} k^2\, dk \tag{2.48}$$

where factor 2 on the left-hand side accounts for two spin states as each state is occupied by two electrons (spin up and spin down). When $k = (2m_n^*/\hbar^2)^{1/2}(E - E_{co})^{1/2}$ is used $k^2\, dk = (2m^*/\hbar^2)^{3/2} 1/2 (E - E_{co})^{1/2}\, dE$. The DOS, $D(E)$ (per unit volume per unit energy), is given by

$$D_{3e}(E) \equiv \frac{\Delta n_x \Delta n_y \Delta n_z}{L_x L_y L_z\, dE} = 2\frac{4\pi}{(2\pi)^3}\left(\frac{2m_n^*}{\hbar^2}\right)^{3/2} \frac{1}{2}(E - E_{co})^{1/2}$$

$$= \frac{1}{2\pi^2}\left(\frac{2m_n^*}{\hbar^2}\right)^{3/2} (E - E_{co})^{1/2} \tag{2.49}$$

The DOS for the valence band can be similarly derived

$$D_{3h}(E) = \frac{1}{2\pi^2}\left(\frac{2m_p^*}{\hbar^2}\right)^{3/2} (E_{vo} - E)^{1/2} \tag{2.50}$$

$D_{3e}(E)$ for the free electron mass ($m_e^* = m_o$) is given by

$$D_3(E) = 1.06 \times 10^{56}\ \text{J}^{-3/2}\ \text{m}^{-3}(E_{Joules} - E_{co})^{1/2} \tag{2.51}$$

In terms of conventional units of eV for energy and m^{-3} for volume, Equation 2.51 transforms for a semiconductor with an effective mass m^* to

$$D_3(E) = 6.8 \times 10^{27}\ \text{eV}^{-3/2}\ \text{m}^{-3}\left(\frac{m^*}{m_o}\right)^{3/2}\left(E_{\text{eV}} - E_{co,\text{eV}}\right)^{1/2} \tag{2.52}$$

This derivation demonstrates well how an analog 3D world is derived from the digital world where spacing between the energy levels is very small because of the large size in each of the three dimensions. The quantum energy levels are packed closely together and their density per unit energy per unit volume is what appears in Equation 2.49.

2.12 2D Quantum Well

As one direction (say z-direction), for example, is squeezed so that $L_z < \lambda_D$ while the other two $L_{x,y} \gg \lambda_D$ maintain their bulky character, a quasi-two-dimensional (Q2D) quantum well emerges. In this Q2D quantum well, analog-type levels appear only in two dimensions, while a third goes quantum:

$$E_{nk} = E_{co} + \frac{\hbar^2 (k_x^2 + k_y^2)}{2\, m_n^*} + n_z^2 E_{1ez} = E_c + \frac{\hbar^2 (k_x^2 + k_y^2)}{2\, m_n^*} \tag{2.53}$$

with

$$E_c = E_{co} + n_z^2\, E_{1ez},\quad n_z = 1, 2, 3, \ldots \tag{2.54}$$

In the extreme case ($n_z = 1$), the spacing between the levels in the z-direction is so large that it is impossible for states with larger quantum number to be populated. The conduction band is lifted by the quantum energy E_1 to give $E_c = E_{co} + E_{1ez}$. Q2D layer becomes truly 2D in this extreme. The conduction band is lifted by zero-point energy E_{1ez} (minimum energy in a quantum wave). Similarly, the valence band $E_v = E_{vo} - E_{1hz}$ drops by zero-point energy E_{1hz} of holes. The effective bandgap now increases to $E_g = E_{go} + E_{1ez} + E_{1hz}$.

Following the same pattern as before (with a spin 2 factor) and noting that $\Delta k_x \Delta k_y = k\, dk\, d\theta$ in circular coordinates with $\theta = 0 \rightarrow 2\pi$ (going full circle), the DOS $D_{2e(h)}$ for electrons e (or for holes h) is given by

$$D_{2e(h)}(E) \equiv \frac{\Delta n_x \Delta n_y}{L_x L_y\, dE} = \frac{2\pi}{(2\pi)^2}\left(\frac{2m_{n(p)}^*}{\hbar^2}\right) = \frac{m_{n(p)}^*}{\pi\hbar^2} \tag{2.55}$$

DOS in this case are the number of quantum states per unit area per unit energy. These can be numerically shown to lead to

$$D_{2e(h)}(E) = \frac{m^*_{n(p)}}{m_o} \times 2.6 \times 10^{37}\ \mathrm{J^{-1}m^{-2}} = \frac{m^*_{n(p)}}{m_o} \times 4.17 \times 10^{18}\ \mathrm{eV^{-1}m^{-2}} \tag{2.56}$$

2.13 1D Quantum Well

As two (say, y- and z-directions) are squeezed so that L_y and $L_z < \lambda_D$, while the remaining one direction $L_x \gg \lambda_D$ maintains its analog character, a quasi-one-dimensional (Q1D) quantum well or nanowire emerges. In this Q1D quantum well, analog-type levels appear only in one dimension, while the other two go quantum:

$$E_{nk} = E_{co} + \frac{\hbar^2 k_x^2}{2m_n^*} + n_y^2 E_{1ey} + n_z^2 E_{1ez} = E_c + \frac{\hbar^2 k_x^2}{2\, m_n^*}, \quad n_{y,z} = 1, 2, 3, \ldots \tag{2.57}$$

with

$$E_c = E_{co} + n_y^2 E_{1ey} + n_z^2 E_{1ez} \tag{2.58}$$

In the extreme case ($n_{y,z} = 1$), the spacing between the levels in the y- and z-directions is so large that it is impossible for higher states to be populated. The lifted conduction band edge is now given by $E_c = E_{co} + E_{1ey} + E_{1ez}$. In this extreme, the Q1D layer becomes truly 1D (a nanowire). The conduction band is lifted by what is known as zero-point energy of electrons $E_{1ey} + E_{1ez}$. Similarly, the valence band $E_v = E_{vo} - E_{1hy} - E_{1hz}$ drops by zero-point energy of the holes $E_{1hy} + E_{1hz}$. The effective bandgap now increases to $E_g = E_{go} + E_{1ey} + E_{1hy} + E_{1ez} + E_{1hz}$.

Following the same pattern as before (with spin 2 factor) and noting that $\Delta k_x = 2k_x$ (positive and negative directions), the DOS $D_{1e(h)}(E)$ for electrons and for holes is given by

$$D_{1e}(E) \equiv \frac{\Delta n_x}{L_x\, dE} = 2\frac{2}{(2\pi)}\left(\frac{2m_n^*}{\hbar^2}\right)^{1/2}\frac{1}{2}(E - E_c)^{-1/2} = \frac{1}{\pi}\left(\frac{2m_n^*}{\hbar^2}\right)^{1/2}(E - E_c)^{-1/2} \tag{2.59}$$

$$D_{1h}(E) \equiv \frac{\Delta n_x}{L_x\, dE} = 2\frac{2}{(2\pi)}\left(\frac{2m_p^*}{\hbar^2}\right)^{1/2}\frac{1}{2}(E_h - E)^{-1/2} = \frac{1}{\pi}\left(\frac{2m_p^*}{\hbar^2}\right)^{1/2}(E_h - E)^{-1/2} \tag{2.60}$$

The factor of 2 is due to spin and the other factor of 2 is for $\pm k_x$. Numerically, it can be put in the form

$$D_1(E) = 4.07 \times 10^{18}\ \mathrm{J}^{-1/2}\ \mathrm{m}^{-1}\left(\frac{m^*}{m_o}\right)^{1/2}(E_J - E_c)^{-1/2}$$

$$= 1.63 \times 10^{9}\left(\frac{m^*}{m_o}\right)^{1/2}\mathrm{eV}^{-1/2}\ \mathrm{m}^{-1}(E_{\mathrm{eV}} - E_c)^{-1/2} \quad (2.61)$$

2.14 Quantum Dots: QOD Systems

In a quantum dot, there are no densely packed quantum levels. The DOS concept is not valid. As all three Cartesian directions are squeezed so that L_x, L_y, and $L_z < \lambda_D$, a quantum dot or box emerges without any analog-like energy levels. The quantized energy is given by

$$E_{n_x,n_y,n_z} = E_{co} + n_z^2\,E_{1ez} + n_y^2\,E_{1ey} + n_x^2\,E_{1ex}, \quad n_{x,y,z} = 1, 2, 3, \ldots \quad (2.62)$$

For a 10-nm GaAs cube (dot), the energy spacing between the lowest and next higher level is $\Delta E = E_{211} - E_{111} = 0.168$ eV. For a spherical dot of radius R, the wavefunction is the spherical Bessel function. Hence, the process of getting energy is a bit involved. However, in its approximate form, it can be obtained as

$$E_n = n^2\frac{\pi^2\hbar^2}{2m^*R^2} = n^2E_1 \quad (2.63)$$

For an $R = 6.2$ nm radius sphere, which has the same volume as 10-nm cube, $\Delta E = E_2 - E_1 = 0.438$ eV for GaAs. $R = 0.62\,L$ when the volume of a sphere $(4/3)\pi R^3 = L^3$ is equated to volume of a quantum dot (cube) of length L. On the other hand, R = 10 nm can be equated to $L = 16.1$ nm for which $\Delta E = E_2 - E_1 = 0.169$ eV.

For a particle bound in a spherical well with infinite boundaries, the energy levels are given by

$$E_n = \alpha_{n,\ell}^2\frac{\hbar^2}{2m^*R^2} = n^2E_1 \quad (2.64)$$

where $\alpha_{1,(0-4)} = 3.142$, 4.493, 5.763, 6.988, and 8.183 and $\alpha_{2,(0-4)} = 6.283$, 7.725, 9.095, 10.417, and 11.705. $\alpha_{1,0} \approx \pi$. The wave functions are $\psi_{n,\ell,m}(r, \theta, \phi) = Aj_n(\alpha_{n,\ell}r)Y_\ell^m(\theta, \phi)$ where $j_n(\alpha_{n,\ell}r)$ is the Bessel function with $j_n(\alpha_{n,\ell}R) = 0$ and $Y_\ell^m(\theta, \phi)$ as spherical harmonics for spherical symmetric system.

The comparison of DOS in 3D, 2D, and 1D is for GaAs ($m^*/m_o = 0.067$) and for a quantum well with $L_z = 10$ nm (2D) and $L_y = L_z = 10$ nm (1D). The curves are normalized, so DOS is in common units $eV^{-1} nm^{-3}$.

2.15 Generalized DOS

DOS can also be calculated by considering that the probability density is high where energy levels exist. In that scenario,

$$D(E) = \frac{1}{L_x L_y L_z} \sum_{\alpha} \delta(E - E_\alpha) \tag{2.65}$$

where α is the set of quantum numbers depending on the dimensionality. For bulk (3D), Q2D, and Q1D nanostructures, the DOS are compared in Figure 2.29. The DOS using Equation 2.65 is calculated in Appendix 2A.

It is possible to obtain a general formula for the DOS that applies to 3D, 2D, and 1D (with dimensionality $d = 3$, 2, and 1). As indicated above, the DOS is a function of energy only. Later on, we will define effective density of states (EDOS), a function of temperature. The EDOS for dimensionality $d = 3, 2, 1$ is given by

$$N_{cd} = 2\left(\frac{m^* k_B T}{2\pi\hbar^2}\right)^{d/2} \tag{2.66}$$

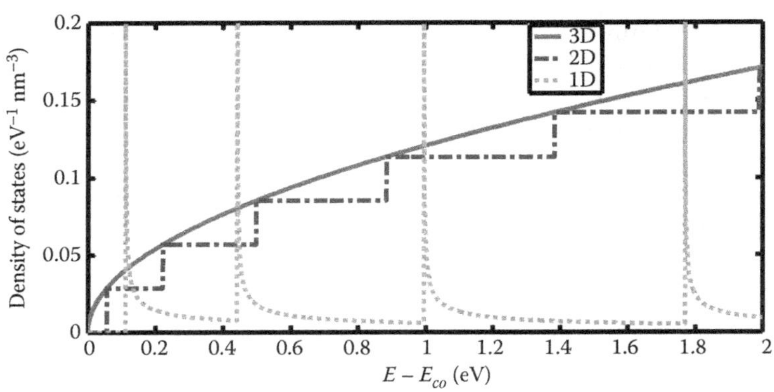

FIGURE 2.29
Comparison of the DOS for 3D, 2D, and 1D for GaAs ($m^*/m_o = 0.067$) and for a quantum well with $L_z = 10$ nm (2D) and $L_y = L_z = 10$ nm (1D).

Differential DOS can be expressed in terms of N_{cd} if the normalized energy x in terms of thermal energy k_BT is defined as follows:

$$x = \frac{E - E_c}{k_B T} \tag{2.67}$$

$D_d(E)\,dE$ then transforms to be

$$D_d(x)\,dx = N_{cd}\frac{1}{\Gamma(d/2)}x^{(d-2)/2}\,dx \tag{2.68}$$

Here, $\Gamma(d/2)$ is a Gamma function whose value for $d = 3$ is $\sqrt{\pi}/2$, for $d = 2$ is 1, and for $d = 1$ is $\sqrt{\pi}$. In terms of $D_d(E)\,dE$, it is given by

$$D_d(E)\,dE = \left(\frac{2m^*}{\hbar^2}\right)^{d/2}\frac{1}{2^{d-1}\pi^{d/2}\Gamma(d/2)}(E - E_c)^{(d-2)/2}\,dE \tag{2.69}$$

The DOS, when higher states in 2D and 1D are also occupied, is given by

$$D_2(E)\,dE = \left(\frac{m^*}{\pi\hbar^2}\right)\mathrm{Int}\left[\sqrt{\frac{E}{\varepsilon_{oz}}}\right]dE \tag{2.70}$$

$$D_1(E)\,dE = \frac{1}{\pi}\left(\frac{2m^*}{\hbar^2}\right)^{1/2}(E - E_{cn_zn_y})^{-1/2}S(E - E_{cn_zny})dE \tag{2.71}$$

Int[.] stands for integer value of the argument and $S(.)$ stands for step function which indicates the expression in the square root is always positive. $E_{cn_zny} = E_{co} + n^2\varepsilon_{oz} + m^2\varepsilon_{oy}$ is the lifted conduction band in a 1D nanowire.

The probability of occupation of a quantum state is the Fermi–Dirac distribution given by

$$f(E) = \frac{1}{e^{(E-E_F)/k_BT} + 1} \tag{2.72}$$

This distribution function for electrons (holes) is discussed in Chapter 3. Here, the Fermi energy E_F is the energy at which probability of occupation is one-half. The differential density of carriers $dn_d = f(x)D_d(x)dx$ for dimensionality d with $f(x) = 1/[\exp(x - \eta_d) + 1]$ obtained from Equation 2.72 with $\eta_d = (E_F - E_c)_d/k_BT$ when integrated from $x = 0(E = E_c)$ $x = \infty$ to $(E = \infty)$ results in

$$n_d = N_{cd}\Im_{(d-2)/2}(\eta_d) \tag{2.73}$$

where $\Im_j(\eta)$ is a Fermi–Dirac integral (FDI) whose properties are discussed in Appendix F and its value tabulated in Appendix G. $\eta_d = (E_F - E_c)_d/k_BT$ is the normalized Fermi energy with respect to the conduction (or valence) band and is a key factor in deciding the degeneracy of the material that itself depends on the carrier density with respect to the EDOS. The details of carier statistics are discussed in Chapter 3.

2.16 Ellipsoidal Conduction Band Valleys

The band structure of real semiconductors is much more complex than the one stated above with the effective mass the same in all three directions. For an isotropic effective mass, the E–k relationship is parabolic:

$$E_k^{c(v)} = E_{c(v)} \pm \frac{\hbar^2 k^2}{2m^*_{n(p)}} \tag{2.74}$$

with

$$k^2 = k_x^2 + k_y^2 + k_z^2 \tag{2.75}$$

The energy surface when plotted as a function of k (or k_x, k_y, and k_z) is a parabola. For a free space parabola shown in Figure 2.30 with the apex at energy E_c, the kinetic energy $E_k - E_c$ is a quadratic function of k. All the discussion stated above centered on this isotropic effective mass. However, the minimum (or apex) of the parabola is shifted to $\pm k_{ox,y,z}$ in the band structure

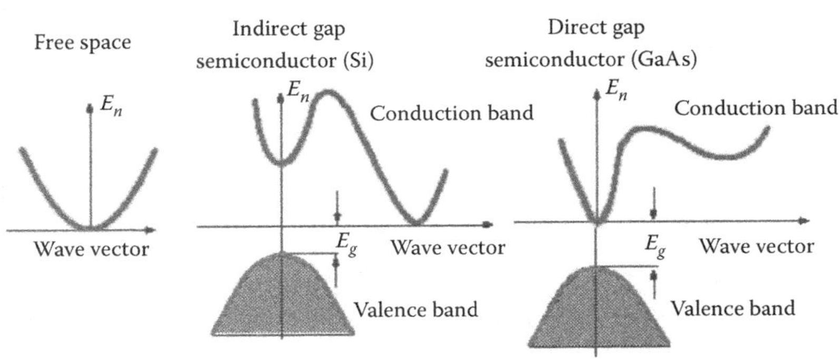

FIGURE 2.30
Description of parabolic energy band in free space centered at $k = 0$ (left), silicon conduction band with minimum at $k \neq 0$ and valence band at $k = 0$, and GaAs conduction and valence bands both at $k = 0$.

of silicon. The inverted parabola of the valence band is at $k = 0$. This misalignment of the conduction and valence band minimum makes silicon less suited for optical transitions as any such transition will necessarily involve a shift in the momentum. The photon with momentum $k = 0$ is highly unlikely to be emitted. However, such transitions do involve phonons of the lattice vibrations allowing energy to be emitted or absorbed by a phonon. GaAs offers a distinct advantage as the minimum of the conduction band and maximum of the valence band are aligned at $k = 0$, enhancing the probability of photon emission or absorption.

When $\in = E_k - E_c$ is plotted in k-space with axes along $k_{x,y,z}$, the constant energy surface looks spherical. However, for silicon, as Figure 2.31 shows, each of the six valleys is ellipsoidal. For a valley with different effective masses m_x, m_y, and m_z in three directions, the kinetic energy $\in$ is given as

$$\in_k = \frac{\hbar^2 k_x^2}{2m_x^*} + \frac{\hbar^2 k_y^2}{2m_y^*} + \frac{\hbar^2 k_z^2}{2m_z^*} \tag{2.76}$$

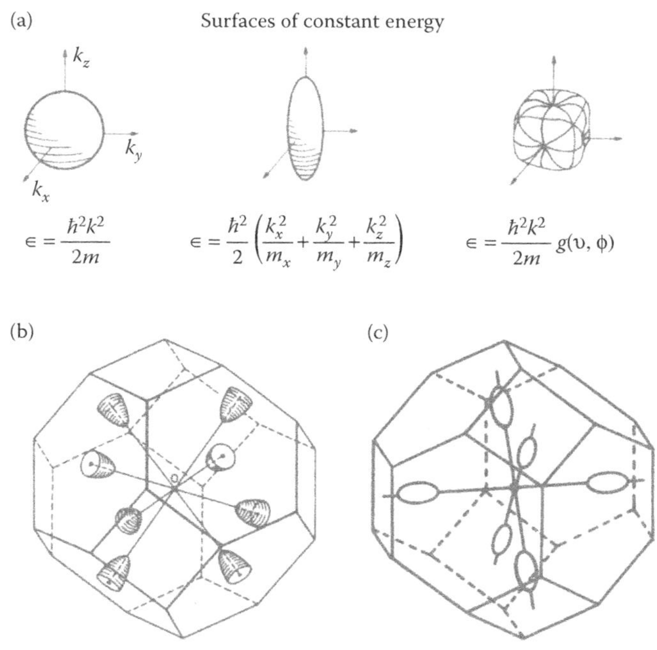

FIGURE 2.31
(a) The constant energy $\in = E_k - E_c$ surfaces for spherical (isotropic effective mass), ellipsoidal (anisotropic effective mass), and warped shapes. (b) Multivalley band structure of germanium and silicon in a Brillouin zone with eight ellipsoidal half-valleys (four full valleys) centered along the eight diagonals Ge. (c) Six silicon ellipsoidal valleys centered at $\pm k_{ox,y,z}$ in the Brillouin zone.

The valley is an ellipsoid of revolution with an elongated major axis in the $\pm z$-direction and minor axes in the x- and y-directions. $m^*_{x,y} = m^*_t$ is the transverse effective mass that is the same in the x- and y-directions. In the z-direction, the mass is longitudinal $m^*_z = m^*_\ell$. The energy surface of Equation 2.76 can be made spherical if momentum components $k_{x,y,z}$ are scaled to define $K_{x,y,z} = k_{x,y,z}/(m^*_{x,y,z}/m_o)^{1/2}$:

$$\epsilon_{K(k)} = \frac{\hbar^2 K_x^2}{2m_o} + \frac{\hbar^2 K_y^2}{2m_o} + \frac{\hbar^2 K_z^2}{2m_o} = \frac{\hbar^2 K^2}{2m_o} \tag{2.77}$$

Equation 2.47 in K-space transforms to

$$\Delta n_x \Delta n_y \Delta n_z = \frac{L_x L_y L_z}{(2\pi)^3}\left(\frac{m_x^* m_y^* m_z^*}{m_o^3}\right)^{1/2} \Delta K_x \Delta K_y \Delta K_z \tag{2.78}$$

With this transformation, Equation 2.78 with analogy of earlier derivation becomes

$$\frac{\Delta n_x \Delta n_y \Delta n_z}{L_x L_y L_z} = 2\frac{1}{(2\pi)^3}\left(\frac{m_x^* m_y^* m_z^*}{m_o^3}\right)^{1/2} K^2\, dk \int_0^{\pi} \sin\theta\, d\theta \int_0^{2\pi} d\phi \tag{2.79}$$

As integration is carried out, the DOS becomes

$$D_{3e}(E) = \frac{\Delta n_x \Delta n_y \Delta n_z}{L_x L_y L_z} = \frac{4\pi}{(2\pi)^3}\left(\frac{m_x^* m_y^* m_z^*}{m_o^3}\right)^{1/2} K^2\, dK \tag{2.80}$$

With the conversion of K to energy $\epsilon = E_k - E_c$, the DOS is given by

$$D_{3e}(E) = \frac{1}{2\pi^2}\left(\frac{m_x^* m_y^* m_z^*}{m_o^3}\right)^{1/2}\left(\frac{2m_o}{\hbar^2}\right)^{3/2}(E - E_{co})^{1/2}\, dE \tag{2.81}$$

When multiplied by the number of valleys g_v ($g_v = 6$ for Si), Equation 2.79 gives for the DOS the expression

$$D_{3e}(E) = \frac{1}{2\pi^2}\left(\frac{2m^*_{dse}}{\hbar^2}\right)^{3/2}(E - E_{co})^{1/2} \tag{2.82}$$

with

$$m_{dsn}^{*3/2} = g_v\left(m_x^* m_y^* m_z^*\right)^{1/2} \Rightarrow m^*_{dsn} = g_v^{2/3}\left(m_t^{*2} m_\ell^*\right)^{1/3} \tag{2.83}$$

The experimentally obtained values of longitudinal and transverse effective mass for silicon is $m_\ell^* = 0.92m_o$ and $m_t^* = 0.198m_o$. With these values, the DOS effective mass is

$$m_{dsn}^* = 6^{2/3}(0.198^2 \times 0.92)^{1/3} m_o = 1.08\, m_o \tag{2.84}$$

When an electric field is applied in the z-direction, two of the valleys in silicon will have longitudinal axis in the direction of the electric field, while the longitudinal axes will be perpendicular to the valleys for the other four valleys. Therefore, the conduction velocity is $v_z = \hbar k_z/m_\ell^*$ for the two longitudinal valleys and $v_z = \hbar k_z/m_t^*$ for the other four transverse valleys. The average for all six valleys then gives the conductivity effective mass

$$\frac{1}{m_{ce}^*} = \frac{1}{6}\left(\frac{2}{m_\ell^*} + \frac{4}{m_t^*}\right) = \frac{1}{6}\left(\frac{2}{0.98} + \frac{4}{0.19}\right)\frac{1}{m_o} = \frac{3.75}{m_o} \Rightarrow m_{ce}^* = 0.26m_o \tag{2.85}$$

The DOS for Q2D and Q1D systems will also transform. For example, in a MOSFET (a Q2D nanostructure) with the electric field applied in the (100) crystalline direction, only two valleys with transverse direction in the direction of the electric field are appreciably populated. In that case, $D_v = 2$ and both the DOS and conductivity effective masses are m_t^*. Similar exercise can be considered for the nanowire carrier statistics.

2.17 Heavy/Light Holes

There are three parabolic valence bands as shown in Figure 2.32. Following the scheme in the DOS for the highest subbands, the DOS effective mass for holes is given by

$$m_{dsp}^* = \left[m_{hh}^{*3/2} + m_{\ell h}^{*3/2}\right]^{2/3} \tag{2.86}$$

In Equation 2.86, the contribution to effective mass of a split-off band is neglected. The split-off band has a parabolic band mass m_{so}^* and can be taken into account in the EDOS when carrier statistics are considered in Chapter 3. In silicon, the split-off band is only $\Delta = 0.044$ eV and hence may influence the occupation of the DOS. It is appropriate, in that case, to calculate the DOS effective mass for carrier statistics as follows:

$$m_{dsh}^* = \left[m_{hh}^{*3/2} + m_{\ell h}^{*3/2} + m_{so}^{*3/2}\, e^{-\Delta/k_BT}\right]^{2/3} \tag{2.87}$$

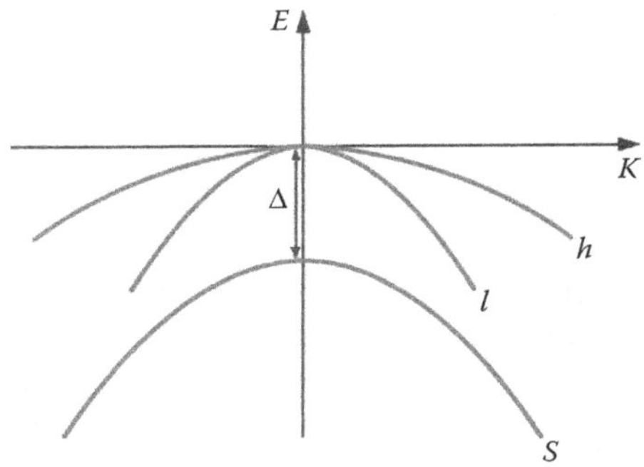

FIGURE 2.32
Light-hole, heavy-hole, and split-off valence bands.

The valence band with quantum effects considered in a Q2D quantum well will be populated only by the heavy holes as a light-hole valley moves further away from the conduction band edge and hence not appreciably populated. Therefore, the conductivity and DOS effective mass will be m_{hh}^*.

2.18 Electrons in a Magnetic Field

As nanowires are being explored for a variety of applications, it is good to have a limited study of electrons in a magnetic field as this configuration offers many similarities to nanowires and other Q1D system being explored.

An electron traveling at an angle to a magnetic field makes a spiral motion with the axis of the spiral in the direction of the magnetic field, as shown in Figure 2.33. Only the velocity component perpendicular to the magnetic field is affected by the magnetic field $\vec{B}$. The component parallel to the electric field is unaffected. The velocity $v_x = \hbar k_x / m_e^*$ is therefore of analog type. However, the perpendicular component follows the periodic circular motion. When the radius of the spiral is comparable to the de Broglie wavelength, the quantum effects are important and cannot be ignored. The magnetic force $qv_{\perp}B$ provides the centripetal force in the circular orbit:

$$qv_{\perp}B = \frac{m_e^* v_{\perp}^2}{R} \Rightarrow v_{\perp} = \frac{qRB}{m_e^*} \tag{2.88}$$

As an orbit contains only the number of de Broglie waves

$$2\pi R_n = n\frac{h}{m_e^* v_{\perp}} \Rightarrow v_{\perp} = n\frac{\hbar}{m_e^* R_n} \tag{2.89}$$

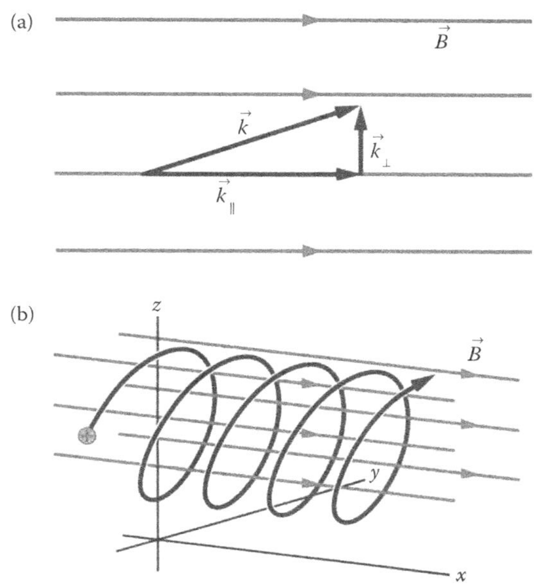

FIGURE 2.33
(a) The velocity vector in a magnetic field applied in the x-direction being decomposed into components parallel and perpendicular to the magnetic field, (b) perpendicular motion is affected by the magnetic field resulting in a circular orbit, the circle being carried away by parallel component resulting in a spiral.

Elimination of $v_\perp$ from Equations 2.88 and 2.89 yields the radius for the quantized orbit

$$R_n = \left(\frac{n\hbar}{qB}\right)^{1/2} = n^{1/2}R_1, \quad R_1 = \left(\frac{\hbar}{qB}\right)^{1/2} \tag{2.90}$$

The quantized kinetic energy $E_{\perp n}$ is given by

$$E_{\perp n} = \frac{1}{2}m_e^* v_{\perp n}^2 = n\hbar\omega_c, \quad \omega_c = \frac{qB}{m_e^*} \tag{2.91}$$

Here, ω_c is known as the cyclotron frequency of the orbit. In a cyclotron resonance experiment, the orbit resonates with the external radio frequency from which the effective mass is extracted. In addition to the kinetic energy, the harmonic oscillations at either y- or z-axis give potential energy (known as zero point energy) of $(1/2)\hbar\omega_c$ in addition to unperturbed kinetic energy parallel to the electric field. The complete energy spectrum with respect to the conduction band edge is given by

$$E_{nk_z} = E_{co} + \left(n + \frac{1}{2}\right)\hbar\omega_c + \frac{\hbar^2 k_x^2}{2m_e^*}, \quad n = 0, 1, 2, \ldots \tag{2.92}$$

$$D_{1e}(E) = \frac{1}{\pi}\left(\frac{2m_e^*}{\hbar^2}\right)^{1/2} (E - E_{cn})^{-1/2}, \quad E_{cn} = E_{co} + (n + \frac{1}{2})\hbar\omega_c \tag{2.93}$$

2.19 Triangular Quantum Well

Another form of quantum confinement that is critical in understanding modern solid-state devices such as the MOSFET is the triangular potential well. Carriers in MOSFETs are confined in an approximately triangular quantum well (TQW) created by the gate electric field.

When an electron is subjected to a potential that has linear proportionality with distance and an infinite barrier existing at $x = 0$, a triangular well is formed. The Schrodinger wave equation for the electron in this well is

$$\frac{d^2\psi}{dz^2} + \frac{2m^*}{\hbar^2}[E_n - q\phi(z)]\psi = 0, \quad n = 1, 2, 3, \ldots \tag{2.94}$$

where E_n are the eigenvalues and

$$\phi(z) = \mathcal{E}z, \quad z > 0 \tag{2.95}$$

is the trapping potential, and $\mathcal{E}$ is a constant electric field. With the linear potential of Equation 2.95, the solution of Equation 2.94 is well known. The solution is the Airy functions $Ai(z)$, which in the normalized form is given by

$$\psi_n(z) = \frac{1}{Ai'(-\xi_n)z_\circ^{1/2}} Ai\left(\frac{z}{z_\circ} - \xi_n\right) \tag{2.96}$$

where $Ai'(-\xi_n)$ is the first derivatives of the Airy functions. ξ_n is the nth zero of the Airy function with the first three values given by

$$\xi_1 = 2.338111, \quad \xi_2 = 4.07895, \quad \xi_3 = 5.52056 \tag{2.97}$$

The first three derivatives of the Airy function at the above ξ values are

$$Ai'(-\xi_1) = 0.701211, \quad Ai'(-\xi_2) = 0.803111, \quad Ai'(-\xi_3) = 0.865204 \tag{2.98}$$

The eigenvalues expressed in terms of the energy parameter E_o are

$$E_n = \xi_n E_o \tag{2.99}$$

with

$$E_o = \left(\frac{\hbar^2 q^2 \mathcal{E}^2}{2m^*}\right)^{1/3} \tag{2.100}$$

The turning point parameter of the well is given by

$$z_o = \frac{E_o}{q\mathcal{E}} \tag{2.101}$$

The Wentzel–Kramers–Brillouin (WKB) approximation is used sometimes to obtain the approximate eigenfunctions and eigenvalues. In this approximation, ξ_n is replaced by α_n:

$$\alpha_n = \left[\frac{3\pi}{2}\left(n - \frac{1}{4}\right)\right]^{2/3}, \quad n = 1, 2, 3, \ldots \tag{2.102}$$

The first three values are: $\alpha_1 = 2.320251$, $\alpha_2 = 4.081810$, and $\alpha_3 = 5.517164$. These are very close to the exact values of ξ_n in Equation 2.97. The energy levels are given by

$$E_n = \left(\frac{\hbar^2}{2m^*}\right)^{\frac{1}{3}} \left[\frac{2}{3}\pi\mathcal{E}\left(n - \frac{1}{4}\right)\right]^{2/3}, \quad n = 1, 2, 3, \ldots \tag{2.103}$$

Because of the difficulty of evaluating integrals with Airy functions, we will consider the approximate eigenfunctions. The eigenfunctions of a particle in a triangular well and the eigenfunctions of a particle in an infinite well are quite similar except for the effective width. The triangular well can be considered an infinite well with an effective width

$$L_n = \left(\frac{2}{a_n^2}\right) z_o \tag{2.104}$$

with

$$a_n = \frac{0.53556}{Ai'(-\xi_n)} \tag{2.105}$$

Here, 0.53556 is the maximum of the Airy function occurring at an argument value of $\xi = 1.019$. The approximate eigenfunctions are

$$\psi_n(z) = \sqrt{\frac{2}{L_n}} \sin\frac{n\pi z}{L_n}, \quad 0 \le z \le L_n \tag{2.106}$$

The eigenvalues can be obtained from

$$E_n = \frac{n^2\hbar^2\pi^2}{2m^* L_n^2} \tag{2.107}$$

The approximate (sinusoidal) wavefunctions from Equation 2.106 and the exact wavefunctions from Equation 2.96 are plotted in Figure 2.34 for the three lowest energy levels.

Because of the simple form of the approximate eigenfunction and eigenvalues, these are used for the inversion layer carriers in MOSFET when a strong gate electric field is applied to the surface of the MOSFET. The carriers subjected to a triangular well that will quantize the energy in the direction perpendicular to the semiconductor–insulator interface.

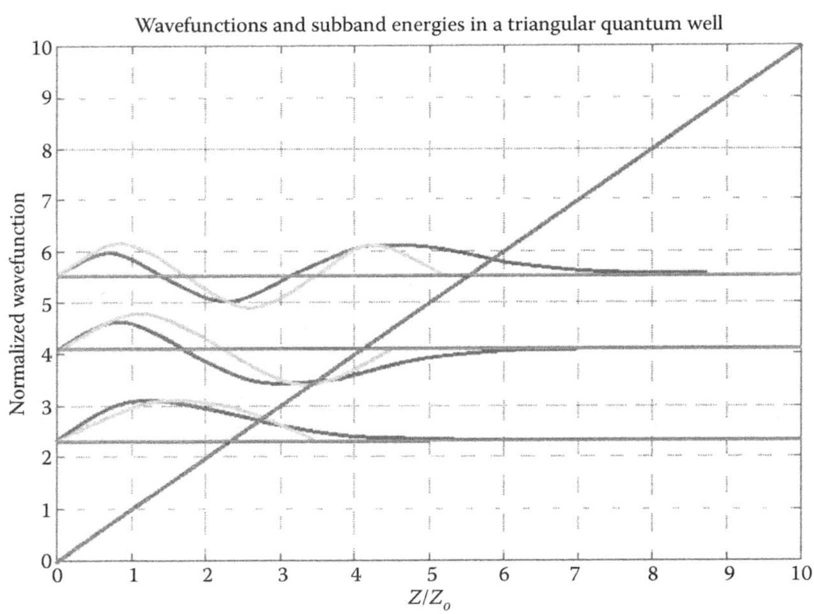

FIGURE 2.34
Eigenfunctions for a particle in a triangular well, the solid lines represent the first three Airy functions. Dashed lines represent the approximated sine wavefunctions.

Yet another way of expressing TQW in terms of a square well is to obtain effective equivalent width L_{neff} by replacing L in the quantum formula $L/2$ due to the reason that the triangular well is one-sided. The vanishing of the wavefunction at $z = 0$ assures only odd parity states are permissible for a square well giving

$$\left(\frac{\hbar^2}{2m^*}\right)^{1/3}\left[\frac{2}{3}\pi\mathcal{E}\left(n-\frac{1}{4}\right)\right]^{2/3} = \frac{\pi^2\hbar^2}{2m^*(L_n/2)^2} \tag{2.108}$$

Another form of approximation which is sometimes useful in obtaining the ground-state $\psi_1(z)$ in a TQW is given by Stern and Howard [5] as

$$\psi_1(z) = 2\beta^{3/2}\frac{1}{\sqrt{z_o}}\frac{z}{z_o}\exp\left(-\beta\frac{z}{z_o}\right) \tag{2.109}$$

where $\beta = 1.077576$. The first excited state was obtained in similar form to be

$$\psi_2(z) = (12\beta^3)^{1/2}\frac{1}{\sqrt{z_o}}\frac{z}{z_o}\left(1-\frac{2}{3}\beta\frac{z}{z_o}\right)\exp\left(-\beta\frac{z}{z_o}\right) \tag{2.110}$$

The eigenfunctions are generally not easy to obtain because of the difficulty of evaluating integrals involving Airy functions. Although the approximation given by Stern and Howard tends to avoid these functions, the approximated wave functions are still complicated in nature. The degree of complexity also increases when eigenfunctions pertaining to higher excited states are required.

2.20 Q2D Electrons in a MOSFET

MOSFETs are fabricated on (100)-oriented substrate to reduce the dangling bonds at the Si/SiO_2 interface. The channel in a nanoscale MOSFET (or otherwise) is a quantum-mechanical one that is constrained by the gate electric field that forms an approximately linear quantum well. The energy spectrum is digital (quantum) in the z-direction, perpendicular to the gate, while other two Cartesian directions (x and y) are analog (or classical). This confinement makes the channel quasi-two-dimensional (Q2D) [6]:

$$\varepsilon_{ki} = \frac{\hbar^2 k_x^2}{2m_1^*} + \frac{\hbar^2 k_y^2}{2m_2^*} + \varepsilon_i \tag{2.111}$$

with

$$\varepsilon_i = \xi_i E_o \approx \left[\frac{\hbar^2}{2m_3}\right]^{1/3} \left[\frac{3\pi q}{2} \mathcal{E}_t \left(i + \frac{3}{4}\right)\right]^{2/3}, \quad i = 0, 1, 2, 3 \tag{2.112}$$

$$E_o = \left(\frac{\hbar^2 q^2 \mathcal{E}_t^2}{2m_3^*}\right)^{1/3} \tag{2.113}$$

where $k_{x,y}$ are the momentum vectors in the analog 2D x–y-plane and ε_i is the quantized energy in the digitized z-direction. $m_{1,2}^*$ is the effective mass in the x–y-plane of the Q2D channel and m_3^* is the effective mass in the z-direction for a given conduction valley. ξ_i are the zeros of the Airy function ($Ai(-\xi_i) = 0$) with $\xi_0 = 2.33811$, $\xi_1 = 4.08795$, and $\xi_2 = 5.52056$.

As shown in Figure 2.35, the conduction band energy surfaces are six ellipsoids with longitudinal direction along $\pm x, y, z$ in k-space [6] for ⟨100⟩-oriented Si MOSFET. The two valleys have $m_3^* = m_\ell = 0.98m_o$ and four valleys have $m_3^* = m_t = 0.19m_o$. $\varepsilon_o = \xi_0 E_o$ is the ground-state energy corresponding to $i = 0$ with $m_3 = 0.916m_o$ for the two valleys. Other four valleys are not occupied in the quantum limit when all electrons are in the lowest energy state. $m_{1,2}^* = 0.19\ m_o$ is the conductivity effective mass in the x–y-plane of the Q2D channel for lower two valleys. Electrons stay in the lowest quantized state appropriate for the two valleys that have a higher mass in the direction of confinement. In the x–y-plane in which electrons are itinerant, the effective mass for these two valleys is the transverse effective mass $m_t = 0.19m_o$ both for the DOS as well as for the mobility. $\mathcal{E}_t$ is the electric field generated by the gate, which in the strong inversion regime is given by [7]

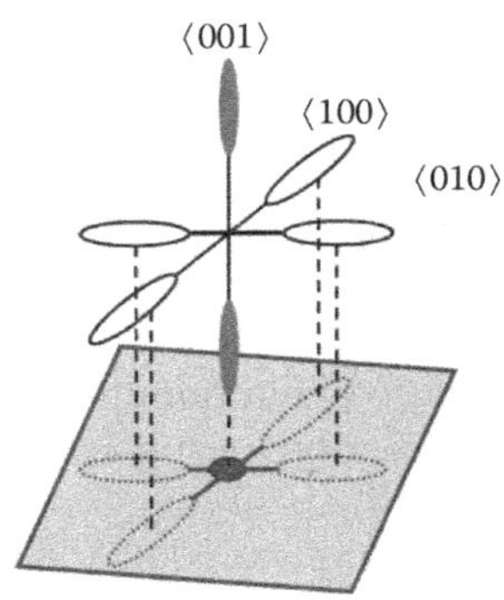

FIGURE 2.35
The populated valley (filled) with quantum confinement in the z-direction with projection on the x–y-plane with isotropic effective mass $m_t = 0.19m_o$.

$$\mathcal{E}_t = \frac{V_{GS} + V_T}{6t_{ox}} = \frac{V_{GT} + 2V_T}{6t_{ox}} \tag{2.114}$$

Here, $V_{GT} = V_{GS} - V_T$ is the gate voltage above the threshold voltage V_T and t_{ox} is the thickness of the gate oxide. The eigenfunctions corresponding to the eigenvalues of Equation 2.111 are given by

$$\psi(x,y,z) = \frac{1}{\sqrt{L_x L_y}} e^{j(k_x x + k_y y)} Z_i(z) \tag{2.115}$$

$$Z_i(z) = \frac{1}{Ai'(-\xi_i) z_o^{1/2}} Ai\left(\frac{z}{z_o} - \xi_i\right) \tag{2.116}$$

$$z_o = \frac{E_o}{q\mathcal{E}_t} \tag{2.117}$$

The average distance z_{QM} of the electrons from the interface ($z = 0$) for electrons in the ground state is [5,8]

$$z_{QM} = \frac{2}{3} z_o \tag{2.118}$$

The wavefunction vanishes at the interface and the probability of existence of electrons at the interface is zero. The distance z_{QM} of the electrons, which is a few nm, is significant when compared to thickness of the gate oxide ($t_{ox} = 1.59$ nm in our case). The effective oxide thickness of the gate after correction for difference in permittivity of the SiO_2 and Si is given by

$$t_{oxeff} = t_{ox} + \frac{\varepsilon_{ox}}{\varepsilon_{Si}} z_{QM} \approx t_{ox} + \frac{1}{3} z_{QM} \tag{2.119}$$

The gate capacitance C_G is lower than C_{ox} due to oxide thickness t_{ox} and is given by [4,9]

$$C_G = \frac{\varepsilon_{ox}}{t_{oxeff}} = \frac{C_{ox}}{1 + \frac{1}{3}\frac{z_{QM}}{t_{ox}}} \tag{2.120}$$

Sometimes quantum capacitance due to z_{QM} is confused with depletion layer capacitance that is also in series with the gate capacitance. However, the gate capacitance C_G given by Equation 2.120 is only responsible for the inversion charge [4]. Depletion layer capacitance affects the threshold voltage and may induce transient currents in the body of the MOSFET. Some of the deleterious body effects are being ameliorated by using the SOI substrate.

2.21 Carbon Allotropes

Allotropes are compounds that exist in forms with different chemical structures. Carbon has allotropes in the form of diamond, graphite, fullerenes, and carbon nanotubes (CNTs). The specific hybridization of carbon and its bonding to surrounding atoms determines which allotrope carbon assumes. Carbon with sp^2 hybridization will form either graphite buckminster-fullerene (60 carbon atoms forming a sphere), or CNTs—depending on the conditions in which it is formed. Though no practical application for buckyballs has been developed yet, scientists are extremely excited about the potential uses of CNTs. These structures have a diameter between 1 and 10 nm, yet are 50 times stronger than steel. CNTs are also structurally perfect, and this property gives rise to a whole host of other unique properties, such as unique electrical properties and high thermal conductivity. Graphene and graphene nanoribbons (GNRs) are other allotropes that are finding potential applications and elucidate a new phenomenon both in physics and chemistry of carbon-based materials.

The natural starting point for describing carbon allotropes is graphene. Graphene is a rapidly rising star on the horizon of materials science and condensed-matter physics [10]. This strictly 2D material exhibits exceptionally high crystal and electronic quality, and, despite its short history, has already revealed an abundance of new physics and potential applications. Although high hopes are placed on entrepreneurial capabilities of graphene, no further proof is required for its importance in exploring fundamental sciences, including biology. More generally, graphene forms the foundation of a conceptually new class of materials that are only one atom thick, and, on this basis offers new inroads into low-dimensional physics that has never ceased to surprise and continues to provide a fertile ground for applications. Graphene is a planar allotrope of carbon where all the carbon atoms form covalent bonds in a single plane. The planar honeycomb structure of graphene has been observed experimentally and is shown in Figure 2.36. Graphene can be considered the mother of three carbon allotropes. As illustrated in Figure 2.36, wrapping graphene into a sphere produces buckyballs, folding into a cylinder produces nanotubes, and stacking several sheets of graphene leads to graphite. Furthermore, cutting graphene into a small ribbon results in GNRs, which are the subject of contemporary research. As a result, understanding quantum processes that form the properties of graphene is of paramount importance in designing the future of nanoelectronics and its many applications.

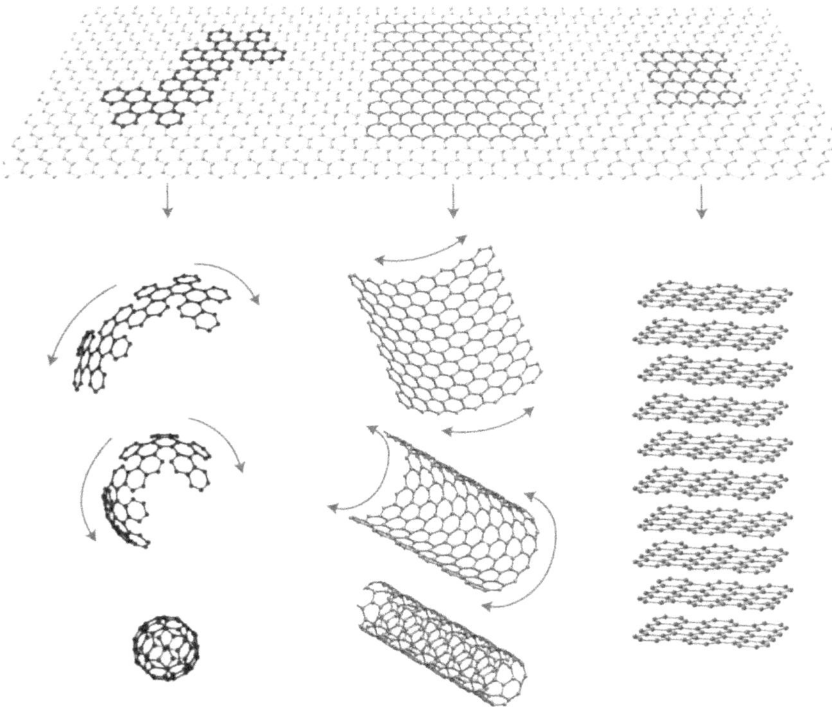

FIGURE 2.36
Carbon allotropes arising from graphene sheet to form zero-dimensional (0D) buckyball, 1D CNT, and 3D graphite. Each layer of 2D graphite can be converted into a 1D GNR by making width smaller. Copyright Macmillan Publishers Limited. (Adapted from A. K. Geim and K. S. Novoselov, *Nature Materials*, 6, 183–191, Mar 2007.)

Graphene is a planar allotrope of carbon with atoms arranged in a regular hexagonal honeycomb lattice one-atom thick, as shown in Figure 2.37. Figure 2.37a shows the role of a chiral vector into a CNT. Figure 2.37b shows the sp^2 bonding of a graphene layer with three σ bonds and one π bond. σ-bond gives graphene the strength with high modulus. The carbon–carbon bond length in graphene is $a_{CC} = 0.142$ nm. The electronic configuration of carbon (C) with atomic number $Z = 6$ is $1s^2 2s^2 2p^2$. Four electrons in $n = 2$ shell make it tetravalent just like silicon (Si) with $1s^2 2s^2 2p^6 3s^2 3p^2$. Both C and Si are from Group IV of the periodic table. Just like Si, carbon has four valence electrons, as shown in Figure 2.37b, which tend to interact with each other to form a crystal. Three sp^2 orbitals form σ-bond residing in the graphene plane. These are pretty strong bonds that demonstrate superior electronic properties. $2p_z$ orbital forms a weakly bound π-bond, making electron delocalized. These delocalized electrons form the conducting properties of graphene and CNTs.

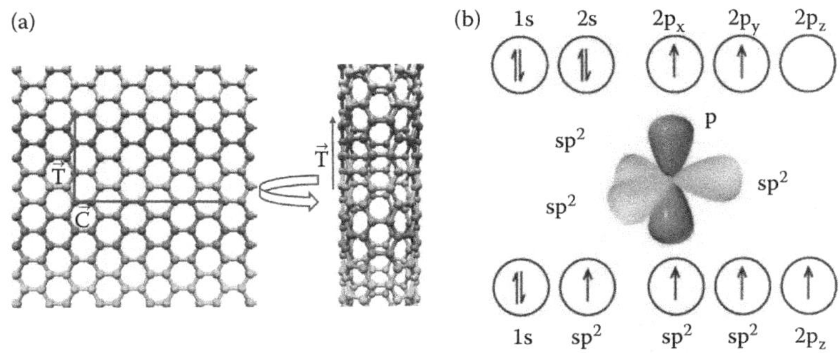

FIGURE 2.37
(a) The rolled up graphene sheet into a CNT. Carbon atoms arranged in a honeycomb lattice of one-layer thick graphene. sp^2 hybrid orbitals shown in (b) result from three σ bonds sp^2 and one π bond (p). (b) Electronic configuration of C (top) with four reconfigured electrons in $n = 2$ shell; hybridization (middle) in a carbon crystal with sp^2 electronic configuration (bottom).

Monoatomic layers makes graphene a perfect 2D material. As a 2D nanolayer, graphene sheet has some semblance to a MOSFET. As graphene is a relatively new material with a variety of allotropes, the landscape of its applications is an open arena. Each atom contributes one-third of the π-electron to a hexagon. With six atoms forming corners of a hexagon, each hexagon has two π-electrons. The areal electronic density of π-electrons is $n_g = 2/A_h$ with $A_h = 3\sqrt{3}a_{CC}^2/2$ the area of the haxagon. The areal density is $n_g = 3.82 \times 10^{19}\ \text{m}^{-2}$ with $a_{CC} = 0.142$ nm. The intrinsic line density n_{CNT} is expected to be a function of diameter $n_{CNT} = 1.2 \times 10^{11} d_t\ (\text{nm})\text{m}^{-1}$ when rolled into a CNT.

CNTs, originally discovered by Iijima, have opened a new vista for design of compact nanostructures and their applications. Some of these applications are the building blocks of nano-VLSI circuit design, including reliable interconnects [3]. It is now well known that a CNT is rolled-up cylinder of graphene, a one-atom-thick allotrope of carbon, whose properties depend on chirality of the roll up. The unzipping of a CNT into graphene has been recently demonstrated by a number of labs. The fabrication of single- and multiple-wall nanotubes is now possible with a wide variety of procedures. The carrier mobility of the CNT nanostructure is not affected by processing and roughness scattering due to the chemical stability and perfection, as it is in the conventional semiconducting channel. The fact that there are no dangling bond states at the surface of the CNT allows for a wide choice of gate insulators in designing a FET. The outer wall in a double-wall CNT can also be designed as a gate that will control the charge on the inner CNT and hence the flexibility of making it n- or p-type. It is no surprise that the CNTs are being explored as viable candidates for high-speed applications. Castro Neto et al. [11] review the basic theoretical aspects of graphene as unusual 2D Dirac-like electronic excitations. The Dirac electrons can be controlled by application of external electromagnetic fields or by altering sample geometry, chirality, and/or topology (sheet or tubular structure).

2.22 Graphene to CNT

Chirality is key to understanding the rollover of a graphene sheet into a CNT. A chiral nanostructure is not superimposable on its mirror image. The achiral (not chiral) nanostructure is identical to its mirror image. As is seen below, zigzag carbon nanotubes (ZCNTs) and armchair carbon nanotubes (ACNTs) are achiral. Chirality is described by a pair index (n, m). The key to understanding chirality is the basis vectors $\vec{a}_1 = (\sqrt{3}/2)a,\ (a/2)$ and $\vec{a}_2 = (\sqrt{3}/2)a,\ (-a/2)$ both with magnitude $a = |\vec{a}_1| = |\vec{a}_2| = a_{CC}\sqrt{3}$, where $a_{CC} = 0.142$ is a carbon-to-carbon bond length giving $a = 0.142\ \text{nm} \cdot \sqrt{3} = 0.246\ \text{nm}$. The chiral vector C_h whose absolute value is equal to the circumference of the nanotube is defined as

$$C_h = n a_1 + m a_2 \equiv (n, m) \quad \text{with} \quad 0 \le |m| \le n \tag{2.121}$$

Figure 2.38 shows basis vector $\vec{a}_1$ and $\vec{a}_2$. Chiral vector $\vec{C}_h$ is obtained in vector sum of $n\vec{a}_1$ and $m\vec{a}_2$. The angle between $\vec{a}_1$ and $\vec{a}_2$ is 60°.

The diameter of the CNT, as calculated from the chiral vector, is given by

$$d_{CNT} = 2\mathbf{R}_{CNT} = \frac{|C_h|}{\pi} = \frac{1}{\pi}\sqrt{C_h \cdot C_h} = \frac{a}{\pi} \cdot \sqrt{n^2 + m^2 + nm} \tag{2.122}$$

The length of the chiral vector $\vec{C}_h = n\vec{a}_1 + m\vec{a}_2$ depends on n units in the direction of $\vec{a}_1$ and m units in the direction of $\vec{a}_2$. The unit cell is spanned by the moving two vectors $\vec{a}_1$ and $\vec{a}_2$ to shift, so they bisect the carbon–carbon bond and form a parallelogram containing two carbon atoms, one with bonds toward the left and the other toward the right. The parallelogram forms a Bravais unit cell that can be replicated to form the complete graphene structure. The area of the unit cell is $|a_1 \times a_2| = a^2 \cdot \sin 60° = 0.05387\ \text{nm}^2$ giving the density of carbon atoms as $2/0.05387\ \text{nm}^2 = 37.13\ \text{nm}^{-2}$.

Graphene lattice is not a Bravais lattice as the two adjacent atoms shown in Figure 2.38 do not replicate. However, if these two atoms are grouped together, the Bravais lattice is formed by repeating the compound structure of two atoms. The configuration shown in Figure 2.38 is one possible way to depict the periodicity of the crystal. There are other models that are used as well. All give the same end result as far as periodic boundary conditions are concerned. Chirality is always pegged with the basis vectors $\vec{a}_1$ and $\vec{a}_2$, as shown in Equation 2.121.

Starting with a flat graphene sheet, a cylinder is made when the two endpoints of the sheet, as shown in Figure 2.39, are superimposed. The diameter

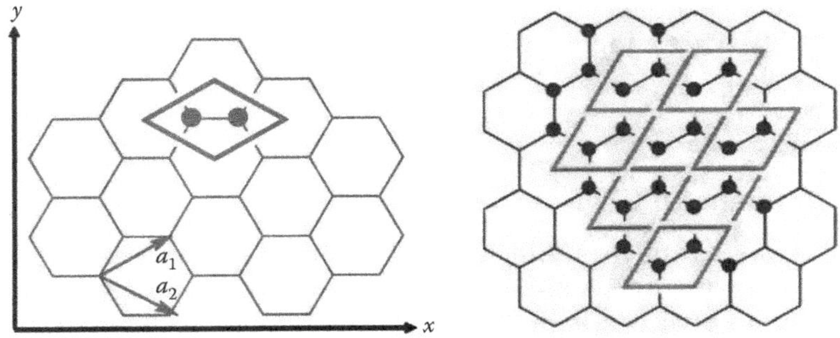

FIGURE 2.38
The basis unit vector in a graphene sheet forming a lattice unit cell containing two carbon atoms when basis vectors are shifted to form the basis of a parallelogram that intersects two carbon bonds. This parallelogram unit cell forms the complete graphene lattice when replicated throughout the lattice.

$d_{CNT} = C_h/\pi$, where $\vec{C}_h$ is the chiral vector going from one edge to the other edge of the graphene sheet that is rolled into a cylinder.

Figure 2.40 shows the rollover direction of ZCNTs, ACNTs, and chiral CNTs. All ZCNTs have chiral vector $(n, 0)$ and the edges of the tube is zigzag in nature, as seen in left of Figure 2.39. All ACNTs having chiral vector (n, n) when rolled show edges that look like a sofa seat with arms, hence, the name armchair (middle of Figure 2.39). Chiral CNTs are inbetween the two extremes (right of Figure 2.39).

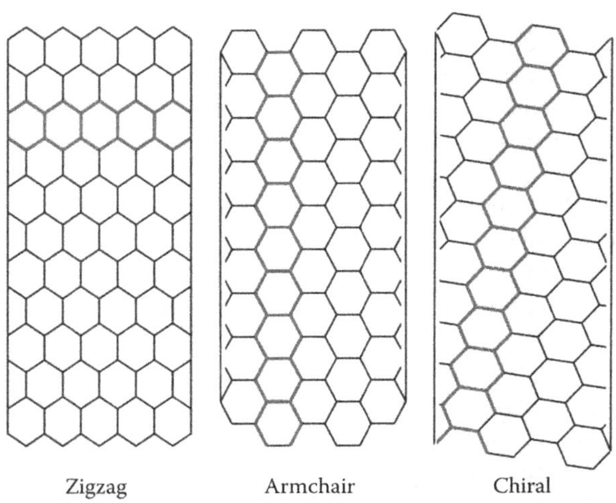

FIGURE 2.39
Rollover of graphene sheet into a CNT with zigzag edges (left), armchair edges (middle), and chiral edges (right).

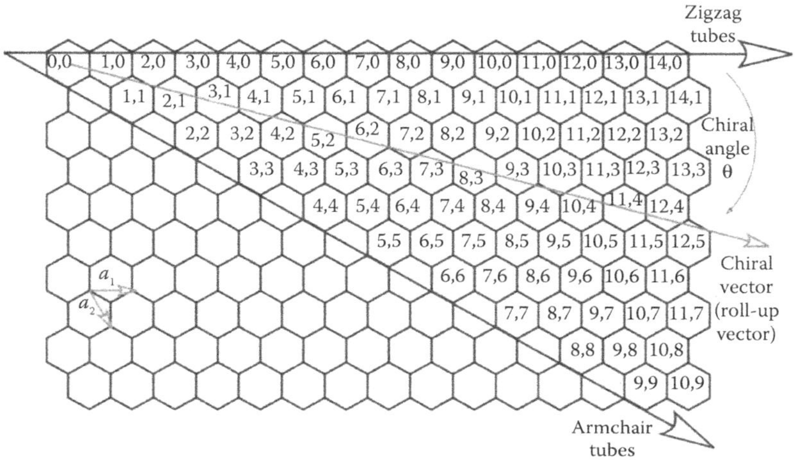

FIGURE 2.40
Rollover in (n, n) direction results in ACNT; in $(n, 0)$ direction results in ZCNT; and all other directions are chiral.

2.23 Bandgap Engineering of Carbon Allotropes

The band structure of graphene, as obtained from tight-binding approximation of electrons to carbon atoms, is very complex. Figure 2.41 shows simplified E–k Dirac cone [12]. Once 6-fold Dirac cones are obtained centered at K and K' points, bandgap engineering comes easily.

The theoretical development of electronic structure in a graphene nanostructure is simplified by starting from the linear E–k relation with zero effective mass of a Dirac fermion near the Fermi points K and K', as shown in Figure 2.41:

$$E = E_{Fo} \pm \hbar v_{Fo}|k| = E_{Fo} \pm \hbar v_{Fo}\sqrt{k_x^2 + k_y^2} \tag{2.123}$$

where k is the momentum vector which in circular coordinates has components $k_x = k\cos\theta$ and $k_y = k\sin\theta$. $v_{Fo} = (1/\hbar)dE/dk$ is constant due to a linear rise of energy E with momentum vector k and hence, constant slope and the intrinsic Fermi velocity $v_{Fo} \approx 10^6$ m/s. $\hbar v_{Fo}$ is the gradient of E–k dispersion. The linear dispersion of the Dirac cone is confirmed up to ±0.6 eV. $E_F - E_{Fo} = 0$ for intrinsic graphene with $E_{Fo} = 0$ as the reference level. Here, subscript "o" is added to emphasize the fact these are intrinsic properties in the metallic state when E–k relation is linear. The Fermi velocity vectors are randomly oriented in the graphene sheet in equilibrium and, hence, are stochastic. The DOS for a 2D graphene nanolayer, as calculated in the Appendix 2A, is given by

$$D_g(E) = D_{go}|E - E_{Fo}| \tag{2.124}$$

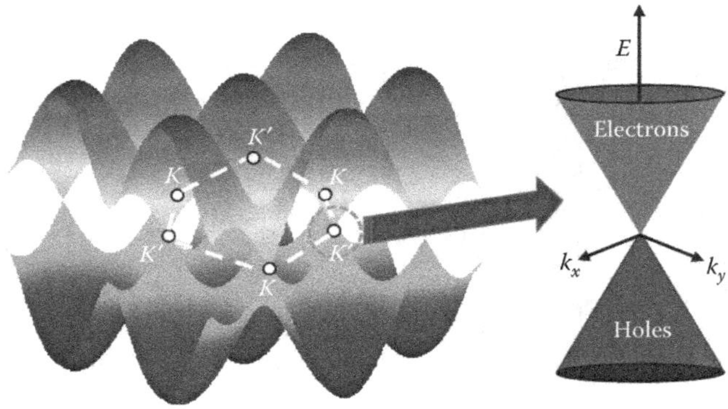

FIGURE 2.41
Full band dispersion over the whole Brillouin zone for π (lower surface or valence band) and π* (upper surface or conduction band) of graphene. There are 3*K* and 3*K'* points in the Brillouin zone, each *K* Fermi point at the apex of the Dirac cone.

where

$$D_{go} = \frac{2}{\pi\hbar^2 v_{Fo}^2} = 1.5\times10^{14}\ \text{eV}^{-2}\ \text{cm}^{-2} = 1.5\times10^{6}\ \text{eV}^{-2}\ \mu\text{m}^{-2} \quad (2.125)$$

The electronic quantum transport in a CNT is sensitive to the precise arrangement of carbon atoms. There are two families of CNTs: single-walled carbon nanotubes (SWCNTs) and multiple-walled carbon nanotubes (MWCNTs). The diameter of SWCNTs spans a range of 0.5–5 nm. The lengths can exceed several micrometers and can be as large as a cm. The MWCNT is a cluster of multiply nested or concentric SWCNTs. The focus here is on SWCNTs. A roll up of Figure 2.40 can lead to either a semiconducting or metallic state depending on the chirality. When the arrangement of carbon atoms is changed by mechanical stretching, a CNT is expected to change from semiconducting to metallic or vice versa. Several unique properties result from the cylindrical shape and the carbon–carbon bonding geometry of a CNT. Wong and Akinwande [12] are vivacious in connecting physics and the technology of a graphene nanolayer to that of a CNT with splendid outcomes. The CNT band structure arising out of sixfold Dirac *K*-points with equivalence of *K* and *K'* points can lead to simplified mathematics. Once nearest-neighbor tight binding (NNTB) formalism is applied, the resulting Dirac cone, as revealed in Figure 2.41, gives useful information for a variety of chirality directions. In fact, the *K*-points offer much simplicity for quantum transport applications. In the metallic state *E*–*k* relation is linear. However, the Fermi energy and associated velocity are different in the semiconducting state. Intrinsic Fermi energy $E_{Fo} = 0$ is applicable for undoped or uninduced carrier concentration. The induction of carriers will move the Fermi level in the conduction (n-type) or valence band (p-type). Once carriers are induced, the Fermi energy E_F will shift from the intrinsic Fermi level E_{Fo}. A positive $E_F - E_{Fo}$ will result in an n-type CNT and negative $E_{Fo} - E_F$ will result in p-type CNT. The distinction between E_F and E_{Fo} is of great significance as CNTs transform to semiconducting state, starting from metallic state.

Equation 2.123 now can be written in terms of k_t, the momentum vector in the longitudinal direction of the tube, and k_c, the momentum vector in the direction of roll up. The new description transforms Equation 2.123 to

$$E = E_{Fo} \pm \hbar v_F |k| = E_{Fo} \pm \hbar v_F \sqrt{k_t^2 + k_c^2} \tag{2.126}$$

The K-point degeneracy is crucial to the profound understanding of the symmetries of graphene folding into a CNT in a given chiral direction. Symmetry arguments indicate two distinct sets of K points satisfying the relationship $\vec{K} = -\vec{K}'$, confirming the opposite phase with the same energy. There are three K and three K' points, each K (or K') rotated from the other by $2\pi/3$. In addition to spin degeneracy $g_s = 2$, the zone degeneracy $g_K = 2$ is based on two distinct sets of K and K' points. There are six equivalent K points, and each K point is shared by three hexagons; hence, $g_K = 2$ is for graphene as well as for a rolled-up CNT. The phase k_cC_h of the propagating wave $e^{\pm ik_c C_h}$ in the chiral direction results in a rolled-up CNT to satisfy the boundary condition

$$k_cC_h = \nu\left(\frac{2\pi}{3}\right) \tag{2.127}$$

where $\nu = (n - m)\text{mod}3 = 0, 1, 2$ is the band index. $(n - m)\text{mod}3$ is an abbreviated form of $(n - m)$ modulo 3 that is the remainder of the Euclidean division of $(n - m)$ by 3. The quantization condition transforms to $k_c = \nu(2/3d_t)$ when $C_h = \pi d_t$ is used for a CNT's circular parameter. Using $k_c = \nu(2/3d_t)$ in Equation 2.126 yields the band structure as given by

$$E = E_{Fo} \pm \hbar v_{Fo} |k| = E_{Fo} \pm \hbar v_{Fo} \sqrt{k_t^2 + \left(\nu \frac{2}{3d_t}\right)^2} \tag{2.128}$$

The equation can be rewritten as

$$E = E_{Fo} \pm \hbar v_{Fo} \left(\nu \frac{2}{3d_t}\right) \sqrt{1 + \left(\frac{3d_tk_t}{2\nu}\right)^2} = E_{Fo} \pm \frac{E_g}{2} \sqrt{1 + \left(\frac{3d_tk_t}{2\nu}\right)^2} \tag{2.129}$$

where E_g is defined below when the square root term is expanded into a binomial expression near $k_t = 0$. It is parabolic in character, as shown in Figure 2.42, for which effective mass is defined.

Figure 2.42 displays the $E_\nu - k_t$ relationship for $\nu = 0$ metallic and $\nu = 1(2)$ semiconducting (SC1(2)) states. ν is the band index confined to these three values only, as $\nu = 3$ is equivalent to $\nu = 0$ and the pattern repeats itself in

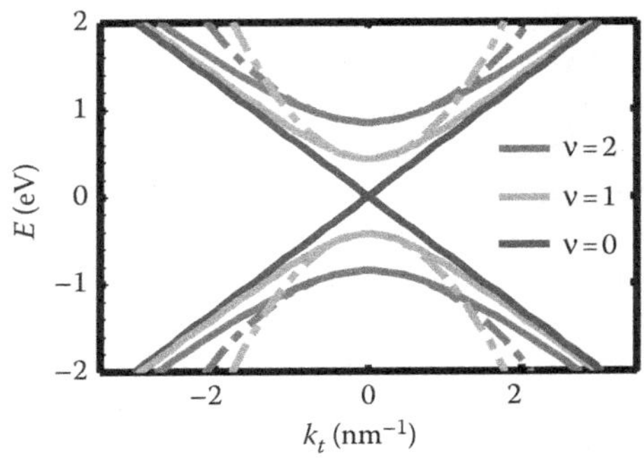

FIGURE 2.42
E_ν versus k_t graph with chirality (10, 4) for ν = 0, (13, 0) for ν = 1, and (10, 5) for ν = 2, all with diameter $d_t \approx 1.0$ nm. Solid line is used for exact formulation (Equation 2.129) and dash-dot line showing parabolic approximation (Equation 2.130).

these three modes. The diameter is $d_t \approx 1.0$ nm for the chosen chirality directions: (10, 4) for ν = 0, (13, 0) for ν = 1, and (10, 5) for ν = 2. Assuming that each of these configurations are equally likely, about one-third of the CNTs are metallic and two-thirds semiconducting with bandgap that varies with chirality. As seen in Figure 2.42, the curvature near $k_t = 0$ is parabolic, making it possible to define the effective mass that also depends on chirality. Figure 2.43 shows the bandgap of CNTs with different chiral configurations, covering metallic (ν = 0) and two semiconducting (ν = 1, 2) states. As expected, the bandgap is zero for the metallic state, in agreement with Figures 2.42 and 2.43. SC2 bandgap is twice as large as that of SC1. Figure 2.43 also shows chirality leading to ν = 0, 1, or 2. The NNTB bandgap is likewise shown. It also exhibits the wide bandgap nature of SC2. The implications of the boundary condition of Equation 2.127 are self-evident from Figures 2.42 through 2.44.

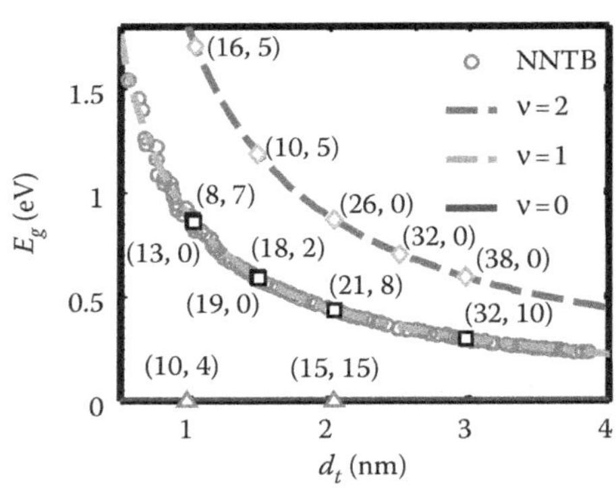

FIGURE 2.43
Calculated bandgap as a function of CNT diameter showing an agreement with NNTB calculation. Chiral vectors are indicated against corresponding points.

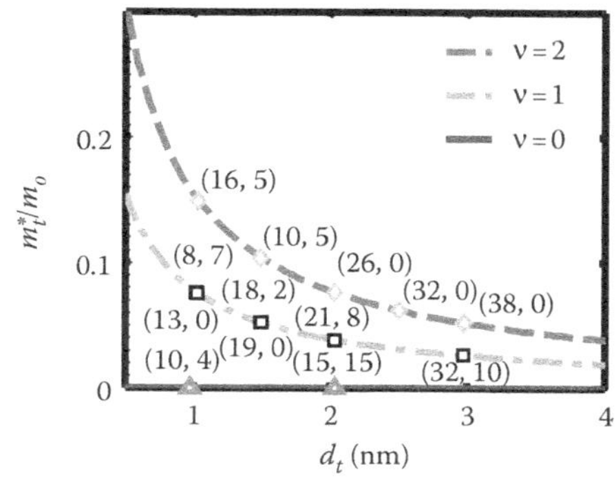

FIGURE 2.44
The CNT-normalized effective mass as a function of diameter for ν = 1 and ν = 2.

With the binomial expansion $(1 + x)^{1/2} \approx (1 + 1/2)x$ and keeping first-order term transforms Equation 2.129 to

$$E = E_{Fo} \pm \left(\frac{E_g}{2} \pm \frac{\hbar^2 k_t^2}{2m_t^*} \right) \tag{2.130}$$

with

$$E_g = \nu \frac{4}{3d_t} \hbar v_{Fo} = \nu \frac{0.88 \text{ eV} \cdot \text{nm}}{d_t} \quad \text{and} \quad \frac{m_t^*}{m_o} = \nu \frac{2}{3d_t} \frac{\hbar}{v_{Fo} m_o} = \nu \frac{0.077 \text{ nm}}{d_t} \tag{2.131}$$

Here, $|v_F| = 3a_{cc}\gamma/2\hbar = 10^6$ m/s with $\gamma = 3.1$ eV, the C–C bond strength.

The normalized effective mass as a function of diameter is shown in Figure 2.44. $m_t^* = 0.051 m_o$ for $\nu = 1$ for a (19, 0) CNT with a 1.5 nm diameter. The effective mass for $\nu = 2$ is twice as large as that of $\nu = 1$ both for electrons and holes. Table 2.3 lists the bandgap and effective mass of each state for representative chirality configurations. The table is arranged, so a comparison of the same diameter for various chirality directions is easily made. As diameter increases, the bandgap decreases.

Figure 2.45 shows the relative effective mass m_t^*/m_o as a function of bandgap represented by

$$\frac{m_t^*}{m_o} = \frac{E_g}{2m_o v_{Fo}^2} = \frac{E_g(\text{eV})}{11.37\,\text{eV}} \tag{2.132}$$

The markers are for (10, 4), (13, 0), and (10, 5) chirality indicating a universal nature as bandgap increases so does the effective mass. A similar relation is found for a compound semiconductor [13] with $m_t^*/m_o = g_v E_g/22.74$ eV, where

TABLE 2.3
Band Index, Diameter, Bandgap, and Effective Mass for a Number of Chiralities

Chirality	$(n-m)$ mod3ν	Diameter $d_t(nm)$	Bandgap E_{gv}(eV)	Effective Mass m^*/m_o
(10, 4)	0	0.98	0	0
(8, 7)	1	1.02	0.84	0.08
(13, 0)	1	1.02	0.84	0.08
(12, 2)	1	1.03	0.86	0.08
(10, 5)	2	1.04	1.72	0.15
(19, 0)	1	1.49	0.58	0.05
(18, 2)	1	1.49	0.58	0.05
(16, 5)	2	1.49	1.18	0.10
(15, 15)	0	2.03	0	0
(21, 8)	1	2.03	0.43	0.04
(26, 0)	2	2.04	0.88	0.08
(16, 14)	2	2.04	0.86	0.08
(32, 0)	2	2.51	0.70	0.06
(32, 10)	1	2.98	0.30	0.03
(38, 0)	2	2.98	0.60	0.05

g_v is the valley degeneracy that is the same as $g_K = 2$ in a CNT, but 1 for compound semiconductors with central Γ valley. This comparative observation indicates that the Fermi velocity v_{Fo} plays a distinctive role in NNTB models for compound semiconductors as well as graphene-based configurations. The narrow bandgap may offer a distinct advantage for enhanced mobility as is obvious in a CNT. Another observation that $E_{g/2} = m_t^* v_{Fo}^2$ may bring us closer to Einstein's $E = mc^2$, where c is the speed of light. The comparative study may unravel the mystery of v_{Fo} being the limited velocity for a Dirac Fermion. Figure 2.42

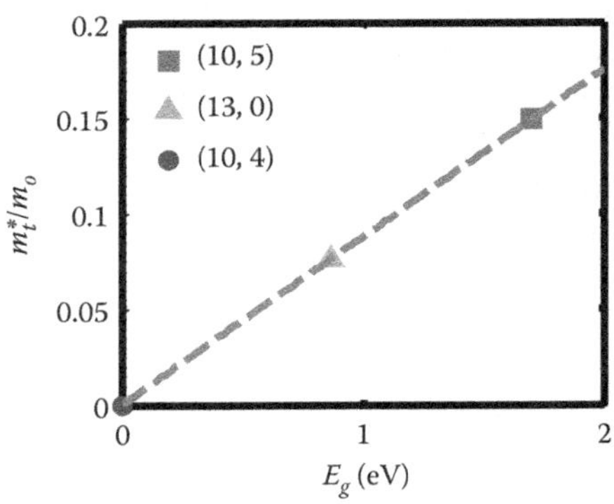

FIGURE 2.45
Normalized effective mass m_t^*/m_o as a function of bandgap. The markers are for chirality (10, 4) with ν = 0, (13, 0) with ν = 1, and (10, 5) with ν = 2.

also indicates that there are only three bands possible. $v = 0$ is always a metallic state. $v = 1$ and $v = 2$ are for semiconducting state. The phase $\pm k_cC_h$ may result in the same two semiconducting states when rotated from clockwise or counter-clockwise direction, creating a two-fold degeneracy. In that perspective, two subband structure will exist for all semiconducting states.

The DOS and probability distribution function form the backbone of carrier statistics that will take the Fermi level E_F, based on induced carrier concentration, higher or lower than that of the intrinsic grapheme $E_{Fo} = 0$. Continuous analog states arise from the digital quantum states with minuscule spacing, so packing of these closely spaced quantum states establishes the DOS as the density of condensation in k_t-space. $k_t = (2\pi/L_t)n_t$ is quantized in the direction of CNT length L_t, where n_t is an integer. In a one-sided quantum well, as discussed before $k_t = (\pi/L_t)n_t$. However, when the complete space spanning $-\infty < L_t < \infty$ is considered $k_t = (2\pi/L_t)n_t$. This condition can also be obtained from the cyclic boundary condition of propagating wave $\exp(jk_tx)$ with $k_tL = 2\pi n_t$ along the length of the tube. The number of quantum states within Δk_t are $\Delta n_t = (L_t/2\pi)\Delta k_t$. The number of states per unit length is then given by

$$\mathbb{N}_c(E_v) = \frac{g_s g_K}{L_t}\sum_{\substack{\vec{k}_t \\ E_{n_t} < E_v}} 1 = \frac{g_s g_K}{L_t}\frac{L_t}{2\pi} 2\int_0^{k_t(E_v)} dk_t \tag{2.133}$$

Here, $g_s = 2$ is the spin degeneracy and $g_K = 2$ is K degeneracy with two sets of K and K' points. The prefactor 2 arises from the integral limits from $-k_t$ to $+k_t$ as energy is an even function of k_t. $k_t(E)$, as obtained from Equation 2.129 with $E_{Fo} = 0$, is given by

$$k_t(E_v) = \frac{8v}{3d_tE_g}\left[E_v^2 - E_{cv}^2\right]^{1/2}, \quad E_{cv} = \frac{E_{gv}}{2} \tag{2.134}$$

The number of quantum states per unit length is now given by

$$\mathbb{N}_{CNT}(E_v) = \frac{16v}{3\pi d_tE_{gv}}\left[E_v^2 - E_{cv}^2\right]^{1/2} \tag{2.135}$$

The differential DOS per unit length per unit energy now naturally follows from Equation 2.135. The resulting DOS is obtained as

$$D_{CNT}(E) = \frac{d\mathbb{N}_c}{dE} = D_0\frac{|E|}{\left[E^2 - E_c^2\right]^{1/2}} \tag{2.136}$$

with

$$D_o = \frac{4}{\pi\hbar v_F} = 1.93\,\text{nm}^{-1}\,\text{eV}^{-1} \tag{2.137}$$

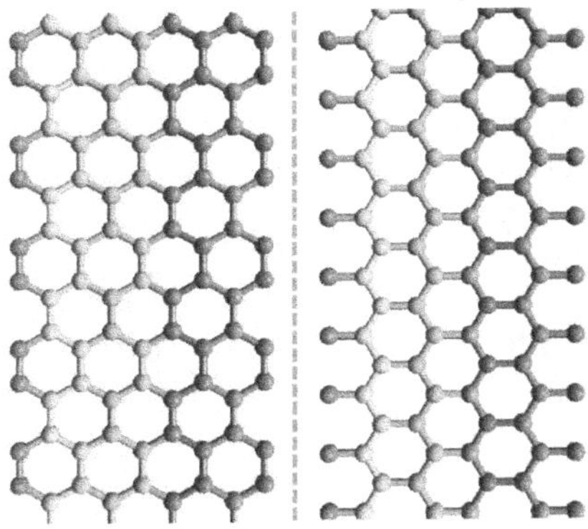

FIGURE 2.46
AGNRs and ZGNRs where edges look like armchair and zigzag, respectively.

Band index ν from energy E is now dropped, as it is redundant for a given ν. This DOS is universal for all configurations $\nu = 0, 1, 2$. The DOS $D_o = 1.93$ nm^{-1} eV^{-1} is constant for metallic CNTs ($\nu = 0$). However, it does show Van Hove singularity at the bandedge for $\nu = 1, 2$.

All CNTs with chirality (n, n) are ACNTs. ACNTs are metallic with zero bandgap since $\nu = (n - m)/3 = 0$. Those with chirality $(n, 0)$ are ZCNTs. Since $\nu = 1$ and 2 lead to bandgap, these are semiconducting. Assuming equal distribution of chirality among $\nu = 0$, 1, and 2, roughly one-third of CNTs are metallic and another two-thirds semiconducting. The diameter, as obtained from Equation 2.122, is $d_t = 3na_{cc}/\pi$ for a ACNT and $d_t = \sqrt{3}na_{cc}/\pi$ for a ZCNT. Small CNTs with a diameter less than 1 nm have a large curvature, which may lead to small bandgap for metallic CNTs widely referred to as quasi-metallic. Research, both in the experimental domain as well as in the theoretical domain, tend to show bandgap that changes quadratically with the tube diameter ($E_g \sim 1/d_t^2$). There is expected to be carbon–carbon bond length asymmetry and σ–π bond hybridization. Additionally, CNT properties may be affected by the substrate on which these are supported because of the presence of substrate phonons. There may be additional piezoelectric, capacitive, and an induced field effect when CNT are on a substrate. This diversity of CNT properties makes them attractive for a variety of applications including interconnects, transistors, and sensors.

CNTs can be unzipped to create graphene nanoribbons (GNRs) of Figure 2.46. GNRs, as shown in Figure 2.46, are chiral as their mirror image is superimposable. A GNR as a cut-out from a graphene sheet with narrow width W appears to have superior transport properties with the potential of novel applications in carbon-based nanoelectronics. The carbon atoms on the edge of GNRs have armchair and zigzag shapes. In a recent paper [14], CNTs were shown as arising from a rolled-up graphene sheet (in reality, GNR of narrow

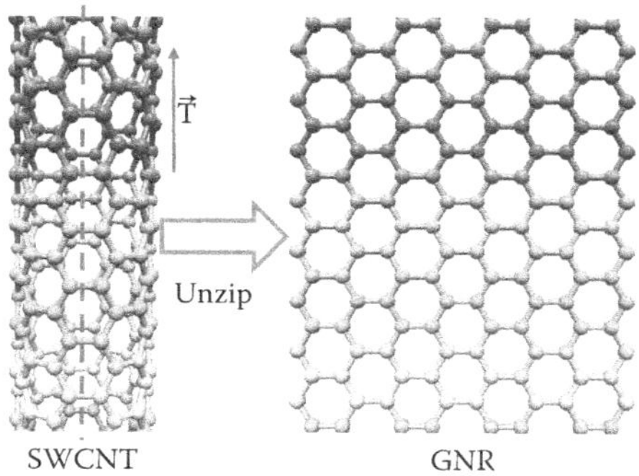

FIGURE 2.47
Single-walled ACNTs unzipping into a ZGNR.

width) of chiral vector with width $C = \pi d_t$. The opposite can also be true: A CNT can be unzipped into a GNR of chain index N_Z or N_A, as shown in Figure 2.47. The example in Figure 2.47 shows the ACNT unzipping to a zigzag-edged graphene nanoribbons (ZGNRs).

Because of a high length-to-width aspect ratio, GNR are 1D conductors. Armchair graphene nanoribbons (AGNRs) have an armchair cross-section at the edges. Zigzag one has zigzag cross-section. The GNR's width (W_A for armchair or W_Z for zigzag) is determined by the number of armchair (N_A) or zigzag (N_Z) chains, as shown in Figure 2.48.

Raza and Kan [15] report electronic structure and electric-field modulation calculations in the width direction for AGNRs using a semiempirical extended Hückel theory. One surprising aspect that emerged from this study

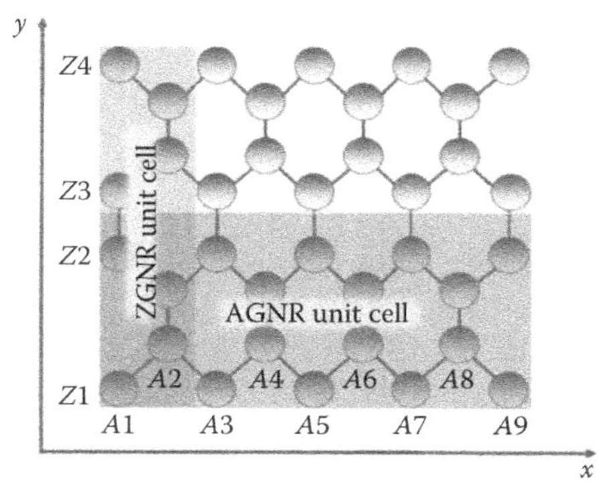

FIGURE 2.48
Schematic representation of the GNRs. The translation of the unit cells (shaded regions) along the x- and y-axes results in zigzag and armchair GNRs, respectively. The "Z" and "A" indices specify the width of zigzag and armchair GNRs, respectively. (S. Dutta and S. K. Pati, *Journal of Materials Chemistry*, 20, 8207–8223, 2010. Reproduced by permission of The Royal Society of Chemistry.)

is the absence of metallic AGNRs with zero bandgap, indicative of the fact that all bandgaps are semiconducting. The α, β, and γ categories used by them is similar to the band index $\nu = 0$, 1, and 2 of Arora and Bhattacharyya [14]. α-AGNRs are identified with $N_A = 8, 11, 14, \ldots$; β-AGNRs with $N_A = 9, 12, 15, \ldots$; and γ-AGNRs with $N_A = 10, 13, 16, \ldots$. The bandgap $E_g = \delta_A/W_A(\text{nm})$ in each category is inversely proportional to the width W with $\delta_A = 0.04$, 0.86, and 1.04 eV·nm for α, β, and γ, respectively. The corresponding effective mass ratio $m^*/m_o = \varepsilon_A/W_A(\text{nm})$ with $\varepsilon_A = 0.005$, 0.091, and 0.160 has been identified. Raza [16] revisited his formalism for ZGNRs and reported $E_g = \delta_Z/W_Z(\text{nm})$ with $\delta_Z = 1.65$ eV·nm, which is considerably larger than that in an AGNR. Son et al. [17], based on the first-principles approach, present scaling rules for the bandgaps of GNRs as a function of their widths. The GNRs considered have either armchair- or zigzag-shaped edges on both sides with hydrogen passivation. Their ab initio calculations show that the origin of energy gaps for GNRs with armchair-shaped edges arises from both the quantum confinement and crucial effect of the edges. No empirical relation indicating the bandgap dependence on GNR width is given.

In terms of the number of chains, the width is given by

$$W_A = \frac{N_A - 1}{2}a, \quad W_Z = \frac{3N_Z - 2}{2\sqrt{3}}a \tag{2.138}$$

where $a = \sqrt{3}a_{cc} = 0.246$ nm is the lattice constant. CNTs are being unzipped to create GNR [19]. Thin, elongated strips of GNRs that possess straight edges gradually transform from semiconductors to semimetals as their width increases and represents a particularly versatile variety of graphene. So, there is naturally a connection between the width W of GNR and the chiral vector $C_z(n, n)$ for ZGNR from ACNT (n, n) and $C_A(n, 0)$ for AGNR from ZCNTs $(n, 0)$. The width so obtained are

$$W_A = an = \sqrt{3}\,na_{CC}, \quad W_Z = \sqrt{3}\,an = 3na_{CC} \tag{2.139}$$

where $a = \sqrt{3}a_{CC} = 0.246$ nm obtained from $a_{CC} = 0.142$ nm. When comparison of Equation 2.138 is made to Equation 2.139, $N_A = (2n + 1)$ and $N_Z = (6n + 2)/3$. In reality, no CNT is made by rolling a GNR into a tubular shape. However, this roll-up allows one to apply the boundary condition in terms of equivalent Dirac K points that must meet in phase for electrons to form standing waves. In the same spirit, the reverse process of a CNT being unzipped to form a GNR with standing electron waves is designed [19]. In fact, a number of attempts have been made to unzip a CNT into a GNR [19,20]. The electron standing waves are formed similar to sound waves in an open pipe on both ends, as standing waves are restricted to a few nanometers width. The spectrum breaks into a set of subbands, and electron energy along the confined direction is discretized in terms of bandgap index of the CNT from which it is rolled out into a GNR. As in a CNT, bandgap is found to be inversely proportional to the GNR width W. Both theory and experiments differ in their outcomes giving metallic or semiconducting GNR, depending on the band index similar to what was used for a CNT [14].

In momentum k-space, there are bonding and antibonding wavefunctions. As seen in Figure 2.49, the valence and conduction states meet at K (or K') points. Dispersion around these points is conical. A hexagon formed by the $3K$ and $3K'$ points defines the graphene unit cell in k-space; beyond this unit cell, the dispersion relation repeats itself. As in CNTs, the wavefunction is given as $\exp[j\nu(2\pi/3)]$. This wavefunction maps onto itself by a rotation of $2\pi/3$. There are two further symmetries of graphene that are important for understanding CNT being flattened into a GNR. The first is between k and $-k$ states. In the absence of a magnetic field, forward k and backward $-k$ moving states have identical eigenenergies as is well known both for parabolic semiconductors as well as for graphene with $E = \hbar v_{Fo}|k|$. This degeneracy occurs both at K and K' with $\vec{K}' = -\vec{K}$. The total phase change as one starts from one of the three K points to other two and returning to the same one is π. The angular spacing between K and K' in k-space is $(2\nu + 1)\pi/6$, where $\nu = 0, 1, 2$. $\nu = 3$ is equivalent to $\nu = 0$ repeating the pattern. This gives $k_W W = (2\nu + 1)\pi/6$ with $\nu = 0, 1$, and 2 for equivalent K or K' points, where k_W is the momentum vector along the width and k_L is along the length of the nanoribbon. The small width of GNRs can lead to quantum confinement of carriers that can be modeled analogous to the standing waves in a pipe open at both ends, as shown in Figure 2.49.

The dispersion for a GNR is then given by

$$E = E_{Fo} \pm \hbar v_F|k| = E_{Fo} \pm \hbar v_F \sqrt{k_L^2 + \left((2\nu + 1)\frac{\pi}{6W}\right)^2} \quad (2.140)$$

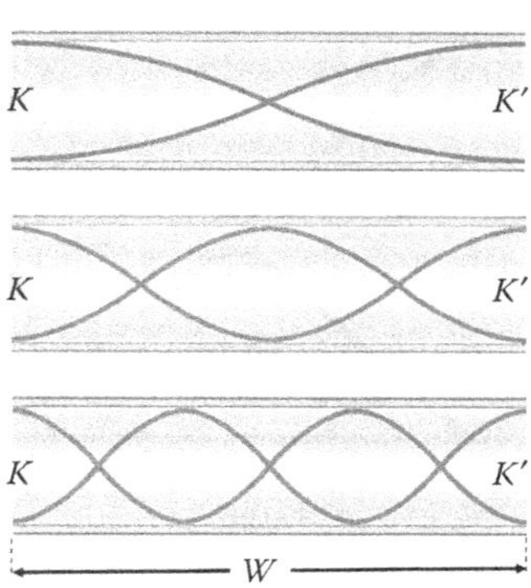

FIGURE 2.49
Standing waves in a GNR with different modes. The propagating wave along the width of the tube have phase $k_y W = (2\nu + 1)\pi/3$ as K space is segmented from K to other $3K'$ points.

This leads to bandgap equation

$$E = E_{Fo} \pm \frac{E_g}{2}\sqrt{1+\left(\frac{3Wk_L}{\pi(2\nu+1)}\right)^2} \tag{2.141}$$

In the parabolic approximation as for CNTs, the bandgap and effective mass are given by

$$E_g = (2\nu+1)\frac{\pi}{3W}\hbar v_F = (2\nu+1)\frac{0.69\ \text{eV}\cdot\text{nm}}{W} \tag{2.142}$$

and

$$\frac{m_L^*}{m_o} = \nu\frac{\pi}{6W}\frac{\hbar}{v_F m_o} = (2\nu+1)\frac{0.06\ \text{nm}}{W} \tag{2.143}$$

Figure 2.50 shows the complete $E-k_L$ spectrum with exemplary $N_A=7$ taken for an AGNR and $N_Z=8$ for ZGNR. The effective chiral index is $n_A+1=4$ for an AGNR and $n_Z=4$ for a ZGNR. The bandgap index is $\nu=1$ in each of these two cases. The linear dotted-dash line passing through the origin in Figure 2.50 is the representation of asymptotic behavior as $k_LW \gg 1$. In this extreme, Equation 2.140 collapses to

$$E_\nu = E_{Fo} \pm \hbar v_{Fo} k_L \tag{2.144}$$

All $E-k_L$ curves are semiconducting in nature with a finite bandgap. The curvature near $k_L=0$ is parabolic, represented by an effective mass. In the large $k_L \to \infty$ limit, the dispersion is linear with Equation 2.144 forming an asymptote.

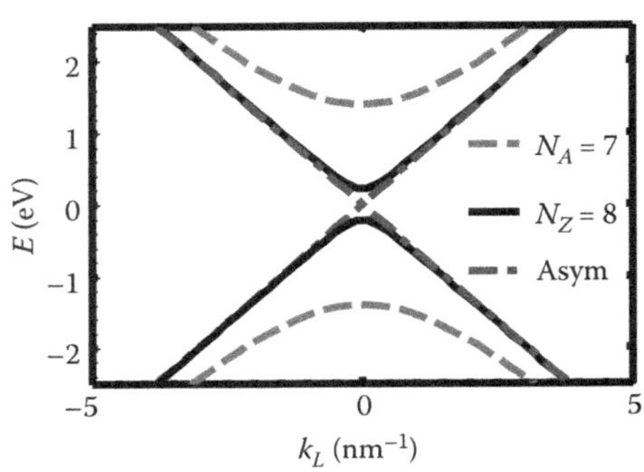

FIGURE 2.50
E versus k_L graph for two different armchair and zigzag nanoribbons having similar number of chains ($N_A=7$ with $n_A+1=4$ and $N_Z=8$ with $n_Z=4$) associated with $v=1$. Asymptotic value for large k_L is also shown with dotted-dash line.

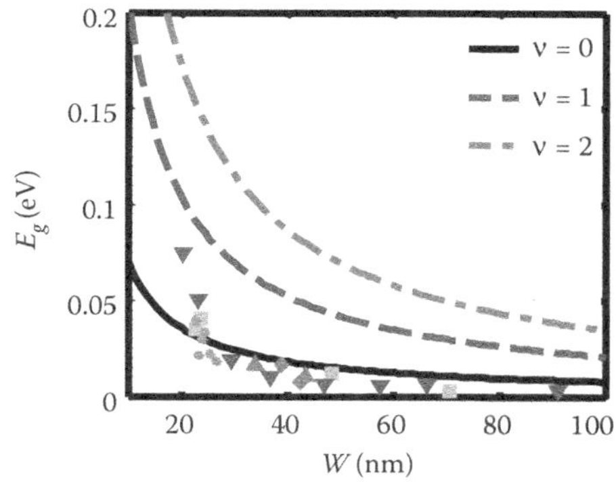

FIGURE 2.51
The GNR bandgap as a function of width *W*. Markers are experimental data. (Adapted from M. Y. Han et al. *Physical Review Letters*, 98, 206805, 2007.)

The bandgap E_g as a function of GNR width W is shown in Figure 2.51. From the experimental data, it appears that the effective thickness is larger by ΔW, indicative of the fact that edge effect plays a dominant role. Also, the experimental data for smaller widths tend to follow $\nu = 2$ curve, indicating the transition to $\nu = 1$ as width is increased, ultimately transitioning to $\nu = 0$ as GNR transforms to layered graphene for width exceeding 100 nm.

As the width of the GNR goes beyond 100 nm, a return to a normal graphene structure is expected. The DOS for the GNR follows the same pattern as for the CNT. The quantum confinement used in the above argument is akin to that in a pipe, open on both ends as in sound waves. The maximum (antinode) on the open edge can be outside the edge to account for edge effect. That makes effective thickness larger by ΔW, estimated to be around 1.5 nm. This edge effect depends on preparation of the sample or on the unzipping of the CNT into GNR. A commonly accepted empirical formula for bandgap as a function of width is given by

$$E_g = \frac{\alpha}{W + \Delta W} \tag{2.145}$$

The simplistic model noted above predicts $\alpha = (2\nu + 1)0.69$ eV·nm. α can be 0.69 eV for GNR with $\nu = 0$, 2.07 eV for $\nu = 1$, and 3.45 for $\nu = 3$. Ideally $\Delta W = 0$, but the passivation of the edges can make antinode of the standing waves further from the edge of GNR. The values reported for E_g in the literature have wide variations from as low as 0.2 eV and rising to more than 2.0 eV. The periodic symmetry model predicts small bandgap GNR with $\nu = 0$, intermediate bandgap with $\nu = 1$, and large bandgap GNR with $\nu = 2$. Complexity also arises from the nature of GNR (armchair or zigzag). For ZGNR, α and W are found to be the same for all values of N_A, the number of armchair chains. This shows

ZGNR have a unique value $\nu = 0$ leading to the smallest bandgap. However, for AGNR, $\nu = 0$, 1, and 2 correspond to three types of semiconducting states. As one can imagine, the dangling bonds at the surface change the value of E_g and ΔW and hence, the difficulty of identifying their values uniquely.

The effective mass along the length of GNR is shown in Figure 2.52. As in CNT, the mass is larger for $\nu = 2$. Once again, as the width increases, the electron reaches massless Dirac Fermion in a graphene nanolayer.

A mental map of the GNRs being unzipped from CNTs makes it easier for us to connect bandgap index ν to chirality (n, m) and GNR index N_A or N_Z. In that case, the width can be related to that of a chial vector $C_Z = a\sqrt{3}n$ for ZGNR unzipped from ACNT. Similarly, $C_A = an = \sqrt{3}a_{CC}n$ for AGNR unzipped from ZCNT. In this setup, $\nu = 0$ for ZGNR and $\nu = n \bmod 3$ for AGNR. There is no metallic GNR with a zero bandgap. The bandgap is a function of the width with $(2\nu + 1) = 1, 3,$ or 5, giving three states SC0, SC1, and SC2 and related effective masses $m^*_{W0,1,2}$. Figure 2.53 shows AGNR is a result of unzipping ZCNT of $(n_A, 0)$ chirality, where subscript A refers to armchair of AGNR. Similarly, ZGNR is a result of unzipping ACNT of (n_Z, n_Z) chirality. Because of the high length-to-width aspect ratio, GNRs are 1D nanoconductors. AGNR has an armchair cross-section at the edges. ZGNR has a zigzag cross-section. GNRs width (W_A for armchair or W_Z for zigzag) is determined by the number of armchair (N_A) or zigzag (N_Z) chains.

Width of the AGNR and ZGNR is stated in the literature according to number of chains N_A and N_Z as shown in Figure 2.53. The other way of describing width is in terms of chiral index. W_A of AGNR is related to the chiral length of flattened ZCNT of chirality $(n_A, 0)$ and W_Z of ZGNR is related to the chiral length of the flattened ACNT chirality (n_Z, n_Z). The width of AGNR is the adjusted chiral vector C_{ZCNT} of the ZCNT, as shown in Figure 2.53, giving

$$W_A = C_{ZCNT} + a = n_A a + a = (n_A + 1)a \qquad (2.146)$$

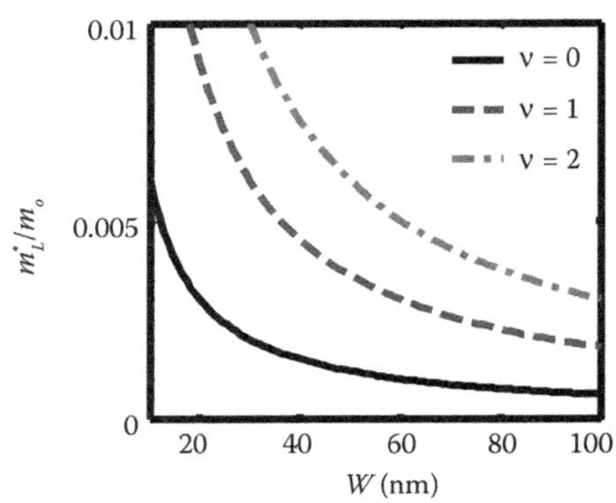

FIGURE 2.52
The GNR-normalized effective mass as a function of its width for $\nu = 0$, 1, and 2.

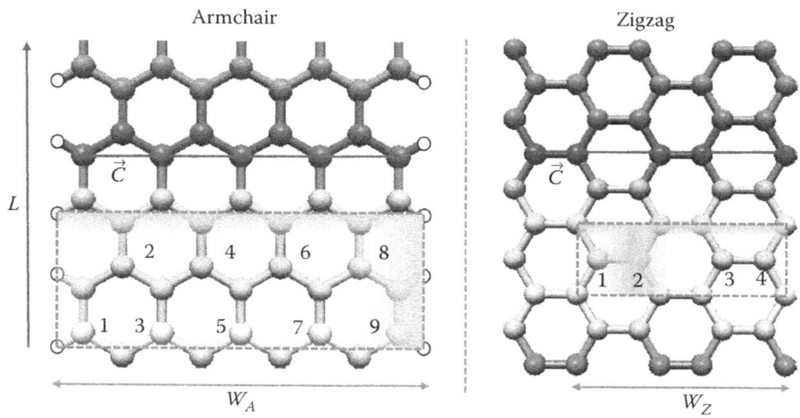

FIGURE 2.53
AGNR and ZGNR where edges look like armchair and zigzag, respectively, with exemplary chain index and chiral vector in each case.

The additional "*a*" in Equation 2.146 exists because of dangling bonds that form the armchair and are normally not included in the total length in counting the number of chains. In terms of chain index, W_A is given by [12]

$$W_A = \frac{N_A - 1}{2}a + a \tag{2.147}$$

Equations 2.146 and 2.147 yield $n_A = (N_A - 1)/2$, which means $N_A = 3, 5, 7, \ldots$ is an odd number. This nomenclature does not permit the even number of chains for AGNR, making comparison difficult with the theoretical framework of Raza and Kan [15]. Table 2.4 lists a few of the chains for which bandgap and effective mass are calculated. Two points for which experimental results are available are noteworthy [21] and included in Table 2.4. Chen et al. [21] fine-tuned the bandgap of AGNR with experimental values reported as 2.5 eV for 7-AGNR with $N_A = 7$ and 1.4 eV for 13-AGNR with $N_A = 13$. This is in reasonable agreement with the corresponding predicted values 2.1 and 1.2 eV, considering that the edge effects and stretching/compressing the chains can change the bandgap.

Figure 2.54 gives the bandgap as a function of width. Circles in Figure 2.54 are the points shown in Table 2.4. Noteworthy in Figure 2.54 is the experimental data of Chen et al. [21] for 7-AGNR and 13-AGNR that follows closely the curve for bandgap index $\nu = 1$ as expected from the corresponding chirality. The local electronic structure of 13-AGNR was characterized by performing measurements on 15 different AGNRs of varying widths from 3 to 11 nm. An asymmetry between the conduction band edge and valence band edge is identified that is not accountable by our theoretical framework.

TABLE 2.4
Band Index, Width, Bandgap, and Effective Mass for a Number of AGNRs

Number of Chains N_A	Chiral Index $n_A = \frac{N_A - 1}{2}$	Band Index $\nu = (n_A + 1) \times \text{mod}3$	Width $W_A = (n_A + 1)a = 0.246(n_A + 1)$ (nm)	Bandgap $E_{g\nu}$ (eV)	Effective Mass m_ν^*/m_o
5	2	0	0.738	0.934	0.082
7	3	1	0.984	2.104	0.185
9	4	2	1.230	1.683	0.247
11	5	0	1.476	0.467	0.041
13	6	1	1.722	1.202	0.106
15	7	2	1.968	1.753	0.154
17	8	0	2.214	0.311	0.027
19	9	1	2.460	0.841	0.074
21	10	2	2.706	1.275	0.112

Perhaps that is due to partially filled valence band that causes this asymmetry. In fact, most intrinsic CNTs have been identified to be p-type and certainly that may be the case with GNR which can be settled by further careful experimentation. This asymmetry is also indicative of the fact that the Fermi energy E_F for a grown GNR is distinct from normally assumed $E_{Fo} = 0$. That is why we made a distinction between E_F and E_{Fo}. The dangling bonds at the edges can enhance the carrier concentration and shift the Fermi energy toward the conduction band.

Figure 2.54 also shows the comparison with the theory of Son et al. Their $N_a = 3p$ and $3p + 1$ curves are well replicated by $\nu = 0$ and 1, respectively.

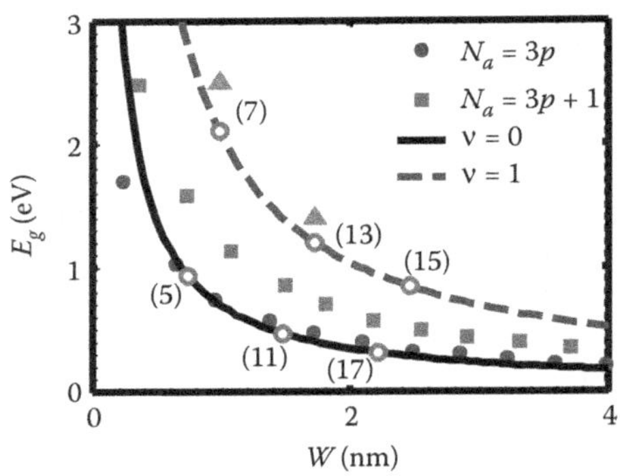

FIGURE 2.54
E_g versus W curve which compares previous theoretical framework and band-index-based formalism (this work) with two distinct ν values ($\nu = 0$, 1) for AGNRs. Available experimental data ($N_A = 7$, 13) agrees well with $\nu = 1$ curve. Open circles represent data available in Table 2.4. (Adapted from Y.-W. Son, M. L. Cohen, and S. G. Louie, *Physical Review Letters*, 97, 216803, 11/22/2006.)

In fact, their outcome differs between TB and LDA calculations. In their note, it is shown that LDA Kohn–Sham gaps in general underestimate the quasiparticle bandgaps of semiconductors as elucidated by Hybertsen and Louie [22]. They attribute the overall increase in the value of the gap by quasiparticle corrections using the GW approximation as discussed by Miyake and Saito [23,24]. The $N_a = 3p + 2$ mode of predicting a small bandgap is incompatible with $\nu = 2$ of our framework, confirming that no metallic or small bandgap state is possible in AGNR. These findings are consistent with the exeprimental data of Chen et al. [21] giving support to the framework presented.

As CNT is unzipped to form a GNR, the width should be connected to the chiral vector. A ZGNR with zigzag edges is a derivative of ACNT of index (n_Z, n_Z). Therefore, the width of ZGNR from the geometry of ACNT is determined to be

$$W_Z = C_{ACNT} - a_{CC} = \sqrt{3}an_Z - a_{CC} = (3n_Z - 1)a_{CC} \tag{2.148}$$

where a_{CC} is the C–C bond length and $a = \sqrt{3}a_{CC}$ is the lattice constant. Subtraction of a_{CC} from C_{ACNT} is necessary to account for the extension of C_{ACNT} into edges. In terms of chain index N_Z, the width W_Z is given by [12]

$$W_Z = \frac{3N_Z - 2}{2\sqrt{3}}a = \frac{3N_Z - 2}{2}a_{CC} \tag{2.149}$$

Equating (2.148) and (2.149) gives $n_Z = N_Z/2$, which means the chain index can take only even values $N_Z = 2, 4, 6, \ldots$. Figure 2.55 shows the bandgap as a function of width for a ZGNR.

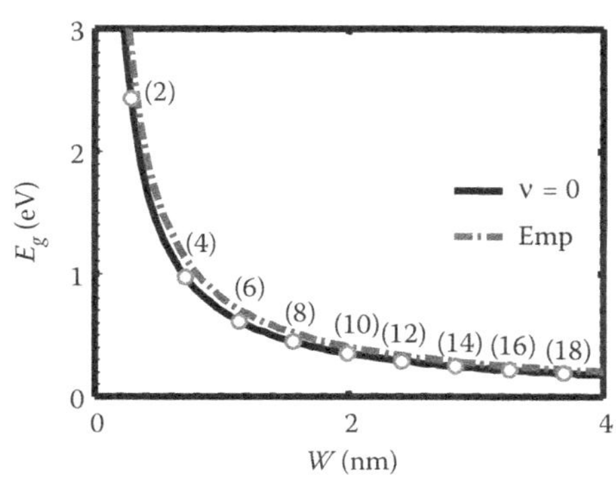

FIGURE 2.55
E_g versus W graph for ZGNR with $\nu = 0$. Calculated bandgap as a function of zigzag chain number is shown with open circles. Empirical relation is also shown by the dash-dotted line. (Adapted from X. Li et al. *Science*, 319, 1229–1232, February 29, 2008.)

Raza [16] studied ZGNR with periodic roughness and reports a significant bandgap opening akin to $\nu = 1$ of AGNR. In fact, he was able to show transitioning of the conduction band to valence band with hole-like effective mass with Γ-point bandgap engineering. The periodic edge roughness is a direct amalgamation of ZGNR, leading to AGNR as periodic roughness progresses to conditions appropriate for an AGNR. His-ball-and-stick model with $N_A = 8$, 9, and 10 gives conditions intermediate between AGNR and ZGNR, which Li et al. [25] describe as mixed edges.

Raza [16] obtained the relation 1.65 eV.nm/W(nm) for amalgamated GNR. Open circles correspond to what is calculated in Table 2.5. The described formalism thus goes beyond what can be achieved by any of the theories in the specialized domain. The electric-field modulation resulting in TQW will lead to a wave function that can be described by Airy functions, as discussed by Fairus and Arora [4], extending the vision of bandgap engineering to uncharted territories. Li et al. [25] are of the opinion that bandgaps result from a staggered sublattice potential from magnetic ordering in ZGNRs. The all-semiconductor nature found in their sub-10-nm GNRs is consistent with the bandgap opening in GNRs with various edge structures suggested theoretically. Bandgap values extracted from the experimental data fall in between the limits of theoretical calculations for ZGNRs and AGNRs with various widths. Our categorization into various modes with band index ν has made this distinction crisp, which makes it easy to interpret the experimental data.

The energy gap in patterned GNRs can be tuned during fabrication with the appropriate choice of ribbon width and associated bandgap index. Theoretical understanding of ribbon dimensions and bandgap index for the electrical properties of graphene structures can be seen as a first step toward the development of graphene-based electronic devices. The most recent evidence of tuning the bandgap based on the formalism presented comes from Chen et al. [21] who developed fabrication tools controlling precisely the edge

TABLE 2.5

Band Index, Width, Bandgap, and Effective Mass for a Number of ZGNRs

Number of Chains N_Z	Chiral Index $n_Z = \frac{N_Z}{2}$	Band Index $\nu = 0$	Width $W_Z = (3n_Z - 1)a_{CC} = (3n_Z - 1)0.142$ (nm)	Bandgap $E_{g\nu}$ (eV)	Effective Mass m_L^*/m_o
2	1	0	0.284	2.430	0.214
4	2	0	0.710	0.972	0.085
6	3	0	1.136	0.607	0.053
8	4	0	1.562	0.442	0.039
10	5	0	1.988	0.347	0.031
12	6	0	2.414	0.285	0.025
14	7	0	2.840	0.243	0.021
16	8	0	3.266	0.211	0.019
18	9	0	3.692	0.187	0.016

geometry as well as the GNR width. There is still a puzzling question that has to be resolved in experimentally determining the bandgap of GNR. GNR is a direct result of making thin layers of graphene where existence of symmetrical Dirac cone with zero bandgap is well established. However, most experimental metallic CNTs tend to indicate a p-type behavior indicative of the fact that valence band is partially full, contrary to what is normally conceived. Also, the presence of dangling bonds can increase the line density of electrons and shift the Fermi energy toward conduction band. Perhaps that is the reason that Chen et al. [21] observed asymmetry in their observation of the bandgap. The conduction band edge of 1.21 eV and valence band edge of −0.15 eV was determined in their work, indicative of the fact that $E_{Fo} \approx 0$ is not quite at the middle of the bandgap. Such asymmetries are noted in a number of other experiments as well. All pervasive data of Li et al. [25] were solution phase-derived, stably suspended in solvents with noncovalent polymer functionalization, and exhibited ultrasmooth edges with possibly well-defined zigzag or armchair-edge structures. Their electrical transport experiments showed that, unlike SWCNTs, all of the sub-10-nanometer GNRs produced were semiconductors and afforded graphene FET with current on–off ratios of about 10^7 at room temperature. These experimental findings are consistent with what is stated above. While CNTs do show a metallic behavior [14], the boundary conditions applicable to GNRs do not show any metallic behavior.

Han et al. [26] did perform a comprehensive experimental measurement in 2007 when GNR band structure theory was not well developed. That is why their experimental data is not categorized into armchair/zigzag, chain index, or equivalent chiral index categories. In fact, controversy on the band structure still continues as we find from our review of the literature. We are not aware of any reference where a distinction of GNR into bandgap index has been made, similar to what has been presented. The closest work to our categorizing the bandgap into three semiconducting states comes from the work of Raza and Kan [15] who have categorized the AGNR bandgap into α, β, and γ categories, based on chain index N_A similar to the categories of Son et al. [17] from the first-principles calculations. In fact, they indicate that a metallic GNR is expected in tight-binding approximation. First-principle work appears to be more consistent with our predictions. However, the exact value of the bandgap does not agree for all categories considered. Dutta and Pati [18] in their review of novel properties of GNR indicate that the passivation of edge carbons by foreign atoms like hydrogen is well captured within the first-principles calculations. In the presence of such passivation, the bonding characteristics change at the edges, leading to the modification of onsite energies or hopping integrals at the edge carbon atoms, in comparison to the bulk. Lack of these considerations within the tight-binding approach results in different observations compared to those of first-principles calculations. In fact, we find the similar problems in our formalism whether or not to include these passivated dangling bonds as we try to make extraction of GNR width from corresponding chiral index. The presence of strains and of applied and induced electric fields from the trapped charges further makes difficult the comparison with theory. Shemella et al. [27] compute the energy gap difference between highest occupied molecular orbital (HOMO) and lowest unoccupied molecular orbital (LUMO) dependence for finite width and length and infinite length for both armchair and zigzag ribbons. They discuss their results in light of the effect of passivation on the electronic properties of graphenes and their impact on nanoelectronic devices based on graphene. Their results are presented in terms of chain index that can be categorized into three categories stated above. The differences do exist in terms of the exact magnitude; the general trend towards bandgap being inversely proportional to width is now well established. We believe that we have given a comprehensive outlook on the bandgap engineering of GNR to encompass all categories which will be useful to experimentalists as well as theorists to make comparisons.

2.24 Tunneling through a Barrier

It is worthwhile to consider the complement of the quantum well (inverted quantum well), the potential barrier of height U_o, and thickness d that can be produced, for example, if the AlGaAs layer is sandwiched in between two GaAs layers. The tunneling is perfect if the energy of the incoming particle resonates with the corresponding energy of a quantum well of depth U_o and well thickness d [28].

Figure 2.56a shows the propagation of the electron waves reflecting in Region I, decaying in Region II, and again propagating in Region III (without reflection).

$$\begin{aligned}
&\text{Region I: } \psi_1(x) = A\,e^{jkx} + B\,e^{-jkx} \Rightarrow k = \sqrt{\frac{2m^*E}{\hbar^2}} \\
&\text{Region II: } \psi_2(x) = C\,e^{-\alpha x} + D\,e^{+\alpha x} \Rightarrow \alpha = \sqrt{\frac{2m^*(U_o - E)}{\hbar^2}} \\
&\text{Region III: } \psi_1(x) = F\,e^{jkx} \Rightarrow k = \sqrt{\frac{2m^*E}{\hbar^2}}
\end{aligned} \tag{2.150}$$

The matching of the wavefunction at the boundaries gives the following set:

$$\begin{aligned}
x = 0 &: \begin{cases} A + B = C + D \\ jkA - jkB = -\alpha C + \alpha D \end{cases} \\
x = d &: \begin{cases} C\,e^{-\alpha d} + D\,e^{+\alpha d} = F\,e^{jkd} \\ -\alpha C\,e^{-\alpha d} + \alpha D\,e^{+\alpha d} = jkF\,e^{jkd} \end{cases}
\end{aligned} \tag{2.151}$$

The tunneling coefficient $T = |F/A|^2$ is obtained as

$$T = \left|\frac{F}{A}\right|^2 = \frac{4}{4\cosh^2(\alpha d) + \left[\frac{\alpha}{k} - \frac{k}{\alpha}\right]^2 \sinh^2(\alpha d)} \qquad E < U_o \tag{2.152}$$

In an alternative form, Equation 2.152 is written as

$$T = \left|\frac{F}{A}\right|^2 = \frac{4E(U_o - E)}{4E(U_o - E) + U_o^2 \sinh^2(\alpha d)} \qquad E < U_o \tag{2.153}$$

The case $E > U_o$ is easily tackled by using $\sin(jx) = j\sinh(x)$ leading to

$$T = \left|\frac{F}{A}\right|^2 = \frac{4E(E - U_o)}{4E(E - U_o) + U_o^2 \sin^2(k_b d)}, \qquad E > U_o \tag{2.154}$$

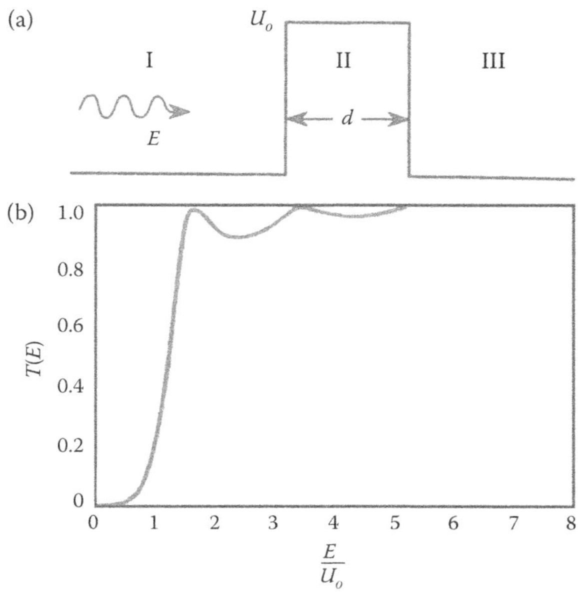

FIGURE 2.56
Figure 2.56 shows a potential barrier of height U_o and thickness d through which electron waves can penetrate. The tunneling is resonant (perfect) when the energy in a corresponding quantum well resonate with the energy of the incident particle.

where

$$k_b = j\alpha = \sqrt{\frac{2m^*(E - U_o)}{\hbar^2}} \tag{2.155}$$

Classically, the particle would be turned back by barrier ($T = 0$) if $E < U_o$ and transmission is perfect ($T = 1$) if $E > U_o$. $T = 1$ for $E > U_o$ is obtained if

$$\sin(k_b d) = 0 \Rightarrow k_b d = n\pi \tag{2.156}$$

This condition is equivalent to the barrier thickness being a half integer multiple of the de Broglie wave in the barrier region $a = n\lambda_D/2$. This is called transmission resonance as shown in Figure 2.56b.

In the case $E < U_o$ and α sufficiently large, Equation 2.153 simplifies to

$$T \approx \frac{16E(U_o - E)}{U_o^2} e^{-2\alpha d} \tag{2.157}$$

One way of looking at this approximation is to the amplitude of the quantum wave decreasing from $C\,e^{-\alpha x}|_{x=0} = 1$ to $C\,e^{-\alpha x}|_{x=d} = C\,e^{-\alpha d}$, since the tunneling coefficient is a ratio of the square of the amplitudes as it emerges from the barrier $T \approx |C\,e^{-\alpha d}/C|^2 = e^{-2\alpha d}$. However, C differs from A by a decrease in the quantum wave being reflected that is $T_o = 16E(U_o - E)/U_o^2$. The net tunneling is then $T \approx T_o\,e^{-2\alpha d}$.

2.25 WKB Approximation

In most cases, the tunneling coefficient is calculated by the WKB approximation with turning points x_1 and x_2 as shown in the Figure 2.57. The turning points x_1 and x_2 can be calculated by equating potential energy to kinetic energy ($E = U(x)$). It is at the turning point the electron comes to rest as its

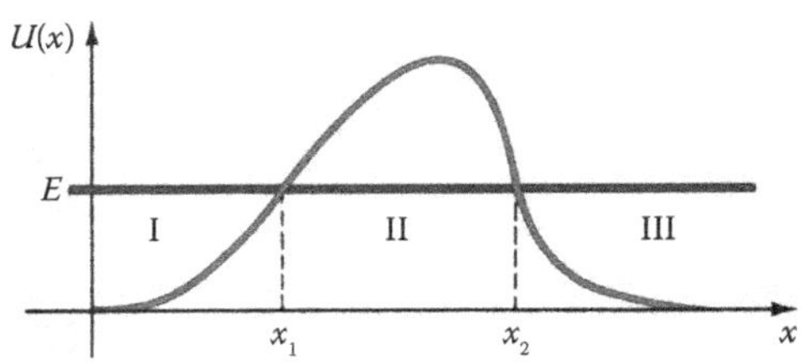

FIGURE 2.57
A potential barrier of energy $qV(x)$ with classical turning point x_1 and x_2.

potential energy is consumed in going up the potential hill and hence all kinetic energy is converted into potential energy. WKB is an acronym for Wentzel–Kramers–Brillouin. WKB approximation is very convenient in finding tunneling coefficient for engineering application.

For this potential barrier, WKB approximation gives the tunneling coefficient given by

$$T \cong \exp\left[-2\int_{x_1}^{x_2} |\alpha(x)\,dx|\right] \tag{2.158}$$

with

$$\alpha(x) = \sqrt{\frac{2m^*}{\hbar^2}(U(x) - E)} \tag{2.159}$$

A modified WKB (MWKB) to make it compatible with the approximation of rectangular barrier can be designed with pre-exponent factor the same as the rectangular barrier as given by

$$T \cong T_o \exp\left[-2\int_{x_1}^{x_2} |\alpha(x)\,dx|\right], \quad T_o \approx \frac{16E(U_o - E)}{U_o^2} \tag{2.160}$$

In T_o, U_o is the peak of the potential barrier. The arbitrary shape potential barrier in this description can be thought of as a series of barriers of differential length dx that can be integrated. When Equation 2.158 is applied to the triangular potential barrier of energy $U_o = qV_o$, as shown in Figure 2.58, the tunneling coefficient is obtained as

$$T = T_o \exp\left\{\frac{-4(2m^*)^{1/2}}{3|q|\mathcal{E}\hbar}(U_o - E)^{3/2}\right\} \tag{2.161}$$

If the electron resides at the Fermi level E_F on the left side of the barrier and the work function is $q\phi$, $U_o = q\phi - E_F$. T_o, as stated earlier, is given by Equation 2.160 with the barrier height the average of the two turning points.

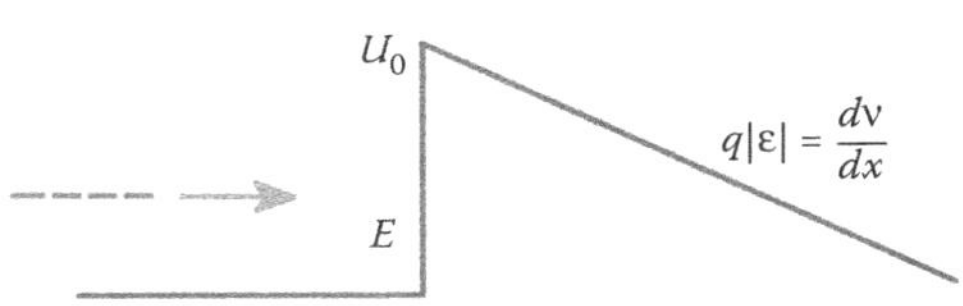

FIGURE 2.58
A triangular potential barrier of energy $U_o = qV_o$ with classical turning point x_1 and x_2.

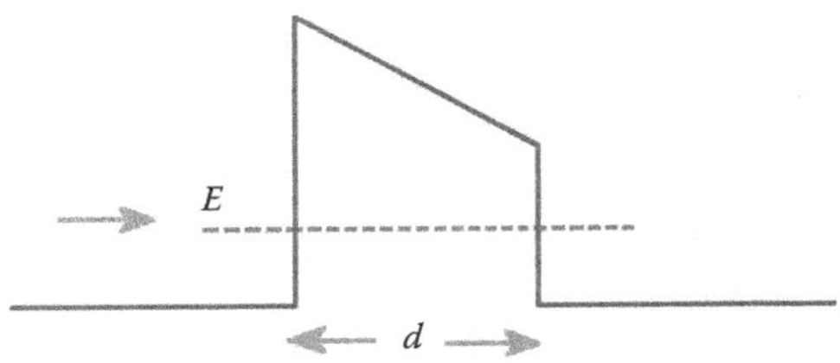

FIGURE 2.59
A trapezoidal potential barrier of energy $U_o = qV_o$ with an applied electric field E.

The arbitrary shape potential barrier in this description can be thought of as a series of barriers of differential length dx that can be integrated to get the WKB formula with correction for reflection coefficient. T_o then is the penetration coefficient as it traverses from barrier-free space to barrier edge.

The tunneling coefficient if the barrier happens to be the trapezoidal barrier as in Figure 2.59 is given by

$$T = T_o \exp\left\{\frac{-4(2m_o)^{1/2}}{3q\mathcal{E}\hbar}\left[(U_o - E)^{3/2} - (U_o - E - q\mathcal{E}d)^{3/2}\right]\right\} \tag{2.162}$$

EXAMPLES

E2.1 Calculate (a) the wavelength of a free electron of energy 1 eV from $\lambda_D = h/p$; (b) the wavelength of a photon of energy 1 eV from $E = hf = hc/\lambda$.

$$E(1\,\text{eV}) = 1.6 \times 10^{-19}\ \text{J}$$

$$\lambda_D = \frac{h}{p} = \frac{h}{\sqrt{2m^*E}} = \frac{6.63 \times 10^{-34}\ \text{J·s}}{\sqrt{2 \times 9.1 \times 10^{-31}\ \text{kg} \times 1.6 \times 10^{-19}\ \text{J}}}$$
$$= 1.23 \times 10^{-9}\ \text{m} = 1.23\,\text{nm}$$

$$\lambda = \frac{hc}{E} = \frac{6.63 \times 10^{-34}\ \text{J} \cdot \text{s} \times 3.0 \times 10^{8}\ \text{m/s}}{1.6 \times 10^{-19}\ \text{J}} = 1.24 \times 10^{-6}\ \text{m}$$
$$= 1.24\ \mu\text{m} = 1240\ \text{nm}$$

E2.2 What is the energy (in J and eV) of a red photon having a wavelength of 750 nm?

$$\lambda = \frac{hc}{E(\text{eV})} = \frac{1240\ \text{nm}}{E(\text{eV})}$$

$$E(\text{eV}) = \frac{hc}{\lambda} = \frac{1240\ \text{nm}}{\lambda} = \frac{1240\ \text{nm}}{750\ \text{nm}} = 1.65\ \text{eV}$$
$$= 1.65 \times 1.6 \times 10^{-19}\ \text{J} = 2.65 \times 10^{-19}\ \text{J}$$

E2.3 The electric vibration in household electricity is 50 Hz, a household microwave is 2.4 GHz, and UV light is 30 THz. Find the energy of quantum in each case.

$$E = hf = 6.63 \times 10^{-34}\ \text{J}\cdot\text{s} \times 60\ \text{s}^{-1} = 3.98 \times 10^{-32}\ \text{J}$$
$$= 2.49 \times 10^{-13}\ \text{eV (elect)}$$
$$= 9.92 \times 10^{-6}\ \text{eV (microwave)}$$
$$= 124.2\ \text{eV (UV light)}$$

Please note that the UV has very high energy that can tear cells.

E2.4 The quantum nature of photon is vividly seen in a photoelectric experiment where a photon takes an electron (photoelectron) from its deep lying state at the Fermi energy to vacuum. The energy required is called the work function. A metal has a work function of $q\phi = 4.0$ eV. The kinetic energy of the emitted photoelectron is 0.3 eV. What is the wavelength of the photon? Assume that the photon is completely absorbed and gives all its energy to the electron.

The energy conservation requires that the energy of a photon is converted into the work done to take an electron out of metal that requires $q\phi = 4.0$ eV. The rest goes in the form of kinetic energy.

$$E_{ph} = q\phi + \text{KE} = 4.0\ \text{eV} + 0.3\ \text{eV} = 4.3\ \text{eV}$$

$$\lambda = \frac{hc}{E(\text{eV})} = \frac{1240\ \text{nm}}{4.3} = 280\ \text{nm}$$

E2.5 Given Bohr's radius of the ground state in the hydrogen atom is $r_1 = 0.053$ nm, calculate the de Broglie wavelength of an electron in the ground state, its momentum, and its velocity.

$$\lambda_D = 2\pi r_1 = 0.333\ \text{nm}$$
$$p = \frac{h}{\lambda_D} = 2.0 \times 10^{-24}\ (\text{kg}\cdot\text{m/s})$$
$$v = \frac{p}{m_o} = \frac{2.0 \times 10^{-24}\ (\text{kg}\cdot\text{m/s})}{9.1 \times 10^{-31}\ \text{kg}} = 2.2 \times 10^{6}\ (\text{m/s})$$

E2.6 Find the wavelength of the first three wavelengths of the light arising from the Balmer series.

$$n_u = 3,\ 4,\ 5\left(E_i = -\frac{13.6\text{ eV}}{n_u^2} = -1.5, -0.85, -0.544\right)$$

$$\Downarrow$$

$$n_\ell = 2\left(E_i = -\frac{13.6\text{ eV}}{2^2} = -3.4\right)$$

$$\lambda = \frac{1240\text{ nm}}{\Delta E_{3\to 2}(\text{eV})} = \frac{1240}{1.9} = 653\text{ nm}$$

$$\lambda = \frac{1240\text{ nm}}{\Delta E_{4\to 2}(\text{eV})} = \frac{1240}{2.55} = 486\text{ nm}$$

$$\lambda = \frac{1240\text{ nm}}{\Delta E_{5\to 2}(\text{eV})} = \frac{1240}{2.856} = 434\text{ nm}$$

E2.7 Find the radius of the $n = 4$ Bohr orbit of a doubly ionized lithium atom Li^{2+} $Z = 3$.

$$r_n = \frac{4\pi\varepsilon_o n^2\hbar^2}{m_o Z q^2} = \frac{n^2 r_1}{Z} r_n$$

$$r_4 = \frac{16(0.053\text{ nm})}{3} = 0.283\text{ nm}$$

E2.8 a. Calculate the first two levels of a free electron confined to a box length of 0.2 nm. Assume infinite boundaries. b. What is the energy of photon emitted if the electron makes a transition from $n = 2$ to $n = 1$? c. Repeat for $L = 1$ cm.

Answer

a. $$E_1 = 1.50 \times 10^{-18}\text{ J} = 9.4\text{ eV}$$
$$E_2 = 6.0 \times 10^{-18}\text{ J} = 37.6\text{ eV}$$
$$\Delta E = 37.6\text{ eV} - 9.4\text{ eV} = 28.2\text{ eV}$$

In general $\Delta E = E_1(2n - 1)$

b. $$E_1 = 6.0 \times 10^{-34}\text{ J} = 3.8 \times 10^{-15}\text{ eV}$$
$$E_2 = 24.0 \times 10^{-18}\text{ J} = 15.2 \times 10^{-15}\text{ eV}$$
$$\Delta E = E_1(2n - 1) = 18.0 \times 10^{-34}\text{ J} = 10.4 \times 10^{-15}\text{ J}$$

The spacing is much less than thermal energy $k_B T = 25.9$ mV at room temperature. Hence, the energy spectrum appears continuous.

E2.9 Exciton, an e–h pair orbiting its center of mass, can be modeled after the hydrogen atom with electron mass being replaced by reduced mass. Calculate for GaAs ($\varepsilon_r = 13.3$) the binding energy and radius of the exciton by using an average of the heavy and light hole masses:

$$m_r^* = \left[\frac{1}{m_n^*} + \frac{1}{m_p^*}\right]^{-1} = \frac{m_n^* m_p^*}{m_n^* + m_p^*}$$

The binding energy is then given by $E = ((-m_r^*/m_o)(1/\varepsilon_r^2))R_Y$ with Rydberg's energy $R_Y = 13.6\,\text{eV}$ and the radius of the exciton is given by

$$r_{ex} = \frac{\varepsilon_r m_o}{m_r^*} r_1 = \frac{13.3}{0.0502} 0.053\text{ nm} = 14.0\text{ nm}$$

E2.10 An AlGaAs/GaAs/AlGaAs quantum well with a width $a = 5$ nm has conduction band discontinuity

$$\Delta E_c = E_c^{\text{AlGaAs}} - E_c^{\text{GaAs}} = 0.238\text{ eV}$$

Find the effective well width $L = a(1 + 1/P)$.

$$P = \frac{a}{2}\sqrt{\frac{2m^*\Delta E_c}{\hbar^2}} = \frac{5\times10^{-9}}{2}\sqrt{\frac{2\times0.067\times9.1\times10^{-31}\times0.238\times1.6\times10^{-19}}{(1.055\times10^{-34})^2}} = 1.62$$

$$L = 5\,nm\left(1 + \frac{1}{1.62}\right) = 8.1\,nm$$

E2.11 A CNT is a rolled-up cylinder of graphene sheet, its radius depending on the chirality (n, m) that determines how the CNT is rolled up.

The radius of the CNT is given by

$$R = \frac{\sqrt{3}}{2\pi} a_{CC}\sqrt{n^2 + nm + m^2}$$

$a_{CC} = 0.142\text{ nm}$ is the interatomic distance from carbon-to-carbon atom.

a. What is the radius of a (19, 0) CNT?
b. What is the radius of a (10, 10) CNT?
c. What is the index n of a ZCNT with chirality $(n, 0)$ of radius 0.3523 nm?

Answer

a. 0.7437 nm b. 0.678 nm c. $n = 9$.

E2.12 A 6 eV electron tunnels through a 2-nm-wide rectangular barrier with the tunneling coefficient of 10^{-8}. What is the height $U_o = qV_o$ of the potential barrier?

Answer

$$U_o = qV_o = 6.858 \text{ eV}$$

E2.13 Calculate the first two energy levels if the gate electric field on a MOSFET with a triangular potential well is 5×10^7 V/m for (a) circular subbands with twofold degeneracy having $m_3^* = m_\ell^* = 0.98m_o$, and (b) elliptic subbands with fourfold degeneracy having $m_3^* = m_t^* = 0.19m_o$.

Answer

a. $$\varepsilon_1 \approx \left[\frac{\hbar^2}{2m_3}\right]^{1/3}\left[\frac{3\pi q}{2}\mathcal{E}_t\left(0+\frac{3}{4}\right)\right]^{2/3}$$

$$\varepsilon_1 = \left[\frac{\left(1.055\times10^{-34}\right)^2}{2\times0.98\times9.11\times10^{-31}}\right]^{1/3}\left[\frac{9\pi\times1.6\times10^{-19}}{8}5\times10^7\right]^{2/3}$$

$$= 113.2 \text{ meV}$$

Through scaling of $(i + 3/4)^{2/3} = (0 + 3/4)^{2/3}$ to (1 + 3/ $4^{2/3}$), 113.2 meV is multipled by $(7/3)^{2/3}$ to obtain $\varepsilon_2 = 113.2 \text{ meV} \times (7/3)^{2/3} = 199.1 \text{ meV}$

b. Similarly, scaling the ε_1 and ε_2 by a changing effective mass by a factor of $(0.98/0.19)^{1/3}$, $\varepsilon_1 = 113.2 \times (0.98/0.19)^{1/3} = 195.4$ meV and $\varepsilon_1 = 199.1 \times (0.98/0.19)^{1/3} = 343.8$ meV

E2.14 In a layered nanostructure of silicon with confinement in the z-direction only two of six valleys with effective mass $m_t^* = 0.198m_o$ are appreciably populated in the quantum limit. One way to define the quantum limit is to ascertain that the spacing between the lowest two quantized levels is k_BT, that is, $\Delta E_{2\to1} = E_2 - E_1 = k_BT$. Assume the quantum-well model with infinite boundaries, so the energy in the confinement direction is given by

$$E_n = n^2\frac{\hbar^2\pi^2}{2m^*L_z^2} \quad \text{with } n = 1, 2, 3, \ldots$$

Find the confinement length L_z for which $\Delta E_{2\to1} = k_BT$ assuming effective mass

$$m_t^* = 0.198m_o$$

$$m_t^* = 0.198m_o$$

$$\Delta E_{2\to 1} = \frac{3\pi^2\hbar^2}{2m^*L^2} = 3E_1 = k_BT \Rightarrow L = \sqrt{\frac{3\pi^2\hbar^2}{2m^*k_BT}}$$

$$= \sqrt{\frac{3\pi^2\left(1.055\times 10^{-34}\right)^2}{2\times 0.198\times 9.1\times 10^{-31}\times 0.0259\times 1.6\times 10^{-19}}}$$

$$= 14.8\times 10^{-9}\text{ m} = 14.8\text{ nm}$$

E2.15 The birth of microelectronics is appreciated by the diameter of a human hair that is typically 100 μm. Now, since CNTs have the diameter in the order of 1.0 nm, estimate the number of tubes that can fit in the space of human hair, ignoring the empty space that exists between the tubes.

The area of a circular cross-section of a human hair is $\pi D^2/4$. With $D = 100$ μm, the cross-section area is 7.85×10^3 μm. The area of the cross-section of a tube is $\pi D_t^2/4$ which is $7.85\times 10^{-1}\text{ nm}^2$. The ratio of the two gives the number of CNTs: 1.00×10^{10} *CNTs*. The scaling arguments can be used to establish the number of tubes that can fit in the same area as that of human hair. Scaling factor $D/D_t = 100\times 10^{-6}/1.0\times 10^{-9} = 100{,}000 = 10^5$. Since area is proportional to the square of diameter, the number of tubes fitting the human-hair cross-section is $(10^5)^2 = 10^{10}$ which is 1010 billion.

PROBLEMS

P2.1 Find the momentum and velocity of a particle with a mass of 9.1×10^{-31} kg and a de Broglie wavelength of 20 nm.

P2.2 Find the frequency f (Hz), radian frequency ω (rad/s), and period T of the infrared radiation with a wavelength of 1 μm. Find the energy of the infrared photon of wavelength 1 μm. Express its energy both in J and eV.

P2.3 The potential (in V) due to a point charge q at a distance r from the charge is given by $V = ((1/4\pi\varepsilon_o)(q/r))$. (a) Evaluate the potential of nuclear charge $+q$ in the hydrogen atom at a distance $r = r_1 = 0.053$ nm (radius of the first Bohr orbit). (b) Obtain the potential energy (PE) $U = -qV$ of the electron with charge $-q$ in the potential of the nucleus. Calculate the kinetic energy $\text{KE} = ((1/2)(1/4\pi\varepsilon_o)(q^2/r_1))$ of the electron revolving in the first orbit. (c) Calculate the total energy $E = \text{KE} + \text{PE}$. Express your answers for PE (U), KE, and total energy E in joules (J) and electron volts (eV). (d) What is the linear momentum and the angular momentum of the electron in the ground state? (e) What is the time required to complete one orbit?

P2.4 Find (a) the longest wavelength in the Lyman series and (b) the shortest wavelength in the Paschen series. What is the energy of the photon in cases (a) and (b). Name the region of the electromagnetic spectrum in which these wavelength are.

P2.5 Calculate the wavelength and energy of a photon in the following transitions of an electron in a hydrogen atom. Which part of electromagnetic spectrum these photons of emitted light reside?

a. $n = 2 \rightarrow n = 1$

b. $n = 5 \rightarrow n = 4$

c. $n = 10 \rightarrow n = 9$

d. $n = 8 \rightarrow n = 2$

e. $n = 12 \rightarrow n = 1$

f. $n = \infty \rightarrow n = 1$

P2.6 (a) Find the de Broglie wavelength of the ground state ($n = 1$) of the hydrogen atom. (b) What is the quantum number n of the hydrogen-atom orbit represented by the following figure. (c) What is the radius of the hydrogen-atom orbit represented by following figure. (d) What is the velocity of the electron in the hydrogen-atom orbit represented by the following figure.

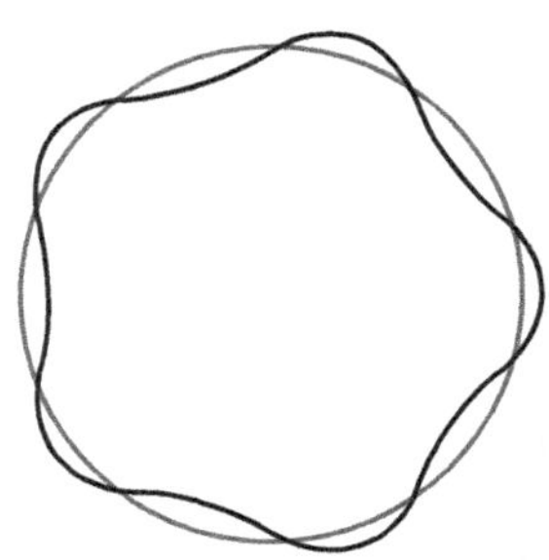

An orbit containing an integer number of de Broglie waves.

P2.7 Hydrogen atom number 1 is known to be in the 4f state. (a) What is the energy of this atom? (b) What is the magnitude of this atom's orbital angular momentum? (c) Hydrogen atom number 2 is in the 5d state. Is this atom's energy greater than, less than, or the same as that of atom 1? Explain. (d) Is the magnitude of the orbital angular momentum of atom 1 greater than, less than, or the same as that of atom 2? Explain.

P2.8 Give the electronic configuration for the ground state of carbon, nitrogen, silicon, phosphorus, boron, gallium, and arsenic.

P2.9 Suppose that the 5d subshell is filled in a certain atom. Write out the 10 sets of four quantum numbers for (n, ℓ, m_ℓ, m_s) the electrons in this subshell.

P2.10 a. In an $Al_xGa_{1-x}As/GaAs/Al_xGa_{1-x}As$ quantum well with electrons, find the value of x for $\Delta E_c = E_c^{AlGaAs} - E_c^{GaAs}$ is 0.238 eV, assuming two-thirds of it goes to the conduction band. The bandgap as a function of x is given by $E_g = 1.426 + 1.247x$.

b. Assume two-thirds of the bandgap difference of $Al_xGa_{1-x}As$ goes to conduction band discontinuity. Since $0.238 \gg k_BT$, it is justified to assume quantum well with infinite boundaries. One way to define quantum limit (only the lowest quantum limit is appreciably populated) is to ascertain that the spacing between the lowest two quantized levels is k_BT. Show that the confinement effects will be important if the length L of the 2D quantum well when

$$L \leq \sqrt{\frac{3\hbar^2\pi^2}{m^* k_B T}}$$

What should be the thickness L of the semiconductor layer to ensure that the difference between the ground (i.e., the lowest) energy level and the first excited level is equal to the thermal energy (k_BT) at room temperature ($T = 300$ K)?

P2.11 Exciton in the semiconductors lessen the bandgap by the exciton energy:

$$E_g' = E_g(\text{eV}) - \frac{m_r^*}{m_o \varepsilon_r^2} 13.6\ \text{eV}$$

a. For GaAs, determine the required photon energy to create an exciton. The reduced effective mass for exciton is $m_r^* = (1/m_n^* + 1/m_p^*)^{-1} = m_n^* m_p^* / m_n^* + m_p^* = 0.0502 m_o$.

b. The application of a dc electric field tends to separate the electron and the hole. Using Coulomb's law, show that the magnitude of the electric field between the electron and hole is

$$|\mathcal{E}| = \frac{1}{4\pi\varepsilon_o\varepsilon_r}\frac{q}{r_{ex}^2}$$

c. For GaAs, determine $|\mathcal{E}|$, the magnitude of an electric field that would break apart the exciton.

P2.12 The conduction band minima in GaP occur right at the first Brillouin zone boundary along ⟨100⟩ directions in *k*-space. Taking the constant energy surface to be ellipsoids with $m_\ell^* = 1.12\ m_o$ and $m_t^* = 0.22\ m_o$, determine the DOS effective mass electrons in GaP. What will be the conductivity effective mass?

P2.13 In Si, what fraction of the holes are heavy holes? How does this fraction change in a 2D quantum well? Discuss and obtain effective DOS effective mass.

P2.14 Estimate tunneling coefficient for an electron in gallium arsenide ($m_e^* = 0.067\ m_o$), tunneling through a rectangular barrier with a barrier height $U_o = 1.0$ eV and a barrier width of 2.0 nm. The electron energy is 0.25 eV.

P2.15 Estimate tunneling coefficient for free electrons with energy of 0.5 eV ($m_o = 9.11 \times 10^{-31}$ kg) incoming onto a parabolic barrier shown in the following figure with parabolic barrier $U_o = f_s x^2$ with $f_s = 4 \times 10^{16}$ eV/m^2 rising from $x = 0$ to 5.0 nm.

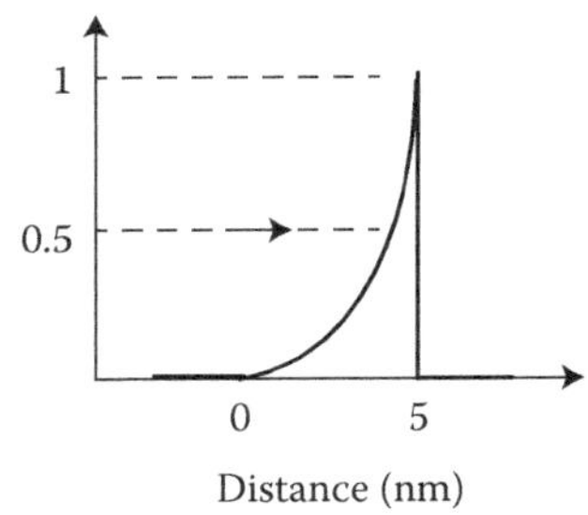

Parabolic potential barrier with peak value 1.0 eV.

P2.16 Obtain an expression for the tunneling coefficient of a rectangular barrier using the WKB approximation and compare with the exact calculations.

P2.17 Calculate the tunneling probability for free electron incoming into the triangular barrier and compare it with the tunneling probability of a rectangular barrier using MWKB approximation. Assume the height in each case is $U_o = 1.0\,\text{eV}$. The triangular barrier is described by $U(x) = U_o(x/L)$ for $0 \le x \le 2.0\,\text{nm}$. The rectangular barrier has the same area as the triangular barrier with $U(x) = U_o$ for $0 \le x \le 1.0\,\text{nm}$. The energy of the incoming electron is $E = 0.5$ eV.

P2.18 The potential energy of electrons in a metal with a surface electric field is shown in the following figure. The electron concentration is 10^{23} cm^{-3}, the electronic effective mass is 9.11×10^{-31}

kg. The velocity of electrons impinging on the metal surface is $1/4(\sqrt{2E/m_o})$ with $E = 4$ eV. (a) Find the strength of the electric field at the surface. (b) Calculate the electric current density $J = nqvT$ (in A/m^2) of electrons escaping the metal.

The strength of the electric field at the surface is $\mathcal{E} = 5/10^{-8}$ V/m $= 5 \times 10^8$ V/m.

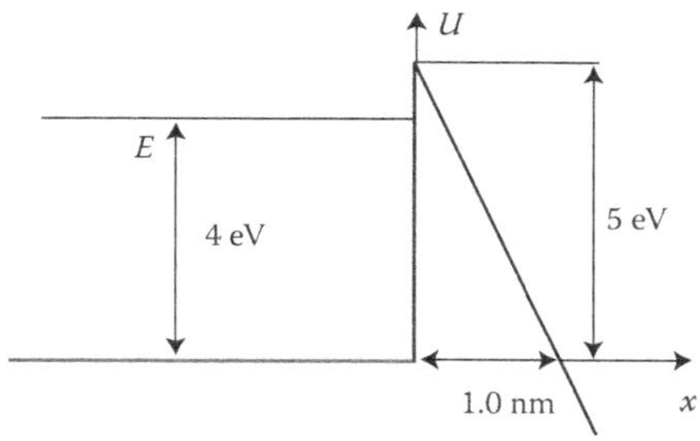

Triangular barrier of 1.0 nm thickness.

P2.19 Al–SiO_2–Al, a metal–insulator–metal heterojunction is a practical example of a rectangular barrier. Its barrier height is $q\phi_b = q\phi - q\chi = 4.1 - 0.9\,\text{eV} = 3.2\,\text{eV}$, where $q\phi = 4.1$ eV is the work function of aluminum and $q\chi = 0.9$ eV is the electron affinity for SiO_2. Determine the tunneling probability if the barrier width is 1 nm and the electron energy is 3.5 eV. State if any approximation used. Repeat for barrier width of 2, 3, and 10 nm.

P2.20 In a GaAs high-electron-mobility transistor (HEMT), the gate electric field with a triangular potential well is 5×10^7 V/m. Calculate the first two energy levels.

P2.21 Silicon crystal has 5×10^{22} atoms/cm^3. It is doped so 1 in 10,000 atoms are replaced by phosphorus. A donor impurity like phosphorus must replace a silicon atom (substitutional impurity) for silicon to become an n-type. The interstitial phosphorus in the empty space between silicon atoms does not donate an electron. Assuming 30% of the implanted phosphorus is substitutional, determine the effective doping level and the electron concentration. Determine the average distance between the phosphorus atoms and between the mobile electrons.

P2.22 An electron leaves a heated cathode with a kinetic energy of 1 eV in free space. Determine the velocity, the wave number, the wavelength, and the frequency of the electron wave. Repeat if the electron acquires an energy of 10 keV while accelerated through a potential drop of 10 kV.

P2.23 Consider the following CNTs with chirality: (4, 3), (4, 4), (5, 1), (6, 6), (9, 0), (9, 9), (12, 4), (10, 10), (11, 5), (12, 12), (12, 0), (9, 3), (10, 2), (11, 7). Make a table of their band index, diameter, bandgap, and effective mass.

P2.24 Consider the following CNTs with chirality: (4, 3), (4, 4), (5, 1), (6, 6), (9, 0), (9, 9), (12, 4), (10, 10), (11, 5), (12, 12), (12, 0), (9, 3), (10, 2), (11, 7). Identify which one is metallic and which ones are SC1 and SC2. Which of these are zigzag, armchair, and chiral? Make a table.

P2.25 Find the number of carbon atoms per m^2 in an intrinsic graphene sheet. Find the electronic density of electrons in a graphene sheet.

P2.26 Considering the graphene rollover into a CNT, what will be the linear density (# of electrons/m) in CNTs of chirality: (4, 3), (4, 4), (5, 1), (6, 6), (9, 0), (9, 9), (12, 4), (10, 10), (11, 5), (12, 12), (12, 0),(9, 3), (10, 2), (11, 7)? Make a table to indicate linear density.

CAD/CAE PROJECTS

C2.1 *Density of States.* Show that Q3D DOS for GaAs ($m^* = 0.067m_o$) are given by

$$D_3(E) = 1.183 \times 10^{20}\left[(E - E_c)(\text{eV})\right]^{1/2} \frac{1}{\text{cm}^3\ \text{eV}}$$

Write similar expressions for Q2D and Q1D quantum wells. Reproduce Figure 2.34.

C2.2 *Tunneling Probability.* Plot the tunneling probability versus electron energy for an electron incident on a triangular potential barrier where $q\phi = 3.0$ eV and the electric field is $\mathcal{E} = 10^9$ V/m. Take energy of the electron $(E - E_F)$ from 0 to 3 eV.

C2.3 *Approximate Wavefunctions in a Rectangular Quantum Well.* By referring to a book on quantum mechanics, calculate the exact function in an $Al_{0.3}Ga_{0.7}As/GaAs/Al_{0.3}Ga_{0.7}As$ quantum well (QW). Compare with the wavefunction obtained from using the effective width in terms of strength of the quantum well. How many levels are there in QW? A good place to start is the paper: Barry I. Barker, Grayson H. Rayborn, Juliette W. Ioup, and George E. Ioup, "Approximating the finite square well with an infinite well: Energies and eigenfunctions," *American Journal of Physics*, vol. 59, pp. 1038–1042, 1991.

C2.4 *Approximate Wavefunctions in a Triangular Quantum Well.* By considering the TQW, compare the exact wavefunction from the Airy function to approximate ones from the infinite quantum well with redefined effective width in terms of electric field or in terms of Stern–Howard exponential function. How good is the fit between the approximate functions and the Airy functions? Good place to start is: A. T. M. Fairus and V. K. Arora, "Quantum engineering of nanoelectronic devices: the role of quantum confinement in mobility degradation," *Microelectronics Journal*, vol. 32, pp. 679–686, 2000.

C2.5 *Exploration Project.* Investigate how tunneling is used in flash memories and describe one commercial application using nanotechnology.

C2.6 *Entrepreneurial Project.* Search and research how field emission is used in displays. As an entrepreneur, propose a number of possible applications using field emission in a product development that will bring out your creativity and innovation.

Appendix 2A: Derivation of the Density of States Using δ-Function

2A.1 3D Parabolic Semiconductors

The DOS, or to be exact, density of quantum states, is given by

$$D(E) = \frac{1}{\Omega}\sum_{n,\vec{k},s} \delta(E - E_{n\vec{k}s}) \tag{2A.1}$$

where $\Omega = L_x L_y L_z$ is the volume of the crystal, n is the band or subband index, $\vec{k} = (k_x, k_y, k_z)$ is the wave vector for 3D (bulk), $\vec{k} = (k_x, k_y)$ for 2D (nanosheet), and $\vec{k} = k_x\hat{x}$ (with $\hat{x}$ the unit vector) for 1D (nanowire). The quantity $D(E)\,dE$ is the number of quantum states between the differential energy interval (E, $E + dE$) per unit volume/area/length for 3/2/1D nanostructure. For nonmagnetic material, spin degeneracy is 2 as energy does not depend upon spin. Therefore,

$$D(E) = 2\sum_{n} D_n(E) \tag{2A.2}$$

with

$$D_n(E) = \frac{1}{\Omega}\sum_{n} \delta(E - E_{n\vec{k}}) \tag{2A.3}$$

Integrating the DOS over energy up to a maximum energy value results in the total number of states per unit volume below this energy

$$\int_{-\infty}^{E} dE'\, D(E') = \frac{2}{\Omega}\int_{-\infty}^{E} dE' \sum_{\substack{n\vec{k} \\ E_{n\vec{k}}<E}} 1 = \mathbb{N}(E) \tag{2A.4}$$

Or, we can write

$$D(E) = \frac{d\mathbb{N}}{dE} \tag{2A.5}$$

For an isotropic parabolic dispersion with effective mass m^*, the energy for a conduction band in 3D material is given by

$$E - E_{co} = \frac{\hbar^2 k^2}{2m^*} = \frac{\hbar^2(k_x^2 + k_y^2 + k_z^2)}{2m^*} \tag{2A.6}$$

The conduction band states are given by

$$\mathbb{N}_c(E) = \frac{2}{\Omega}\sum_{\substack{\vec{k} \\ E_{n\vec{k}}<E}} 1 = \frac{2}{\Omega}\frac{L_x L_y L_z}{(2\pi)^3}\int_0^{k(E)} dk_x\, dk_y\, dk_z \tag{2A.7}$$

In spherical polar coordinates $dk_x\, dk_y\, dk_z = k^2\, dk \sin\theta\, d\theta\, d\phi$. The integral then expands to

$$\mathbb{N}_c(E) = \frac{2}{(2\pi)^3}\int_0^{(E-E_{co})} dk\, k^2 \int_{\theta=0}^{\pi} d\theta \sin\theta \int_{\phi=0}^{2\pi} d\phi \tag{2A.8}$$

which, when Equation 2A.6 is used, reduces to

$$\mathbb{N}_c(E) = \frac{2}{(2\pi)^3} 4\pi \int_0^{(E-E_{co})} dk\, k^2 = \frac{2}{(2\pi)^2}\left(\frac{2m^*}{\hbar^2}\right)^{3/2} \int_0^{E-E_{co}} dE \sqrt{E} \tag{2A.9}$$

Hence, the 3D DOS is

$$D_3(E) = \frac{d\mathbb{N}_c}{dE} = \frac{2}{(2\pi)^2}\left(\frac{2m^*}{\hbar^2}\right)^{3/2} (E - E_{co})^{1/2} \tag{2A.10}$$

2A.2 2D (Nanosheet or Nanolayer) Parabolic Semiconductors

The energy for a conduction band in 3D material is given by

$$E - E_{co} = E_n + \frac{\hbar^2(k_x^2 + k_y^2)}{2m^*} \tag{2A.11}$$

where E_n are the quantized levels for electrons in quantum-confined nanostructure in the z-direction. The DOS is then given by

$$\begin{aligned}\mathbb{N}_c(E) &= \frac{2}{A}\sum_{\substack{k\\ E_{n\vec{k}}<E}} 1 = \frac{2}{A}\frac{L_x L_y}{(2\pi)^2}\int_0^{k(E)} dk_x\, dk_y \\ &= \frac{2}{(2\pi)^2}\int_0^{E-E_{co}-E_n} dk\, k \int_0^{2\pi} d\theta \\ &= \frac{m^*}{\pi\hbar^2}(E - E_{co} - E_n) = \frac{m^*}{\pi\hbar^2}(E - E_{cn})\end{aligned} \tag{2A.12}$$

where $E_{cn} = E_{co} + n^2E_1$, E_1 being the ground-state energy in $n = 1$ level. Of course in strictly 2D case, $n = 1$. The resulting DOS is constant

$$D_2(E) = \frac{d\mathbb{N}_c}{dE} = \frac{m^*}{\pi\hbar^2} \tag{2A.13}$$

In fact in quasi-2D case, the DOS is multiplied by an integer $n = \text{Int}[E/E_1]$.

$$D_2(E) = \frac{d\mathbb{N}_c}{dE} = \frac{m^*}{\pi\hbar^2}\text{Int}\left[\frac{E}{E_1}\right] \tag{2A.14}$$

2A.3 1D (Nanowire) Parabolic Semiconductors

The energy for a conduction band in 1D material is given by

$$E - E_{co} = E_{ny.nz} + \frac{\hbar^2 k_x^2}{2m^*} \tag{2A.15}$$

where $E_{ny.nz}$ are the quantized levels for electrons in a quantum-confined nanostructure in the z-direction. The DOS is then given by

$$\mathbb{N}_c(E) = \frac{2}{L_x}\sum_{\substack{k\\ E_{n\vec{k}}<E}} 1 = \frac{2}{L_x}\frac{L_x}{2\pi}\int_0^{k(E)} dk_x \tag{2A.16}$$

$$\mathbb{N}_c(E) = \frac{2}{2\pi} \int_0^{\left(\frac{2m^*}{\hbar^2}\right)^{1/2}(E-E_{co}-E_{ny,nz})^{1/2}} dk_x$$

$$= \frac{1}{\pi}\left(\frac{2m^*}{\hbar^2}\right)^{1/2} (E - E_{co} - E_{ny,nz})^{1/2} \tag{2A.17}$$

The factor 2 comes from the fact that the energy is the same for $\pm k_x$. The resulting DOS is constant

$$D_{1e}(E) = \frac{d\mathbb{N}_c}{dE} = \frac{1}{2\pi}\left(\frac{2m_n^*}{\hbar^2}\right)^{1/2} \left(E - E_{co} - E_{nynz}\right)^{-1/2}$$

$$= \frac{1}{2\pi}\left(\frac{2m_n^*}{\hbar^2}\right)^{1/2} \left(E - E_{cnynz}\right)^{-1/2} \tag{2A.18}$$

2A.4 DOS in Graphene

The energy spectrum of a Dirac Fermion in a graphene nanostructure is similar to a photon and is given by

$$E(k) = E_{Fo} \pm \hbar|k|v_F \tag{2A.19}$$

The number of quantum states per unit area in the conduction band below the energy E is given by

$$\mathbb{N}_{cg}(E) = \frac{4}{A}\sum_{\substack{k \\ E_{n\vec{k}}<E}} 1 = \frac{4}{A}\frac{L_x L_y}{(2\pi)^2}\int_0^{k(E)} dk_x\, dk_y$$

$$= \frac{4}{(2\pi)^2}\int_0^{E-E_{Fo}} dk\,k\int_0^{2\pi} d\theta = \frac{|E - E_{Fo}|^2}{\pi\hbar^2 v_F^2} \tag{2A.20}$$

The prefactor 4 results for twofold k-space and twofold valley degeneracy (K and K'). The valence band is symmetric to the conduction band. The conduction band in graphene is normally taken to be at zero energy ($E_{Fo} = 0$).

The DOS in graphene is calculated similarly and is given by

$$D_g(E) = \frac{d\mathbb{N}_{cg}}{dE} = \frac{2|E - E_{Fo}|}{\pi\hbar^2 v_F^2} = D_{go}|E - E_{Fo}| \tag{2A.21}$$

where

$$D_{go} = \frac{2}{\pi\hbar^2 v_F^2} = 1.5 \times 10^{14}\ \text{eV}^{-2}\ \text{cm}^{-2} \quad \text{or} \quad D_{go} = 1.5 \times 10^{6}\ \text{eV}^{-2}\ \mu\text{m}^{-2} \tag{2A.22}$$

2A.5 DOS in CNT

The energy for a conduction band in a CNT is given by

$$E_v - E_{Fo} = \pm\frac{E_{gv}}{2}\sqrt{1+\left(\frac{3d_t k_t}{2v}\right)^2} \tag{2A.23}$$

where quantized $k_t = (2\pi/L_t)n_t$, where n_t is an integer, and L_t is the length of the tube. The number of states per unit length is then given by

$$\mathbb{N}_c(E_v) = \frac{g_s g_K}{L_t}\sum_{\substack{\vec{k}_t \\ E_{n_t}<E_v}} 1 = \frac{g_s g_K}{L_t}\frac{L_t}{2\pi}2\int_0^{k_t(E_v)} dk_t \tag{2A.24}$$

Here, $g_s = 2$ is the spin degeneracy and $g_K = 2$ is K degeneracy with two set of K and K' points. The prefactor 2 before the integral arises from twofold $\pm k_t$ degeneracy. With $k_t(E)$, as obtained from Equation (2A.21), is given by

$$k_t(E_v) = \frac{4v}{3d_t E_g}\left[(E_v - E_{Fo})^2 - \left(\frac{E_{gv}}{2}\right)^2\right]^{1/2} \tag{2A.25}$$

The valence band is symmetric to the conduction band. The conduction band in graphene is normally taken to be at zero energy ($E_{Fo} = 0$).

$$\mathbb{N}_{CNT}(E_v) = \frac{8v}{3\pi d_t E_{gv}}\left[(E_v - E_{Fo})^2 - \left(\frac{E_{gv}}{2}\right)^2\right]^{1/2} \tag{2A.26}$$

The resulting DOS is obtained as

$$D_{CNT}(E) = \frac{d\mathbb{N}_c}{dE} = \frac{D_0 g_K}{2}\frac{|E_v - E_{Fo}|}{\left[(E_v - E_{Fo})^2 - \left(\frac{E_{gv}}{2}\right)^2\right]^{1/2}} \tag{2A.27}$$

with

$$\begin{aligned} D_o &= \frac{16v}{3\pi d_t E_{gv}} = \frac{4}{\pi\hbar v_F} = \frac{8}{3\pi a_{cc}\gamma} \\ &= 1.93\times 10^9\ \text{m}^{-1}\,\text{eV}^{-1} = 1.93\ \text{nm}^{-1}\,\text{eV}^{-1} \end{aligned} \tag{2A.28}$$

Here, use is made of $\hbar v_F = 3a_{CC}\gamma/2$, where $a_{CC} = 0.142$ nm is the bond length and $\gamma = 3.1$ eV is the bonding energy. $g_K = 2$ is the valley degeneracy because of K and K' states for which boundary condition apply equally well.

References

1. J. Walker, *Physics*, 4th ed. Upper Saddle River, NJ: Pearson/Prentice-Hall, 2010.
2. B. L. Anderson and R. L. Anderson, *Fundamentals of Semiconductor Devices*. New York, NY: McGraw-Hill, 2005.
3. B. G. Streetman and S. K. Banerjee, *Solid State Electronic Devices*, 6th ed. Upper Saddle River, NJ: Prentice-Hall, 2006.
4. A. T. M. Fairus and V. K. Arora, Quantum engineering of nanoelectronic devices: The role of quantum confinement on mobility degradation, *Microelectronics Journal*, 32, 679–686, 2001.
5. F. Stern and W. E. Howard, Properties of semiconductor surface inversion layers in the electric quantum limit, *Physical Review*, 163, 816, 1967.
6. M. L. P. Tan, V. K. Arora, I. Saad, M. T. Ahmadi, and R. Ismail, The drain velocity overshoot in an 80 nm metal–oxide–semiconductor field-effect transistor, *Journal of Applied Physics*, 105, 074503, 2009.
7. Y. Taur and T. H. Ning, *Fundamentals of Modern VLSI Devices*. Cambridge, UK: Cambridge University Press, 1998.
8. F. Stern, Self-consistent results for n-type Si inversion layers, *Physical Review B*, 5, 4891, 1972.
9. V. K. Arora, Quantum engineering of nanoelectronic devices: The role of quantum emission in limiting drift velocity and diffusion coefficient, *Microelectronics Journal*, 31, 853–859, 2000.
10. A. K. Geim and K. S. Novoselov, The rise of graphene, *Nature Materials*, 6, 183–191, Mar 2007.
11. A. H. Castro Neto, F. Guinea, N. M. R. Peres, K. S. Novoselov, and A. K. Geim, The electronic properties of graphene, *Reviews of Modern Physics*, 81, 109–162, Jan–Mar 2009.
12. P. H. S. Wong and D. Akinwande, *Carbon Nanotube and Graphene Device Physics*. Cambridge: Cambridge University Press, 2011.
13. T. Ihn, *Semiconductor Nanostructures*. Oxford, UK: Oxford University Press, 2010.
14. V. K. Arora and A. Bhattacharyya, Cohesive band structure of carbon nanotubes for applications in quantum transport, *Nanoscale*, 5, 10927–10935, 2013.
15. H. Raza and E. C. Kan, Armchair graphene nanoribbons: Electronic structure and electric-field modulation, *Physical Review B*, 77, 245434, 2008.
16. H. Raza, Zigzag graphene nanoribbons: Bandgap and midgap state modulation, *Journal of Physics: Condensed Matter*, 23, 382203, 2011.
17. Y.-W. Son, M. L. Cohen, and S. G. Louie, Energy gaps in graphene nanoribbons, *Physical Review Letters*, 97, 216803, 11/22/2006.
18. S. Dutta and S. K. Pati, Novel properties of graphene nanoribbons: A review, *Journal of Materials Chemistry*, 20, 8207–8223, 2010.

19. D. V. Kosynkin, A. L. Higginbotham, A. Sinitskii, J. R. Lomeda, A. Dimiev, B. K. Price et al. Longitudinal unzipping of carbon nanotubes to form graphene nanoribbons, *Nature*, 458, 872–876, 2009.
20. H. Santos, L. Chico, and L. Brey, Carbon nanoelectronics: Unzipping tubes into graphene ribbons, *Physical Review Letters*, 103, Aug 21, 2009.
21. Y.-C. Chen, D. G. de Oteyza, Z. Pedramrazi, C. Chen, F. R. Fischer, and M. F. Crommie, Tuning the band gap of graphene nanoribbons synthesized from molecular precursors, *Acs Nano*, 7, 6123–6128, 07/23 2013.
22. M. S. Hybertsen and S. G. Louie, Electron correlation in semiconductors and insulators: Band gaps and quasiparticle energies, *Physical Review B*, 34, 5390–5413, 1986.
23. T. Miyake and S. Saito, Band-gap formation in (n, 0) single-walled carbon nanotubes (n = 9, 12, 15, 18): A first-principles study, *Physical Review B*, 72, 073404, 2005.
24. T. Miyake and S. Saito, Quasiparticle band structure of carbon nanotubes, *Physical Review B*, 68, 155424, 2003.
25. X. Li, X. Wang, L. Zhang, S. Lee, and H. Dai, Chemically derived, ultrasmooth graphene nanoribbon semiconductors, *Science*, 319, 1229–1232, February 29, 2008.
26. M. Y. Han, B. Özyilmaz, Y. Zhang, and P. Kim, Energy band-gap engineering of graphene nanoribbons, *Physical Review Letters*, 98, 206805, 2007.
27. P. Shemella, Y. Zhang, M. Mailman, P. M. Ajayan, and S. K. Nayak, Energy gaps in zero-dimensional graphene nanoribbons, *Applied Physics Letters*, 91, 042101–042101-3, 2007.
28. G. W. Hanson, *Fundamental of Nanoelectronics*. Upper Saddle River, NJ: Pearson/Prentice-Hall, 2008.

3

Carrier Statistics

In Chapter 2, extensive discussion on quantum wells was held and how analog-like continuous energy and momentum states get digitized was discussed. The analog-like continuous spectrum thus can be viewed as a collection of digitized quantum states with step-size miniscule compared to the thermal energy as we go from one energy level to the next higher one. In this chapter, the focus is on the filling of quantum states. The probability that a quantum state is occupied is described by the Fermi–Dirac distribution function. The carrier concentration is a multiplication of the effective density of quantum states multiplied by the effective probability.

3.1 Fermi–Dirac Distribution Function

The probability $f(E)$ that a quantum state of energy E is occupied is given by [1,2]

$$f(E) = \frac{1}{e^{(E-E_F)/k_BT} + 1} \tag{3.1}$$

where E_F is the Fermi energy with the probability of occupation 1/2 at $E = E_F$ (see Appendix 3A for more discussion). This probability function is shown in Figure 3.1. The placement of the Fermi energy with respect to the conduction band edge describes the degeneracy of the system. The degeneracy of a carrier sample depends on the carrier concentration with respect to the DOS. A sample is nondegenerate (ND) if the carrier concentration $n_d < N_{cd}$, where $d = 1, 2, 3$ is the dimensionality of the nanoscale sample and N_{cd} is the effective density of states (EDOS) for which an expression will be found later in the chapter. On the other hand, when $n_d \gg N_{cd}$, a sample is strongly degenerate. The three-dimensional (3D) conduction band edge E_{co} is lifted by the ground-state energy ε_o in a quantum well, so the lifted conduction band edge $E_c = E_{co} + \varepsilon_o$. The minimum energy of occupation is E_{co} or E_c in a quantum well as no quantum states exist in the forbidden bandgap. In the Maxwell–Boltzmann (ND) approximation, the 1 in the denominator of Equation 3.1 is neglected as the Fermi energy E_F is below E_{co} as shown in Figure 3.1. In this case, the reduced Fermi energy $\eta = (E_F - E_{co})/k_BT$ is negative. As a general rule of thumb, the "1" in the denominator of Equation 3.1 is negligible if

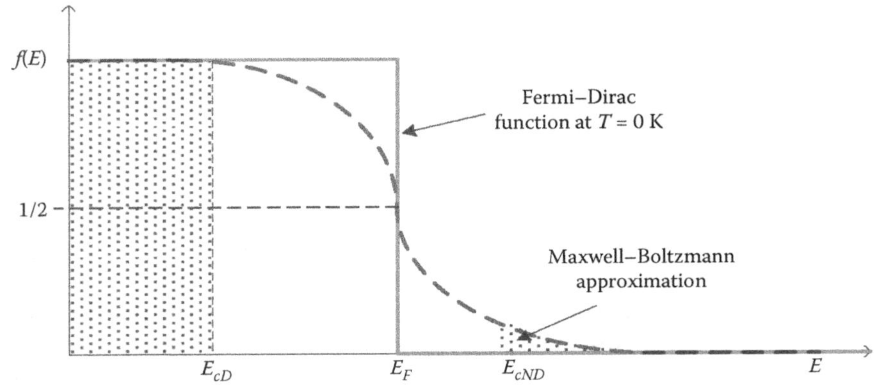

FIGURE 3.1
Fermi–Dirac distribution function as a function of energy. In the ND limit, the Fermi energy is below the conduction band edge making the tail of the distribution ($E > E_{cND}$) Maxwellian. In the degenerate regime ($E_{cD} > E > E_F$), every state is occupied up to the Fermi level and every state is empty above it at 0 K.

$\eta < -3$. Figure 3.1 shows that the conduction band $E_F < E_{cND}$ and hence $\eta < 0$ for ND statistics. In the limit $\eta < -3$, the distribution follows the dashed line of Figure 3.1, its tail fitted with the exponential distribution given by

$$f(E) \approx \frac{1}{e^{(E-E_F)/k_BT}} = e^{-(E-E_F)/k_BT} \tag{3.2}$$

On the other hand, in the strongly degenerate regime, where the Fermi energy is above the conduction band edge, the probability of occupation is 1 up to the Fermi energy level and zero above it. An extreme case exists at $T = 0$ K when $(E - E_F)/k_BT \to \infty$ for $E > E_F$. This limit makes $f(E) = 0$. As $(E - E_F)/k_BT \to -\infty$, the probability of occupation is $f(E) = 1$. Figure 3.1 shows the conduction band $E_F > E_{cD}$ and hence $\eta > 0$ for a degenerate statistics. As shown by the solid line in Figure 3.1, the probability is 1 for every state being occupied below the Fermi level and is 0 above it. Step-like distribution at 0 K near the Fermi energy point smears out at a finite temperature *T*. The degenerate approximation is always valid in case the Fermi energy is well above the conduction band edge.

3.2 Bulk (3D) Carrier Distribution

The carrier concentration is a function of probability $f(E)$ and the DOS between E and $E + dE$. The differential carrier concentration per unit volume dn_3 between E and $E + dE$ is given by

$$dn_3 = f(E)D_3(E)\,dE = \frac{1}{e^{(E-E_F)/k_BT}+1}\frac{1}{2\pi^2}\left(\frac{2m_n^*}{\hbar^2}\right)^{3/2}(E-E_{co})^{1/2}\,dE \tag{3.3}$$

The graphical representation of the distribution function and DOS plotted with energy E on the vertical axis is shown in Figure 3.2. The top of the conduction band is finite, but the probability goes to zero as energy increases away from the conduction band edge because of the exponentially decaying tail. Hence, no appreciable error occurs if integration over energy E is performed from $E = E_{co}$ to $E = E_{top} \approx \infty$. The integration of Equation 3.3 (or the area in the last graph of Figure 3.2) gives for n_3 (carrier concentration with number of carriers per m^3) the expression

$$n_3 = \frac{1}{2\pi^2}\left(\frac{2m_n^*}{\hbar^2}\right)^{3/2}\int_{E_{co}}^{\infty}\frac{1}{e^{(E-E_{co}+E_{co}-E_{F3})/k_BT}+1}(E-E_{co})^{1/2}\,dE \tag{3.4}$$

where E_{co} is the conduction band edge in the bulk semiconductor and n_3 is the carrier concentration with number of carriers per m^3. E_{co} is added and subtracted in the exponent for the convenience of defining $\eta_3 = (E_{F3} - E_{co})/k_BT$, the reduced Fermi energy for a bulk semiconductor. Similarly, the energy variable E is reduced to $x = (E - E_{co})/k_BT$. With this transformation, the carrier concentration n_3 is obtained as

$$n_3 = \frac{1}{2\pi^2}\left(\frac{2m_n^*}{\hbar^2}\right)^{3/2}(k_BT)^{3/2}\int_0^{\infty}\frac{1}{e^{(E-E_{co}+E_{co}-E_{F3})/k_BT}+1}\left(\frac{E-E_{co}}{k_BT}\right)^{1/2}d\left(\frac{E-E_{co}}{k_BT}\right) \tag{3.5}$$

With the substitution of x and η_3, Equation 3.5 becomes

$$n_3 = \left[\frac{1}{2\pi^2}\left(\frac{2m_n^*}{\hbar^2}\right)^{3/2}(k_BT)^{3/2}\Gamma(3/2)\right]\frac{1}{\Gamma(3/2)}\int_0^{\infty}\frac{1}{e^{x-\eta_3}+1}x^{1/2}\,dx \tag{3.6}$$

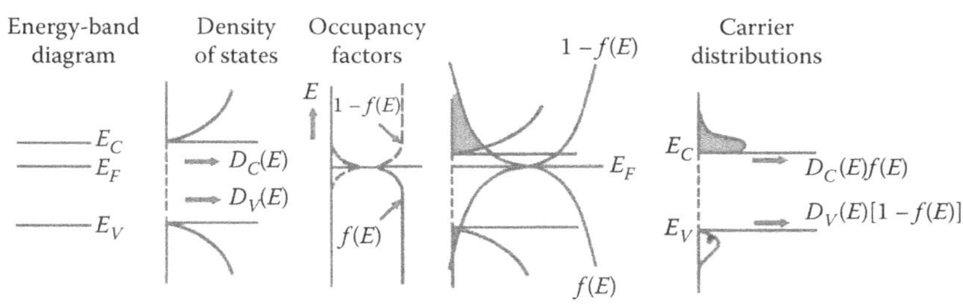

FIGURE 3.2
Band diagram showing the position of the Fermi level (ND), DOS, the distribution function, and carrier distribution for an n-type semiconductor.

Here, $\Gamma(3/2) = \sqrt{\pi}/2$ is the Gamma function of order 3/2. The expression in the bracket of Equation 3.6 is the EDOS N_{c3} for the conduction band in a bulk (3D) sample. In a simplified form N_{c3} is given by

$$N_{c3} = 2\left(\frac{m_n^* k_B T}{2\pi\hbar^2}\right)^{3/2} \tag{3.7}$$

The other factor following the square bracket is the Fermi–Dirac integral (FDI) of order 1/2. FDI of any order j is given by

$$\Im_j(\eta) = \frac{1}{\Gamma(j+1)}\int_0^\infty \frac{x^j}{e^{x-\eta}+1}\,dx \tag{3.8}$$

Equation 3.6 with these substitutions becomes

$$n_3 = N_{c3}\Im_{1/2}(\eta_3) \tag{3.9}$$

In general, $\Gamma(j+1) = \int_0^\infty e^{-x}x^j\,dx = j!$ if j is an integer. $\Gamma(j+1) = j\Gamma(j)$ is the recursion relation. Using this recursion $\Gamma(3/2) = (1/2)\Gamma(1/2) = \sqrt{\pi}/2$ as $\Gamma(1/2) = \sqrt{\pi}$. The properties of FDI are listed in Appendix F and table of FDI values appears in Appendix G.

Equation 3.9 sets the stage for carrier statistics in a bulk semiconductor with all three dimensions larger than the de Broglie wavelength. It is worthwhile to examine the general properties of the FDI in the ND and strongly degenerate regimes. As justified earlier, the ND limit arises from the neglect of 1 in the denominator of Equation 3.1 under the pretext that $\eta < -3$. In this limit, regardless of the value of j, the FDI is always e^η independent of index j:

$$\Im_j(\eta) \approx e^\eta \quad \text{(non-degenerate)} \tag{3.10}$$

Because of this simplicity of the ND limit, it is the most commonly used limit in the analysis of semiconductor devices. However, this limit is not applicable to heavily doped semiconductors especially at low temperatures. For heavily doped semiconductors ($\eta > +3$), the Fermi factor is approximated as

$$\frac{1}{e^{x-\eta}+1} \approx \begin{cases} 1, & x < \eta \\ 0, & x > \eta \end{cases} \quad \text{(strongly degenerate)} \tag{3.11}$$

In this degenerate limit, FDI of Equation 3.8 simplifies to

$$\Im_j(\eta) = \frac{1}{\Gamma(j+1)}\int_0^{\eta} x^j\,dx = \frac{1}{\Gamma(j+1)}\frac{\eta^{j+1}}{j+1} \tag{3.12}$$

The FDI is obtainable in the closed form for $j = 0$ and -1 for all values of degeneracy

$$\Im_0(\eta) = \ln(e^{\eta} + 1) \tag{3.13a}$$

$$\Im_{-1}(\eta) = \frac{d\Im_o}{d\eta} = \frac{e^{\eta}}{(e^{\eta} + 1)} \tag{3.13b}$$

The expression for holes is similarly obtained. In the valence band moving from the valence band edge E_{vo} downwards toward the bottom of the valence band ($E_{bottom} \approx -\infty$), the probability of finding a hole goes to zero as all states are occupied. The probability of finding a hole is $(1 - f(E))$ as it is complement of a missing electron for which probability of occupation is $f(E)$. When these factors are taken into account, the expression for hole concentration p_3 is given by

$$p_3 = \int_{-\infty}^{E_{vo}} (1 - f(E))D_{3v}(E)dE = N_{v3}\Im_{1/2}(\eta_{v3}) \tag{3.14}$$

with

$$N_{v3} = 2\left(\frac{m_p^* k_B T}{2\pi\hbar^2}\right)^{3/2} \tag{3.15}$$

$$\eta_{v3} = \frac{E_{vo} - E_{F3}}{k_B T} \tag{3.16}$$

such that m_p^* is the effective mass of the hole. The graphical display of the integral in Equation 3.14 is displayed in Figure 3.3. A convenient expression for calculating the DOS at all temperatures for conduction or valence band and for a DOS effective mass $m^*_{dsn(p)3}$ for an electron (hole) in a bulk sample ($d = 3$) is given by

$$N_{c(v)3} = 2\left(\frac{m^*_{dsn(p)3} k_B T}{2\pi\hbar^2}\right)^{3/2} = 2.54 \times 10^{25}\left(\frac{m^*_{dsn(p)3}}{m_o}\right)^{3/2}\left(\frac{T}{300}\right)^{3/2} m^{-3} \tag{3.17}$$

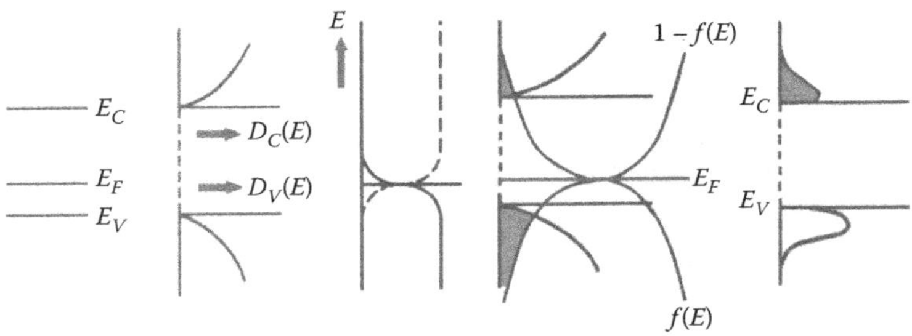

FIGURE 3.3
Band diagram showing the position of the Fermi level (ND), DOS, the distribution function, and carrier distribution in a p-type semiconductor.

3.3 Bulk (3D) ND Approximation

In the ND limit when Equation 3.10 is applicable for approximating FDI, the carrier concentration both for electrons and holes is expressed as

$$n_3 = N_{c3}\, e^{(E_{F3}-E_{co})/k_BT} = N_{c3}\, e^{-(E_{co}-E_{F3})/k_BT} \tag{3.18}$$

$$p_3 = N_{v3}\, e^{(E_{vo}-E_{F3})/k_BT} = N_{v3}\, e^{-(E_{F3}-E_{vo})/k_BT} \tag{3.19}$$

These equations apply for a bulk sample where n_3 is the electron concentration per unit volume and p_3 is the hole concentration per unit volume. $N_{c(v)3}$ is three-dimensional (bulk) effective DOS.

In an extrinsic (doped) semiconductor, either the electrons or holes are created by substitutional impurities replacing the silicon atoms. For example,

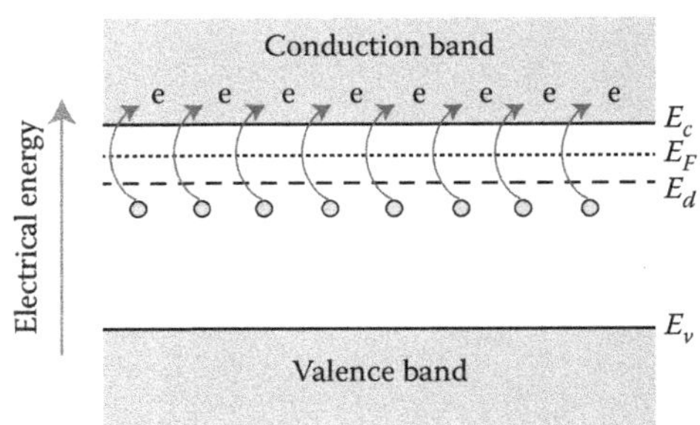

FIGURE 3.4
Electrons supplied to the conduction band by the donors close to the conduction band edge.

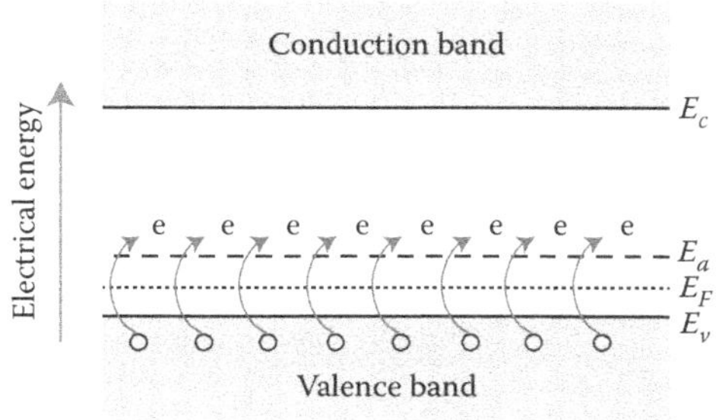

FIGURE 3.5
Holes created in the valence band as electrons are accepted by acceptors close to the valence band edge.

when a tetravalent silicon atom is substituted with a pentavalent phosphorus atom, an extra electron of silicon is loosely bound to the donor phosphorus atom and hence can be easily excited to the conduction band provided the binding energy $E_c - E_d$ is much smaller than the thermal energy k_BT. This thermal hop from the donor atom to the conduction band at room temperature is almost 100%; every donor donates an electron for conduction to take place, as shown in Figure 3.4. Similarly, the acceptors close to the valence band are able to accept electrons that are thermally activated from the valence band, thereby creating holes in the valence band that become itinerant (conducting), as shown in Figure 3.5.

3.4 Intrinsic Carrier Concentration

Intrinsic carriers are inherent to the effect of finite temperature as electrons are excited from otherwise full valence band. At 0 K, electrons are frozen in the valence band and there are no electrons in the conduction band for an intrinsic (undoped Si). As temperature increases, an electron in the valence band may be energetic enough to have energy larger than the bandgap and transfer to the conduction band, leaving behind a hole in the valence band. An electron–hole (e–h) pair is thus generated. An electron is free to hop in the conduction band from one quantum state to the other and hence is termed itinerant (or conduction) electron. On the other hand, the frozen electron in the valence band at a finite temperature can hop to the hole (empty) state giving the impression as the hole has moved to the localized state previously occupied by an electron. So, the hole also becomes itinerant. The holes (missing electrons) are less mobile because of sluggish nature of the localized electrons in the valence band. Figure 3.6 shows how electrons are frozen

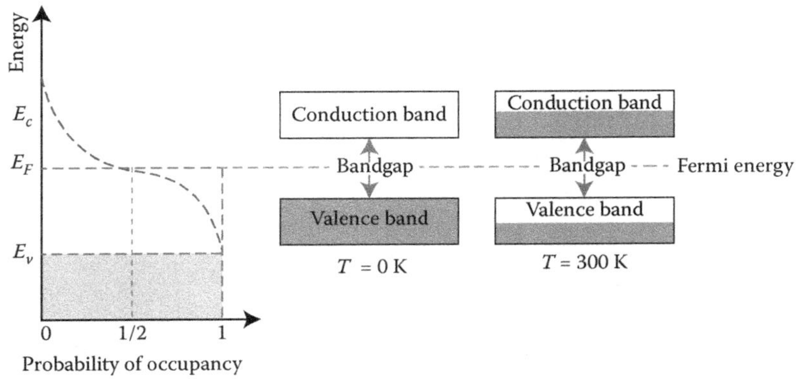

FIGURE 3.6
The Fermi function and corresponding energy band in a semiconductor at $T = 0$ K and 300 K. At 0 K, the valence band is completely full and the conduction band is completely empty. At 300 K, some electrons in the valence band hop over to the conduction band leaving behind hole (empty) states in the valence band. Electrons are itinerant in the conduction band and holes are itinerant in the valence band.

in the valence band at 0 K and excited to create electron–hole pairs at 300 K. The thermal energy arising from the finite temperature allows carriers to be activated.

The concentration of electrons and holes is equal in intrinsic semiconductors ($n_3 = p_3 = n_{i3}$) and $E_{F3} \rightarrow E_{i3}$ which is the intrinsic Fermi energy. Equations 3.18 and 3.19 then transform to

$$n_{i3} = N_{c3}\, e^{-(E_{co}-E_{i3})/k_B T} \tag{3.20}$$

$$n_{i3} = N_{v3}\, e^{-(E_{i3}-E_{vo})/k_B T} \tag{3.21}$$

Equations 3.20 and 3.21 are solved simultaneously to eliminate either n_{i3} or E_{i3}. When multiplied to eliminate E_{i3}, the resulting equation for n_{i3} is

$$n_{i3} = \sqrt{N_{c3}N_{v3}}\, e^{-E_{go}/2k_B T} \tag{3.22}$$

such that $E_{go} = E_{co} - E_{vo}$. A graphical display of the intrinsic semiconductor is depicted in Figure 3.7, with the intrinsic Fermi level ($E_{i3} = E_F$) in the middle of the bandgap. EDOS $N_{c3} = 2.86 \times 10^{19}$ cm^{-3} for electrons and $N_{v3} = 3.10 \times 10^{19}$ cm^{-3} for holes are obtained from Equation 3.17, with the DOS effective

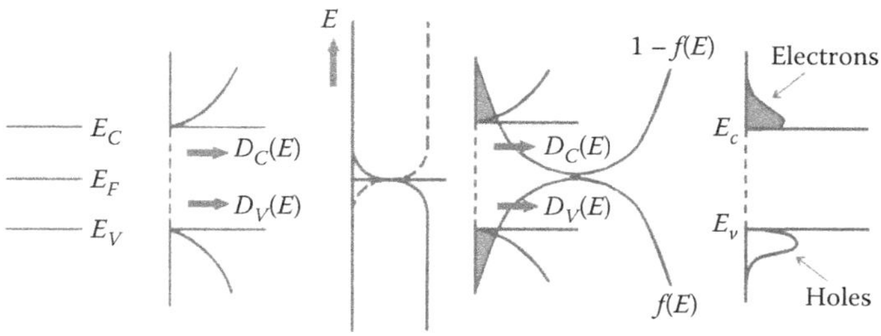

FIGURE 3.7
Band diagram showing the intrinsic position of the Fermi level, DOS, the distribution function, and carrier distribution. Solid line is the distribution $f(E)$ for electrons and dashed line $1 - f(E)$ is for holes.

masses $m^*_{dsp} = 1.15\,\mathrm{m}_o$ and $m^*_{dsn} = 1.09\,\mathrm{m}_o$ for silicon. Equation 3.22 gives $n_{i3} = 1.08 \times 10^{10}$ cm^{-3} with a bandgap $E_{go} = 1.1242$ eV at room temperature.

The intrinsic level E_{i3} is obtained when the right-hand side of Equation 3.20 and that of Equation 3.21 are equated to eliminate n_{i3} which yields

$$E_{i3} = \frac{E_{co} + E_{vo}}{2} + \frac{3}{4} k_B T \ln\left(\frac{m^*_{dsh}}{m^*_{dse}}\right) \approx \frac{E_{co} + E_{vo}}{2} \tag{3.23}$$

The term that includes the effective masses is a direct result of the ratio of the EDOS for conduction and valence bands. EDOS is proportional to 3/2 power of the effective mass: $N_{c(v)} \propto m^{*3/2}_{n(p)}$. Therefore, $\ln(N_v/N_c) = \ln(m^*_n/m^*_p)^{3/2} = 3/2\ln(m^*_n/m^*_p)$. The contribution of this term is very small compared to the bandgap energy. Hence, the intrinsic level lies almost in the middle of the bandgap.

Another way of writing the carrier concentration is in terms of n_{i3} and E_{i3}. The carrier concentration appears in terms of the intrinsic concentration when Equations 3.18 and 3.19 are divided by Equations 3.20 and 3.21, respectively, giving

$$n_3 = n_{i3}\, e^{(E_{F3} - E_{i3})/k_B T} \tag{3.24}$$

$$p_3 = n_{i3}\, e^{(E_{i3} - E_{F3})/k_B T} \tag{3.25}$$

These equations are useful when it is desired to obtain the Fermi energy with respect to the intrinsic level that is almost in the middle of the bandgap.

3.5 Charge Neutrality Compensation

The semiconductor and its atoms are neutral in the normal state. When a donor atom (Group V) donates an electron by ionization (removal from the parent atom), the electron moves to the conduction band and becomes itinerant. The fixed atom left behind now carries a positive charge. Similarly, when an acceptor (Group III) with one deficient electron in tetravalent silicon culture accepts an electron from the valence band, the hole in the valence band become itinerant and the acceptor acquires a negative charge. If both donors and acceptors are present in the same sample, they tend to compensate their charges by sharing among donors and acceptor atoms. A donor gets compensated by giving its extra electron to an acceptor that is deficient by one electron in a tetravalent culture of silicon. If both donors and acceptors are present in equal concentration, there is a perfect compensation and the semiconductor is intrinsic even in the presence of substituted impurities in place of a Si atom. To assess the effect of compensation, assume that N_A^- acceptors atoms per unit volume are implanted with each atom accepting an electron from the valence band with the net hole concentration p_3. Similarly, donor atoms N_D^+ per unit volume are implanted with each donating an electron to the conduction band leaving the net electron concentration n_3. The charge neutrality requires

$$N_D^+ + p_3 = N_A^- + n_3 \tag{3.26}$$

Equations 3.24 and 3.25 when multiplied yield what is known as law of mass action

$$n_3 p_3 = n_{i3}^2 \tag{3.27}$$

The net electron concentration n_3 is obtained after $p_3 = n_{i3}^2/n_3$ from Equation 3.27 is substituted into Equation 3.26 to eliminate p_3 and to solve for n_3. The result is

$$n_3 = \frac{N_D^+ - N_A^-}{2} \pm \sqrt{\left(\frac{N_D^+ - N_A^-}{2}\right)^2 + n_{i3}^2} \tag{3.28}$$

Similarly, an expression for p_3 is obtained as

$$p_3 = \frac{N_A^- - N_D^+}{2} \pm \sqrt{\left(\frac{N_A^- - N_D^+}{2}\right)^2 + n_{i3}^2} \tag{3.29}$$

In case the semiconductor is doped with equal number of donors and acceptors ($N_D = N_A$), a condition called perfect compensation, only carriers remaining are intrinsic ($p_3 = n_3 = n_{i3}$). The perfect compensation eliminates the effect of doping as each donor electron instead of hopping to conduction band drops to open acceptor states because of lower energy. However, partial compensation is used to counterdope a semiconductor to change its polarity from n-type to p-type or vice versa. For example, the resulting semiconductor is n-type with $n_3 = N_D^+ - N_A^-$ if the donor concentration $N_D^+ > N_A^-$ exceeds acceptor concentration and $N_D^+ - N_A^- \gg n_{i3}$. Similarly, the resulting semiconductor is p-type with $p_3 = N_A^- - N_D^+$ if the acceptor concentration $N_A^- > N_D^+$ exceeds the donor concentration and $N_A^- - N_D^+ \gg n_{i3}$.

Figure 3.8 shows the effect of doping for a sample doped with $N_D = 10^{16}$ cm^{-3}. $n_3 \approx N_D = 10^{16}$ cm^{-3} in the temperature range 100–600 K. However, above 600 K, the intrinsic carriers are much larger than the doped carriers. At low temperatures, the fraction of ionized atoms donating the electrons N_D^+/N_D and the fraction of those atoms accepting the electrons giving holes N_A^-/N_A is given by

$$\frac{N_D^+}{N_D} = \frac{1}{1 + g_D\, e^{(E_F - E_D)/k_BT}}, \quad g_D = 2 \text{ for 2 spin states} \tag{3.30}$$

$$\frac{N_A^-}{N_A} = \frac{1}{1 + g_A\, e^{(E_A - E_F)/k_BT}} \tag{3.31}$$

Here, $g_A = 2 \times 2 = 4$ for 2 spin states multiplied by light-heavy hole ($\ell h - hh$) states.

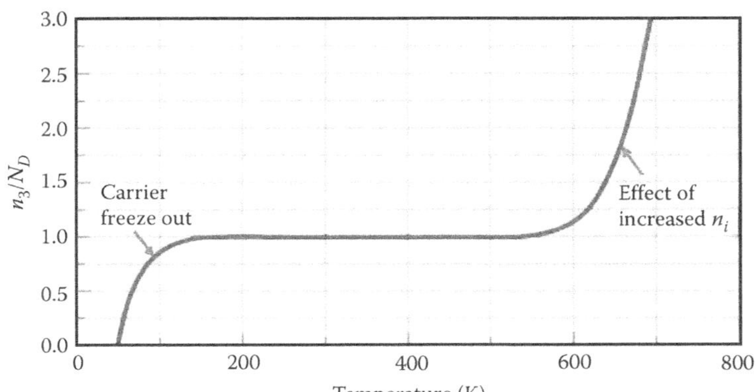

FIGURE 3.8
Description of carrier concentration at room temperature that transforms to a carrier freeze-out at low temperature and an overwhelm of intrinsic carriers at high temperature.

Here, E_D is the donor energy and E_A is the acceptor energy. The Fermi level is thus shifted due to carrier freeze-out. The quantum state degeneracy is 2 for two spin states of electrons. There are light and heavy holes and each can have two spin states giving degeneracy of 4. At a low temperature, $N_D^+ < N_D$, as only a few donor electrons will have thermal energy to hop to the conduction band. However, at an elevated temperature $N_D^+ \approx N_D$ until intrinsic carrier concentration dominates the doped concentration. Above 600 K, the intrinsic carrier concentration is not negligible and becomes dominant as temperature is raised beyond 600 K. Every donor has donated an electron to the conduction band and hence all donors are ionized. The same argument can be given for the acceptor atoms. Electrons are not able to move from the conduction band to the acceptor atom because of lack of thermal energy at 0 K and hence no holes are created. However, as electrons from the valence band become energetic, they tend to move to acceptor atoms leaving behind holes in the valence band.

3.6 Strong 3D (Bulk) Degenerate Limit

In order to enhance the current-carrying capacity of the conducting channel, the semiconductors are heavily doped. A strong degeneracy arises when the Fermi level for electrons crosses over to the conduction band from the forbidden bandgap to a level that is well above the conduction band edge. An onset of transition from ND-to-degenerate state can be defined when the Fermi level coincides with the conduction band edge ($E_{Fdeg} - Ec = 0$ or $\eta_3 = 0$) resulting in the carrier concentration n_{3deg} above which the electrons are degenerate [3]. The cross-over concentration n_{3deg} is given by

$$n_{3deg} = N_{c3}\Im_{1/2}(0) \tag{3.32}$$

which for silicon is $n_{3deg} = N_{c3}\Im_{1/2}(0) = 2.86 \times 10^{19}\text{cm}^{-3} \times 0.76515 = 2.2 \times 10^{19}\text{cm}^{-3}$.

The strong degeneracy occurs when $n_3 \gg n_{3deg}$. It has two major effects. The first is smearing of the conduction band edge into the forbidden zone as radii of the hydrogen-like donor atoms overlap. The conduction band edge is lowered due to merging of quantum states from the donor atoms that are below the conduction band. This merging of the quantum states from donor atoms lowers the conduction band edge and, hence, narrows the bandgap. This bandgap narrowing has been experimentally observed for semiconductor with heavy doping. This narrowing, although small, does influence the intrinsic carrier concentration through the exponential function with argument containing the bandgap E_g. The increase in the carrier concentration due to the bandgap narrowing $\Delta E_g = E_{gdeg} - E_{go}$ may influence the noisy behavior of the devices with strongly degenerate doping.

The other noticeable effect is in the carrier statistics. The Fermi level in the degenerate limit is weakly dependent on temperature and strongly dependent on its carrier concentration. When FDI in the degenerate limit as given by Equation 3.12 is used, the resulting Fermi energy is

$$E_{F3} - E_{co} = \left[\frac{3\sqrt{\pi}}{4}\frac{n_3}{N_{c3}}\right]^{2/3} k_B T = \frac{\hbar^2\left(3\pi^2 n_3\right)^{2/3}}{2m_n^*} \tag{3.33}$$

The Fermi momentum $k_{F3} \approx (3\pi^2 n_3)^{1/3}$ arises naturally as $E_{F3} - E_{co} = \hbar^2 k_{F3}^2/2m_n^*$. This is in fact the equation for the kinetic energy $E_{F3} - E_{co} = (1/2)\, m_n^* v_F^2 = (1/2)\, p_F^2/m_n^* = (1/2)(\hbar^2 k_F^2/m_n^*)$ with $p = mv = \hbar k$ with $k = 2\pi/\lambda$ following de Broglie hypothesis $\lambda = 2\pi/k = 2\pi\hbar/\hbar k = h/p$. Another important feature that arises from this description is the Fermi velocity v_{F3} given by

$$v_{F3} = \frac{\hbar k_{F3}}{m_n^*} = \sqrt{\frac{2(E_{F3} - E_{co})}{m_n^*}} \approx \frac{\hbar\left(3\pi^2 n_3\right)^{1/3}}{m_n^*} \tag{3.34}$$

The Fermi velocity holds a special significance not only in degenerate semiconductors, but also in metals that hold carriers in the degenerate limit. The Fermi energy for strongly degenerate metals is given in Appendix E.

3.7 Carrier Statistics in Low Dimensions

The procedure noted above can be repeated to develop the carrier statistics in low-dimensional materials. Quasi-one-dimensional (Q1D) and quasi-two-dimensional (Q2D) nanostructures discussed in Chapter 2 become truly one-dimensional (1D) (nanowires) and two-dimensional (2D) (layered thin films) in the quantum limit when only the lowest quantized level is appreciably populated. Equation 3.5 when all dimensionalities (d = 1, 2, 3) are considered in the quantum limit transforms to [3,4]

$$n_d = N_{cd}\Im_{d/2-1}(\eta_d), \quad d = 3,\, 2,\, \text{or } 1 \tag{3.35}$$

n_3 and N_{c3} have dimensions per unit volume, n_2 and N_{c2} per unit area, and n_1 and N_{c1} per unit length. The EDOS N_{cd} is given by

$$N_{cd} = 2\left(\frac{m_n^* k_B T}{2\pi\hbar^2}\right)^{d/2} = \frac{2}{\left(2\sqrt{\pi}\lambda_D\right)^d} \tag{3.36}$$

with

$$\eta_d = \frac{(E_F - E_c)_d}{k_B T} \tag{3.37}$$

$$\lambda_D = \frac{\hbar}{\sqrt{2m^* k_B T}} \tag{3.38}$$

$\lambda_D = \lambda/2\pi$, where λ is the electron wavelength. Naturally, n_2 and N_{c2} will be in units of m^{-2} and n_1 and N_{c1} in m^{-1} when SI system is being used. $E_c = E_{co} + \varepsilon_{od}$ with $\varepsilon_{o3} = 0$ is the modification in the conduction band energy as conduction band is lifted due to quantum effects by zero-point energy of the lowest quantized level.

The normalized Fermi energy η_d as a function of normalized carrier concentration is given in Figure 3.9. η_d logarithmically depends on normalized

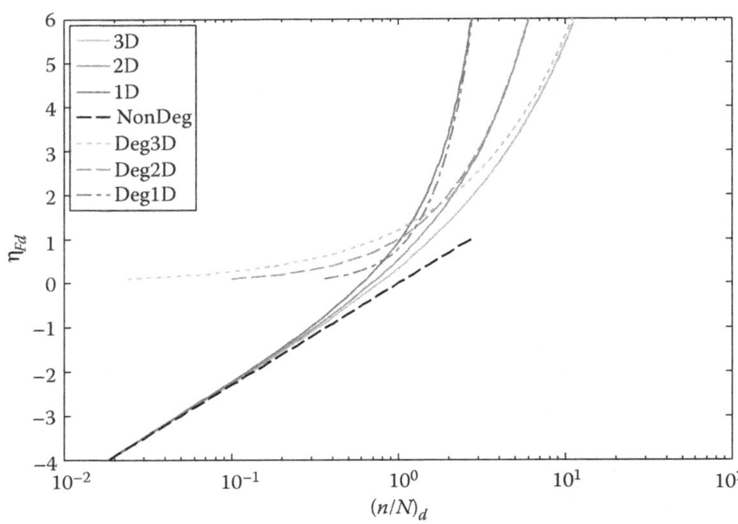

FIGURE 3.9
The normalized Fermi energy for all dimensionalities as a function of normalized concentration. Also shown are their degenerate and ND limit.

carrier concentration for all dimensionalities in the ND regime. Hence, the same normalized curve applies for all dimensionalities:

$$\eta_d = \ln\left(\frac{n}{N_c}\right)_d \tag{3.39}$$

However, the functional dependence on the carrier concentration in the degenerate regime differs as we move from 3D to 1D nanostructure.

In the strong degenerate limit, Equation 3.35 transforms to

$$n_d(Deg) = N_{cd}\Im_{d/2-1}(\eta_d) \approx N_{cd}\frac{1}{\Gamma(d/2)}\frac{\eta_d^{(d/2)}}{(d/2)} \tag{3.40}$$

The Fermi velocity in 2D and 1D semiconductor is given by

$$v_{F2} = \frac{\hbar k_{F2}}{m_n^*} = \sqrt{\frac{2(E_{F2} - E_c)}{m_n^*}} \approx \frac{\hbar(2\pi n_2)^{1/2}}{m_n^*} \tag{3.41}$$

$$v_{F1} = \frac{\hbar k_{F1}}{m_n^*} = \sqrt{\frac{2(E_{F1} - E_c)}{m_n^*}} \approx \frac{\hbar(\pi n_1/2)}{m_n^*} \tag{3.42}$$

The Fermi wavelength is $\lambda_{Fd} = 2\pi/k_{Fd}$ with $k_{F3} = (3\pi^2 n_3)^{1/3}$, $k_{F2} = (2\pi n_2)^{1/2}$, and $k_{F1} = \pi n_1/2$. The Fermi energy, velocity, and wavelength are shown in Appendix E for various metals.

The quantum limit may not be easily realizable in metals as many levels may be filled below the Fermi energy. In a nanowire, the energy spectrum is given by

$$E = E_{co} + n_y^2\frac{\pi^2\hbar^2}{2m^* L_y^2} + n_z^2\frac{\pi^2\hbar^2}{2m^* L_y^2} + \frac{\hbar^2 k_x^2}{2m^*} = E_{c,n_y,n_z} + \frac{\hbar^2 k_x^2}{2m^*} \tag{3.43}$$

Assuming that the quantum wire has a square cross-section with $L_x = L_y = W$, the subband energy E_{c,n_y,n_z} is given by

$$E_{c,n_y,n_z} = E_{co} + n^2\frac{\pi^2\hbar^2}{2m^* W^2} = E_{co} + n^2\frac{\pi^2\hbar^2}{2m^* W^2} \tag{3.44}$$

with

$$n^2 = n_x^2 + n_y^2 \tag{3.45}$$

The number of channels (subbands) that are filled up to the Fermi energy is given by

$$N = \sqrt{\frac{2(E_F - E_{co})m^* W^2}{\pi^2 \hbar^2}} = \frac{W}{(\lambda_F/2)} \tag{3.46}$$

In analogy with the electromagnetic waveguide, as W increases, the number of channels also increases. A new channel appears as the width goes down another half the Fermi wavelength.

3.8 The Velocity and the Energy Averages

In discussing carrier transport, an important unresolved dilemma is the nature and limitations put on the velocity of the carriers as their equilibrium stochastic character transforms to directed velocity vectors in a very high electric field. The magnitude of the randomly oriented (stochastic) carrier velocity vectors plays an important role as unidirectional streamlined velocity vectors or directed moments give the ultimate velocity known as the saturation velocity. The randomly oriented carrier velocity is intrinsic to the sample and exists even in the absence of any external stimulation and hence, it is termed intrinsic velocity. The vector sum of these randomly oriented intrinsic velocity vectors is zero. However, the average magnitude $|v_{n(p)}|$ of each velocity in a random direction is not zero. It is no secret that without any stimulation present, the carriers are moving in random directions. When the time average of a single carrier is taken over a long time, we expect $\Sigma \vec{v}_{n(p)} = 0$. The same is true if the ensemble (group) average of all carriers is taken frozen at any instant. The question often arises: what is the average magnitude of the velocity $|v| = \sqrt{2(E - E_c)/m_n^*}$ as a function of temperature and carrier concentration? An itinerant electron moving at random will encounter atoms or their oscillating energy quanta, the phonons, and change direction. The atoms or their oscillations in the form of phonons as quanta of energy tend to inhibit (scatter is more esoteric to denote the randomness) the motion (both magnitude and direction). The intrinsic velocity (average of the magnitude of randomly oriented velocity vectors) in a nanostructure is obtained by following the same process that led to Equations 3.4 and 3.5, with an added factor $|v| = \sqrt{2(E - E_c)/m_n^*}$ introduced in the integrand and dividing the total velocity by the carrier concentration n_d for dimensionality d. The integration yields the expression for intrinsic velocity v_{id} [3] as

$$v_{id} = v_{thd} \frac{\Im_{(d-1)/2}(\eta_d)}{(n/N_c)_d} = v_{thd} \frac{\Im_{(d-1)/2}(\eta_d)}{\Im_{(d-2)/2}(\eta_d)} \tag{3.47}$$

with

$$v_{thd} = v_{th}\frac{\Gamma\left((d+1)/2\right)}{\Gamma(d/2)} \tag{3.48}$$

$$v_{th} = \sqrt{\frac{2k_BT}{m^*}} \tag{3.49}$$

In the ND limit, the FDI is always an exponential and does not depend on its order. The ND intrinsic velocity is given by

$$v_{idND} = v_{thd} = v_{th}\frac{\Gamma(d+1/2)}{\Gamma(d/2)} \tag{3.50}$$

The ND intrinsic velocity v_{idND} with $d = 1$ (a nanowire), $d = 2$ (a nanosheet), or $d = 3$ (bulk) obtained from Equation 3.50 is given by

$$v_{i1ND} = v_{th1} = \frac{1}{\sqrt{\pi}}v_{th} = 0.564v_{th} \tag{3.51}$$

$$v_{i2ND} = v_{th2} = \frac{\sqrt{\pi}}{2}v_{th} = 0.886v_{th} \tag{3.52}$$

$$v_{i3ND} = v_{th3} = \frac{2}{\sqrt{\pi}}v_{th} = 1.128v_{th} \tag{3.53}$$

The intrinsic velocity in the ND limit does not depend on the carrier concentration and hence has been utilized for various applications even for degenerate semiconductors where ND statistics is not valid. In the degenerate limit of the FDI, the intrinsic velocity is given by

$$v_{idDeg} = \frac{d}{d+1}v_{Fd} = \frac{d}{d+1}\sqrt{\frac{2(E_F - E_c)_d}{m^*}} \tag{3.54}$$

The intrinsic velocity is independent of the carrier temperature and is a function of carrier concentration only. When the degenerate limit of $(E_F - E_c)_d$ is used from Equation 3.40, the intrinsic velocity is given by

$$v_{idDeg} = 2\frac{d}{d+1}\frac{\hbar}{m^*}\sqrt{\pi}\left[\Gamma\left(\frac{d+2}{2}\right)\frac{n_d}{2}\right]^{1/d} \tag{3.55}$$

The intrinsic velocity, when applied to each dimension ($d = 1, 2$, and 3), is obtained as

$$v_{i1Deg} = \frac{\pi\hbar}{4m^*} n_1 \tag{3.56}$$

$$v_{i2Deg} = \frac{\hbar}{m^*}\left[8\pi n_2/9\right]^{1/2} \tag{3.57}$$

$$v_{i3Deg} = \frac{3\hbar}{4m^*}\left[3\pi^2 n_3\right]^{1/3} \tag{3.58}$$

Figure 3.10 demonstrates the normalized intrinsic velocity as a function of normalized carrier concentration. As expected, the intrinsic velocity is constant for low carrier concentrations and is a function of carrier concentration in the degenerate domain rising with the carrier concentration. The rise is dramatic for a nanowire (1D) nanostructure.

Continuing with the same procedure as for $|v|$ with the average kinetic energy $E_{KE} = E - E_c = (1/2)m_n^* v^2$, the mean kinetic energy E_{md} is obtained as

$$E_{md} = \frac{d}{2}k_B T \frac{\Im_{d/2}(\eta_d)}{\Im_{d-2/2}(\eta_d)} = \frac{d}{2}k_B T_{md}, \quad \frac{T_{md}}{T} = \frac{\Im_{d/2}(\eta_d)}{\Im_{d-2/2}(\eta_d)} \tag{3.59}$$

Here, we define elevated temperature T_{md} to determine the elevation of mean energy over and above the lattice temperature due to degeneracy.

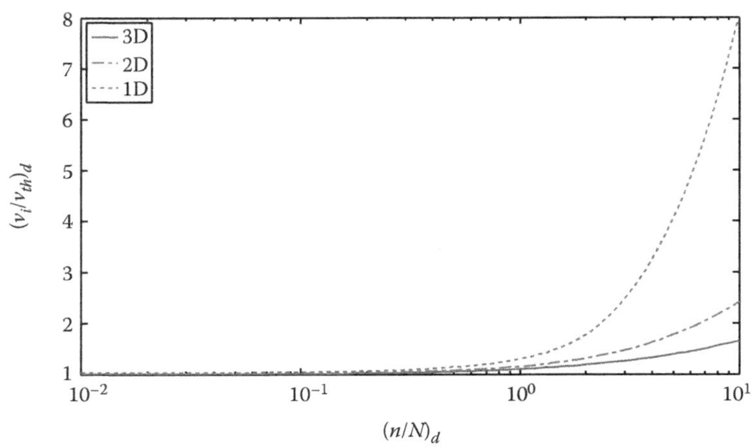

FIGURE 3.10
Normalized intrinsic velocity as a function of normalized concentration.

In the ND regime $T_{md} = T \cdot T_{md}$ will elevate with the carrier concentration as Fermi energy rises in the degenerate regime.

The mean energy is $E_{md} = (1/2)m^* v_{rmsd}^2$, where v_{rmsd} is the root mean square (rms) velocity that is the square root of the average of v^2. This rms velocity $v_{rmsd} = \sqrt{2E_m/m_n^*}$ for dimensionality d, as is obtained from Equation 3.59, is given by

$$v_{rmsd} = \sqrt{\frac{d\,k_B T}{m_n^*}\frac{\Im_{d/2}(\eta_d)}{\Im_{(d/2)-1}(\eta_d)}} = \sqrt{\frac{d\,k_B T_{md}}{m_n^*}} \tag{3.60}$$

In the ND limit, Equation 3.58 gives the familiar law of equipartition of energy that states that the energy per degree of freedom is $(E_{md}/d) = (1/2)k_B T$. This ND intrinsic energy is given by

$$E_{mdND} = \frac{d}{2}k_B T \tag{3.61}$$

The rms velocity in the ND limit is given by

$$v_{rmsdND} = \sqrt{\frac{dk_B T}{m_n^*}} \tag{3.62}$$

The ND intrinsic energy and rms velocity (commonly recognized as the thermal energy and thermal velocity, respectively), for each dimensionality, are given by

$$E_{m1ND} = \frac{1}{2}k_B T, \quad v_{rms1ND} = \sqrt{\frac{k_B T}{m_n^*}} \quad \text{(1D)} \tag{3.63}$$

$$E_{m2ND} = k_B T, \quad v_{rms2ND} = \sqrt{\frac{2\,k_B T}{m_n^*}} \quad \text{(2D)} \tag{3.64}$$

$$E_{m3ND} = \frac{3}{2}k_B T, \quad v_{rms3ND} = \sqrt{\frac{3k_B T}{m_n^*}} \quad \text{(3D)} \tag{3.65}$$

In the degenerate limit, the mean energy is given by

$$E_{mdDeg} = \frac{d}{d+2}(E_F - E_c)_d = \frac{d}{d+2}\frac{\hbar^2}{m^*}2\pi\left[\frac{d}{2}\Gamma\left(\frac{d}{2}\right)n_d/2\right]^{2/d} \tag{3.66}$$

Similarly, the rms velocity in the degenerate limit is given by

$$v_{rmsdDeg} = \sqrt{\frac{2}{m^*}\frac{d}{d+2}(E_F - E_c)_d}$$
$$= \frac{\hbar}{m^*}\left(\frac{d}{d+2}4\pi\right)^{1/2}\left[\frac{d}{2}\Gamma\left(\frac{d}{2}\right)n_d/2\right]^{1/d} \tag{3.67}$$

The mean energy and rms velocity (related to the Fermi velocity that itself depends on carrier concentration) for each individual dimensionality is given by

$$E_{m1Deg} = \frac{\pi^2}{24}\frac{\hbar^2}{m^*}n_1^2, \quad v_{rms1Deg} = \frac{\hbar}{m^*}\frac{\pi\, n_1}{2\sqrt{3}} = \left(\frac{1}{3}\right)^{1/2} v_{F1} \tag{3.68}$$

$$E_{m2Deg} = \frac{\pi}{2}\frac{\hbar^2}{m^*}n_2, \quad v_{rms2Deg} = \frac{\hbar}{m^*}(\pi n_2)^{1/2} = \frac{1}{\sqrt{2}}v_{F2} \tag{3.69}$$

$$E_{m3Deg} = \frac{6\pi}{5}\frac{\hbar^2}{m^*}\left[\frac{3\sqrt{\pi}}{8}n_3\right]^{2/3}, \quad v_{rms3Deg} = \frac{\hbar}{m^*}\left(\frac{12\pi}{5}\right)^{1/2}\left[\frac{3\sqrt{\pi}}{8}n_3\right]^{1/3} = \left(\frac{3}{5}\right)^{1/2} v_{F3} \tag{3.70}$$

The hot electron term is normally used to indicate the elevated energy of carriers in an electric field. The energy is also elevated due to enhanced degeneracy of the sample when electrons change their orientation in a high electric field and become unidirectional as explained in Chapter 4. Additional elevation of degeneracy due to unidirectional character is accounted for in the unidirectional electron temperature given by

$$\frac{T_{ud}}{T} = \frac{T_{id}}{T}\frac{v_{ud}}{v_{id}} \tag{3.71}$$

The first factor is due to the kinetic energy $(1/2)m^*\, v_{id}^2$ rising above its ND counterpart $(1/2)m^*\, v_{thd}^2$ as given by

$$\frac{T_{id}}{T} = \left(\frac{v_{id}}{v_{thd}}\right)^2 \tag{3.72}$$

The second factor arises due to unidirectional velocity rising above the equilibrium Fermi velocity in the strongly degenerate state as randomly oriented concentration n_d becomes unidirectional in a very strong electric field making the Fermi energy appropriate to that of $2n$ electrons in a single direction (antiparallel to electric field) and none in the other (parallel to the electric field).

Figure 3.11 shows the relative temperature (unidirectional and mean for all dimensionalities $d = 3$, 2, and 1). The rise for $d = 1$ in a nanowire is dramatic. In the ND regime, there is no distinction between T and T_{ud} ($T_{ud} = T$). As stated in Equation 3.71, there are two factors that elevate T_{ud} over and above its ND value equal to the lattice temperature T. One factor T_i/T is obvious that arises due to the mean electron energy rising above its ND value as given by Equation 3.72. The other factor v_{ud}/v_{id} is subtler. In the strong degenerate regime, electrons oriented in the direction parallel to the applied electric field change the orientation opposite (antiparallel) to the electric filed. Pauli Exclusion Principle does not allow occupation of already occupied quantum states below the Fermi energy. The converted unidirectional electrons in a very strong electric filed reach unidirectional intrinsic energy appropriate for $2n_d$ electrons filling half the k-space. It is important to emphasize that the density of electrons does not increase as there are $2n_d$ electrons antiparallel to the electric field and none in the parallel direction, giving an average n_d. The unidirectional transformation thus raises the Fermi level that appears to be arising from $2n_d$ electrons. In the ND regime, the intrinsic velocity and unidirectional velocity are equal to the thermal velocity, that is, $v_{ud} = v_{id} = v_{thd}$. However, as the degeneracy level rises, so does the transformation of v_{id} to v_{ud}. As the expressions below show, $v_{u3} = 2^{1/3}v_{i3}$, $v_{u2} = 2^{1/2}v_{i2}$, and $v_{u1} = 2v_{i2}$ in the extreme degenerate case. Thus, the unidirectional character represented by T_{ud} holds special importance in the strong degenerate state and in the strong electric field.

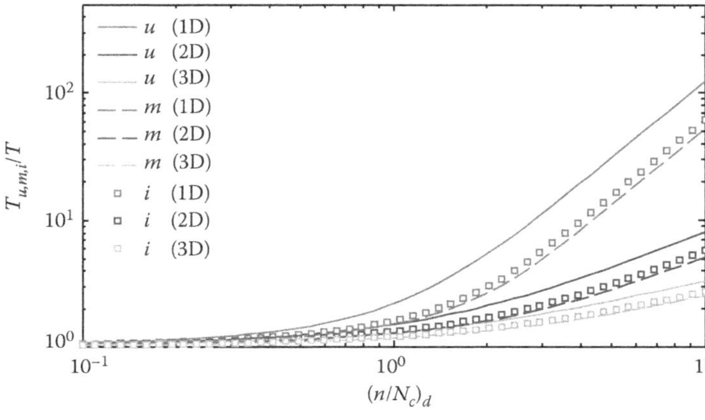

FIGURE 3.11
Normalized temperature T_u/T (unidirectional, solid lines), T_m/T (mean, dashed lines), and intrinsic T_i/T (squares) as a function of normalized concentration n/N_c for $d = 1$ (top), 2 (middle), and 3 (bottom). Top line in each category is for 1D, intermediate for 2D, and bottom for 3D.

All metals are degenerate with effective mass of the electron the same as free electron mass m_o. As evaluated earlier, the Fermi velocity v_F in terms of the carrier concentration for dimensionality d is given by

$$v_{Fd} = \sqrt{\frac{2(E_F - E_c)_d}{m_o}} = \frac{\hbar}{m_o} 2\sqrt{\pi}\left[\frac{d}{2}\Gamma\left(\frac{d}{2}\right)n_d/2\right]^{1/d} \tag{3.73}$$

Specifically, for each dimensionality, the Fermi velocity is given by

$$v_{F3} = \frac{\hbar k_{F3}}{m_o} = \frac{\hbar}{m_o}\left(3\pi^2 n_3\right)^{1/3} \tag{3.74}$$

$$v_{F2} = \frac{\hbar k_{F2}}{m_o} = \frac{\hbar}{m_o}(2\pi n_2)^{1/2} \tag{3.75}$$

$$v_{F1} = \frac{\hbar k_{F1}}{m_o} = \frac{\hbar}{m_o}\frac{\pi n_1}{2} \tag{3.76}$$

The Fermi energy, velocity, and de Broglie wavelength for various metals are given in Appendix E.

3.9 Graphene/CNT Nanostructures

The theoretical development of electronic transport in a graphene nanostructure is complicated due to linear E–k (energy vs. momentum) relation with zero effective mass of a Dirac fermion, as reviewed in a number of notable works [5–7]. As shown in Chapter 2, an ideal graphene is a monoatomic layer of carbon atoms arranged in a honeycomb lattice. A monoatomic layer makes graphene a perfect 2D material. As a 2D nanolayer, the graphene sheet has some semblance to a MOSFET. Wu et al. [5] give an excellent comparison of the linear E–k (energy vs. momentum) relation in a graphene nanolayer to a quadratic one in a nano-MOSFET. Six-valley parabolic band structure in a MOSFET, even though anisotropic, has a finite effective mass [8–10]. As graphene is a relatively new material with a variety of its allotropes, the landscape of electronic structure and applications over the whole range of electric and magnetic fields is in its infancy [11].

The Dirac cone described in Chapter 2 shows a rise in energy with the magnitude of momentum vector:

$$E = E_{Fo} \pm \hbar v_F \,|\, k \,| = E_{Fo} \pm \hbar v_{Fo}\sqrt{k_x^2 + k_y^2} \tag{3.77}$$

where k is the momentum vector which in circular coordinates has components $k_x = k\cos\theta$ and $k_y = k\sin\theta$ and $v_{Fo} = (1/\hbar)dE/dk$ is constant due to the linear rise of energy E with momentum vector k. $\hbar v_{Fo}$ is the gradient of E–k dispersion. The linear dispersion of the Dirac cone is confirmed up to ±0.6 eV [12,13]. $v_{Fo} \approx 10^6$ m/s is the accepted value of Fermi velocity near the Fermi energy that lies at the cone apex $E_F - E_{Fo} = 0$ for intrinsic graphene with $E_{Fo} = 0$ as the reference level. The Fermi velocity vectors are randomly oriented in the graphene sheet. $E_F - E_{Fo}$ defines the degeneracy of the Fermi energy that itself depends on 2D carrier concentration n_g given by [12]

$$n_g = N_g \Im_1(\eta_c) \tag{3.78}$$

with

$$N_g = (2/\pi)\left(\frac{k_B T}{\hbar v_{Fo}}\right)^2 \tag{3.79}$$

$$\eta_c = \frac{E_F - E_{Fo}}{k_B T} \tag{3.80}$$

Equation 3.78 is easily derived by following the same procedure as for semiconductors, integrating the product of DOS with the Fermi-Dirac distribution function.

$\Im_1(\eta)$ is the FDI of order 1 [14,15] with $\Im_1(0) = \pi^2/12$. Hence, the carrier density in intrinsic graphene with $\eta_c = 0$ is given by

$$n_{ig} = \left(\frac{\pi}{6}\right)\left(\frac{k_B T}{\hbar v_{Fo}}\right)^2 \approx 8 \times 10^{10}\left(\frac{T}{300}\right)^2 \text{cm}^{-2} \tag{3.81}$$

In an extrinsically degenerately doped sample $\Im_1(\eta) = \eta^2/2$, giving $E_F - E_{Fo} = \sqrt{\pi n_g}\,\hbar v_{Fo}$. Equation 3.78 transforms in a doped p-type graphene as follows:

$$p_g = N_g \Im_1(\eta_v), \quad \eta_v = \frac{E_{Fo} - E_F}{k_B T} \tag{3.82}$$

with $E_c = E_v = E_{Fo} = 0$ giving zero bandgap $E_g = E_c - E_v = 0$. The Fermi energy is $E_{Fo} - E_F = \sqrt{\pi p_g}\,\hbar v_{Fo}$ in strongly degenerate state of the valence band. In contrast to ND semiconductors where Fermi energy resides in the forbidden bandgap, graphene with zero bandgap has the Fermi energy either in the conduction or valence band guaranteeing a degenerate state, at least partially, even at room temperature.

Graphene and CNTs have been called the "wonder material of the 21st century," "the building blocks for the future of electronics," or possible "replacement for Silicon circuits." While it is debatable whether such grand predictions will come true, CNTs have unarguably generated tremendous interest amongst chemists, physicists, and electrical engineers alike by virtue of their unique properties and potential to offer solutions to several problems as conventional technology approaches fundamental limits.

The Fermi–Dirac distribution function when multiplied by the DOS of CNT and integrated from the conduction band edge to infinity gives carrier density per unit length n_{CNT} as a function of normalized Fermi energy $\eta = (E_F - E_c)/k_BT$ as given by

$$n_{CNT} = N_{CNT}\Im_{CNT}(\eta, e_g) \tag{3.83}$$

$$\Im_{CNT}(\eta, e_g) = \int_0^\infty \frac{x + (e_g/2)}{\sqrt{x^2 + xe_g}}\left(\frac{1}{e^{x-\eta} + 1}\right)dx \tag{3.84}$$

with

$$N_{CNT} = D_o k_B T \tag{3.85}$$

$$e_g = \frac{E_g}{k_B T} \tag{3.86}$$

where T is the ambient temperature. The CNT integral $\Im_{CNT}(\eta, e_g)$ can be evaluated numerically. The reduced Fermi energy η as a function of reduced carrier concentration $u_{CNT} = n_{CNT}/N_{CNT}$ is plotted in Figure 3.12 for three chirality values: (10,4) with $\nu = 0$, (13,0) with $\nu = 1$, and (10,5) with $\nu = 2$. In the ND limit ($\eta < -2$) for low carrier concentration, η appears independent of chirality. However, in the degenerate limit ($\eta > 2$)η depends on chirality. Also

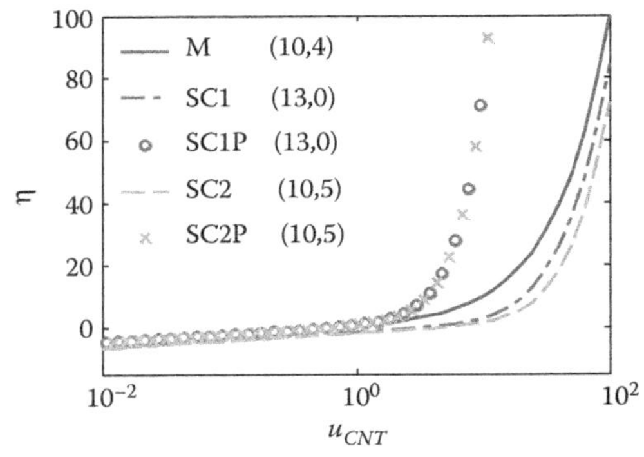

FIGURE 3.12
The reduced Fermi energy $\eta = (E_F - E_c)/k_BT$ as a function of $\Im_{CNT}(\eta, e_g) = u_{CNT}$ for metallic and semiconducting CNTs. Chirality (10,4), (13,0), and (10,5) represent metallic, SC1, and SC2 CNTs with band index $\nu = 0, 1, 2$. Parabolic approximation is compared with the exact form showing an agreement in the ND limit.

shown is the parabolic approximation (Equation 3.35) for which effective mass approximation is valid. As Figure 3.12 shows, the parabolic approximation breaks down in the degenerate regime.

In strong degeneracy limit ($\eta > 2$), Equation 3.83 simplifies to

$$n_{CNT} = N_{CNT}\left[\eta^2 + \eta e_{gv}\right]^{1/2} = \frac{N_{CNT}\left[E_F^2 - E_{cv}^2\right]^{1/2}}{k_B T} \tag{3.87}$$

The ND approximation for CNT is not as straightforward as for parabolic semiconductors. Wong and Akinwande [12] give an expression in terms of Bessel function

$$\begin{aligned} n_{CNT} &= N_{CNT}\frac{e_g}{2}e^{E_F/k_BT}\int_1^{\infty} dt e^{-(e_g/2)t}\frac{t}{\sqrt{t^2-1}} \\ &= N_{CNT}\frac{e_g}{2}e^{E_F/k_BT}K_1\left(\frac{e_g}{2}\right) \end{aligned} \tag{3.88}$$

An asymptotic form of $K_1(z)$ is available [16] for large z and is given by

$$K_1(z) = \sqrt{\frac{\pi}{2z}}\, e^{-z}\left(1 + \frac{3}{8z}\right) \tag{3.89}$$

The ND form simplifies with the use of Equation 3.89. The simplified expression is given by

$$n_{CNT} = N_{CNT}\frac{\sqrt{\pi e_{gv}}}{2}e^{\eta} \tag{3.90}$$

where only the leading term is retained. $\Im_{CNT}(\eta, 0) = \Im_0(\eta)$ for a metallic CNT of zero bandgap ($\nu = 0$), consistent with that in a nanowire [17]. $\Im_0(\eta) = \exp(\eta)$ in the ND limit changes to $\Im_0(\eta) = \eta$ in the strong degeneracy limit. Figure 3.13 shows the approximation for the extreme degenerate and ND limits. It is clear from the figure that the ND approximation is good for $u_{CNT} = n_{CNT}/N_{CNT} \ll 1$, while the degenerate approximation is good for $u_{CNT} = n_{CNT}/N_{CNT} \gg 1$.

It is instructive to compare Equation 3.83 with the parabolic approximation that is always valid in the ND regime as the Fermi level is below the conduction band edge. The parabolic expression for a 1D semiconductor with valley degeneracy $g_K = 2$ is given by

$$n_{CNT} = N_c\Im_{-1/2}(\eta) \tag{3.91}$$

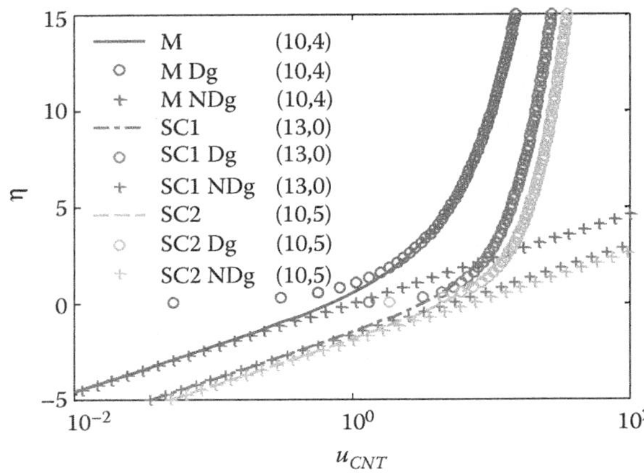

FIGURE 3.13
The plot of η versus $\Im_{CNT}(\eta, e_g) = u_{CNT}$ for chirality (10,4) with $\nu = 0$, (13,0) with $\nu = 1$, and (10,5) with $\nu = 2$. Also revealed are the ND limit both with $K_1(e_g/2)$ and its asymptotic form indexed as NDg. Degenerate approximation is also displayed (Dg).

$$N_{c\nu} = g_K \left(\frac{2m_{t\nu}^* k_B T}{\pi \hbar^2} \right)^{1/2} = N_{CNT} \frac{\sqrt{\pi e_{g\nu}}}{2} \tag{3.92}$$

where $\Im_{-1/2}(\eta) \approx e^{\eta}$ in the ND regime and N_c is the effective DOS in the conduction band. Equation 3.92 is indicative of the fact that the DOS effective mass is $m_{dos}^* = g_K^2 m_t^*$ which is four times the effective mass in a single-valley 1D nanostructure because of valley degeneracy $g_K = 2$. The conductivity effective mass will remain the same as m_t^*.

The intrinsic carrier concentration for a metallic CNT ($\nu = 0$), when both electrons and holes are considered, is given by

$$n_{iM} = N_{CNT} \Im_0(0) = \ln(2) D_o k_B T \tag{3.93}$$

The intrinsic carrier concentration for a metallic CNT rises linearly with temperature. However, for a semiconducting CNT, it is a function of bandgap E_g as given by

$$n_{iSC\nu} = N_{c\nu} e^{-e_{g\nu}/2} \tag{3.94}$$

Obviously, the intrinsic carrier concentration is considerably less in the band with $\nu = 2$ as compared to that of index $\nu = 1$ of the same diameter.

In equilibrium, the velocity vectors are randomly oriented in the tubular direction with half oriented in the positive x-direction and half directed in the negative x-direction for a tubular direction along the x-axis. This makes the sum of velocity vectors equal to zero, as expected. However, the average magnitude of the carrier motion is not zero at a finite temperature. The group average velocity of a carrier in essence informs the speed of a propagating signal. It is also a useful parameter giving information as velocity vectors are re-aligned in the direction of an electric field [17] as it sets the limit at saturation velocity that is the ultimate attainable velocity in any conductor. In a ballistic transport when electrons are injected from the contacts, the Fermi velocity of the contacts plays a predominant role [18,19]. It is often closely associated with the maximum frequency of the signal with which the information is transmitted by the drifting carriers. Formally, the carrier group velocity is defined as

$$v(E) = \frac{1}{\hbar}\left|\frac{dE}{dk}\right| \tag{3.95}$$

The magnitude of the velocity can be related to the DOS by rewriting it as

$$v(E) = \frac{1}{\hbar}\frac{dE}{dN}\frac{dN}{dk} = \frac{g_s}{2\pi\hbar D_{CNT}(E)} \tag{3.96}$$

where $(dN/dk) = (g_s/2\pi)$ (with $g_s = 2$ for spin degeneracy) in the k-space and $D_{CNT}(E) = dN/dE$ is the DOS for a single valley. When multiplied with the DOS and the Fermi–Dirac distribution function and divided by the electron concentration given by Equation 3.91 the magnitude of the velocity vector, the intrinsic velocity v_i, for a CNT is given by

$$v_i = \frac{v_{F0}\,\Im_0(\eta)}{u_{CNT}} \tag{3.97}$$

The name intrinsic is given to this velocity as it is intrinsic to the sample as compared to the drift velocity that is driven by an external field. The intrinsic velocity for arbitrary degeneracy is shown in Figure 3.14 for band index $\nu = 0, 1, 2$. The intrinsic velocity is not equal to the Fermi velocity $v_{Fo} \approx 10^6$ m/s for semiconducting samples approaching v_{Fo} as expected in strong degeneracy. However, for a metallic CNT, the intrinsic velocity is the intrinsic Fermi velocity. The Fermi velocity in the parabolic model is calculable. However, it has no physical meaning as parabolic approximation works only in the ND regime.

Figure 3.14 shows the ratio of the intrinsic velocity to the Fermi velocity as a function of normalized carrier concentration with respect to the EDOS. The saturation velocity in a high electric field is limited to the intrinsic velocity that is affected by the onset of phonon emission [20]. Therefore, the saturation velocity by itself is ballistically unaffected by scattering parameters, but affected only by the energy of an optical phonon. This paradigm is in direct contrast to scattering-limited saturation often quoted in the published literature.

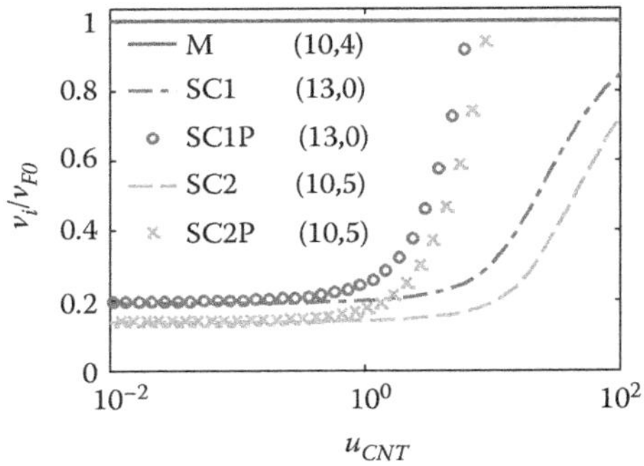

FIGURE 3.14
The normalized intrinsic velocity v_i/v_{F0} as a function of normalized carrier concentration u_{CNT} for CNTs of different chiralties. Exact formulation Equation 3.97 is compared with the parabolic approximation (SC1P and SC2P). Parabolic approximation overestimate the intrinsic velocity.

EXAMPLES

E3.1 a. A GaAs crystal is doped with $N_A = 10^{16}$ cm^{-3} and $N_D = 0$ at room temperature. Sketch its energy band diagram showing the numerical position of the Fermi level with respect to both band edges (E_c and E_v). Include actual calculated values on the diagram. The band diagram for a p-type GaAs doped with $N_A = 10^{16}$ cm^{-3} and $N_D = 0$ at room temperature.

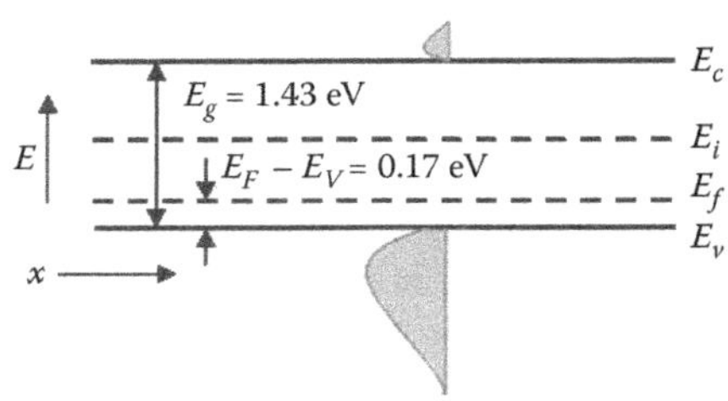

Here, GaAs is a p-type with acceptor concentration $N_A = 10^{16}$ cm^{-3} and zero donor concentration $N_D = 0$. Assuming that every acceptor has accepted an electron from the valence band, the hole concentration is

$$p_3 \approx N_A - N_D = 1.0 \times 10^{16} \text{ cm}^{-3}$$

The hole EDOS for GaAs ($m_p^* = 0.48\, m_o$) from Equation 3.17 is $N_v = 8.3 \times 10^{18}$ cm^{-3}. The Fermi energy $E_{F3} - E_{vo}$ with use of Equation 3.19 is

$$E_{F3} - E_{vo} = k_B T \ln\left(\frac{N_v}{p_3}\right) = 0.0259\text{ eV}\ln\left(\frac{8.30\times10^{18}}{10^{16}}\right) = 0.17\text{ eV}$$

E3.2 a. A silicon crystal is doped with 3.0×10^{16} boron (acceptor) atoms per cm^3. Sketch its energy band diagram showing the numerical position of the Fermi level with respect to both band edges (E_c and E_v). Include actual calculated values on the diagram.

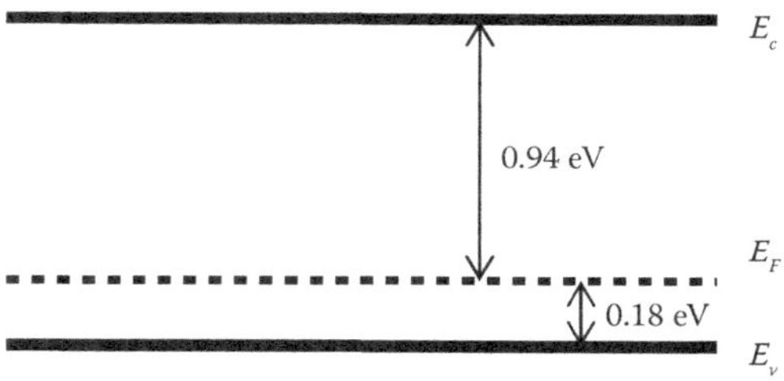

b. The crystal of part (a) is counterdoped with 3.0×10^{16} phosphorus (donors) atoms in addition to 3.0×10^{16} boron atoms already present. Sketch the new energy band diagram. What is the concentration of electrons and holes in the sample so counterdoped?

Answer

a. $p_o = N_v\, e^{-(E_F - E_v)/k_B T}$

$$E_F - E_v = k_B T \ln\left(\frac{N_v}{p_o}\right)$$
$$= 0.0259\text{ eV}\ln\left(\frac{3.1\times10^{19}}{3.0\times10^{16}}\right)$$
$$= 0.18\text{eV}$$

$$E_c - E_F = E_g - (E_F - E_V) = 1.12\text{ eV} - 0.18\text{ eV} = 0.94\text{ eV}$$

$$N_A = N_D \Rightarrow p_o = \frac{N_A - N_D}{2} \pm \sqrt{\left(\frac{N_A - N_D}{2}\right)^2 + n_i^2} = n_i$$
$$= 1.08\times10^{10}\text{ cm}^{-3}$$

b.
$$n_o = \frac{N_D - N_A}{2} \pm \sqrt{\left(\frac{N_D - N_A}{2}\right)^2 + n_i^2} = n_i$$
$$= 1.08\times10^{10}\text{ cm}^{-3}$$

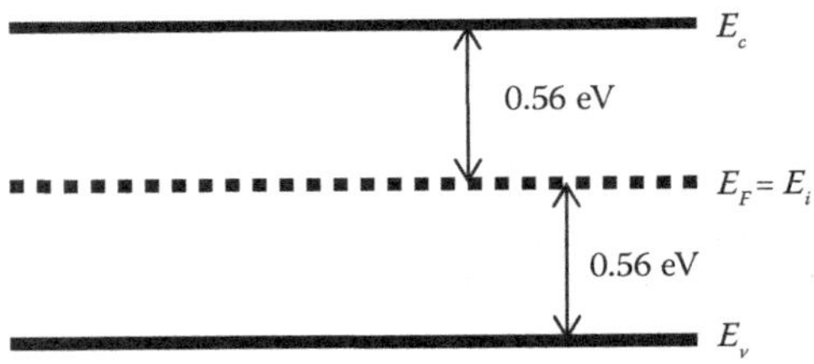

Hence, the semiconductor is intrinsic (perfect compensation).

$$E_F = E_i$$
$$E_c - E_F \approx E_F - E_v = 0.56 \text{ eV}$$

E3.3 a. Silicon is doped with $N_D = 5 \times 10^{15}\text{cm}^{-3}$. Find n_0 and p_0 and locate the Fermi level. Draw the energy band diagram with numerical values indicated with respect to each bandedge (conduction and valence band edge).

b. A new batch of silicon is doped with boron to $N_A = 4 \times 10^{15}$ cm^{-3}. Find n_0 and p_0 and locate the Fermi level. Draw the energy band diagram with numerical values indicated with respect to each bandedge (conduction and valence band edge).

Answer

a. The material will be n-type. Since the number of donors is much greater than the intrinsic concentration, and assuming all dopants to be ionized, then $n_0 = N_D$. Thus, $n_0 = 5 \times 10^{15}\text{cm}^{-3} = N_C e^{-(E_C - E_F/kT)}$ assuming ND material (which we will check later).

So

$$E_C - E_F = -kT \ln\left(\frac{N_D}{N_C}\right) = -0.026 \text{ eV} \ln\left(\frac{5 \times 10^{15}}{2.86 \times 10^{19}}\right) = 0.22 \text{ eV}$$

Since $0.22 > 3\, k_B T$, the assumption of nondegeneracy is verified.

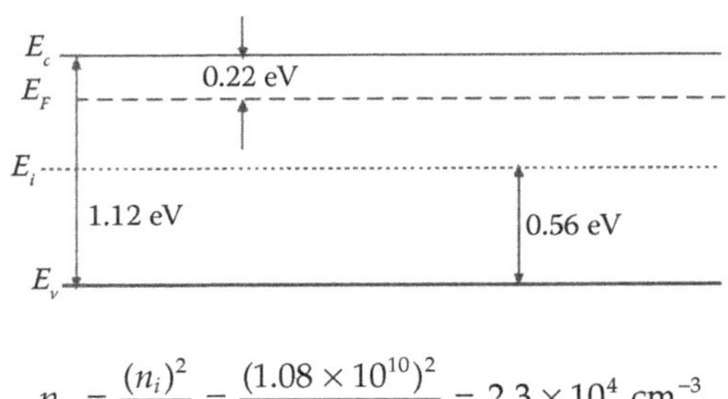

$$p_0 = \frac{(n_i)^2}{n_0} = \frac{(1.08 \times 10^{10})^2}{5 \times 10^{15}} = 2.3 \times 10^4 \text{ cm}^{-3}$$

b. Again we assume ND material and complete ionization. Thus, $p_0 = 4 \times 10^{15}$ cm^{-3}, and

$$n_0 = \frac{(1.08 \times 10^{10}\,\text{cm}^{-3})^2}{4 \times 10^{15}\,\text{cm}^{-3}} = 2.916 \times 10^{4}\,\text{cm}^{-3}$$

$$E_F - E_v = -kT \ln\left(\frac{N_A}{N_V}\right) = -0.026\ \text{eV} \ln\left(\frac{4 \times 10^{15}}{3.1 \times 10^{19}}\right) = 0.23\ \text{eV}$$

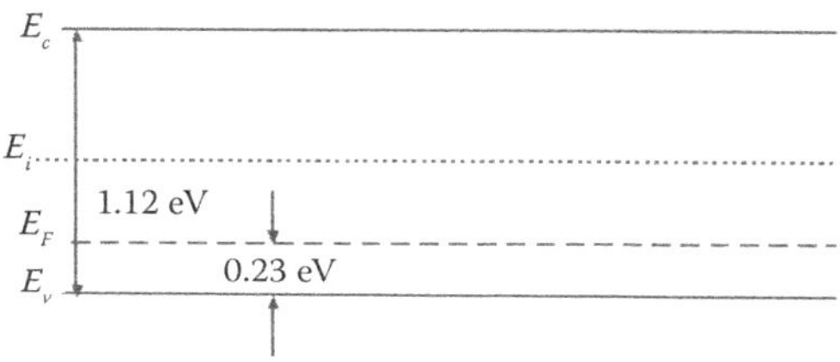

E3.4 The *Fermi energy versus carrier concentration* indicates that electron gas is strongly degenerate when the Fermi energy is 3 k_BT above the edge of the conduction band.

a. Find the probability of finding the electron at the conduction band edge E_c.
b. Find the probability of finding electron at the Fermi energy E_F.
c. What is the carrier concentration (in m^{-3}) of electrons in GaAs at room temperature at the threshold of being strongly degenerate?
d. Sketch a band diagram showing the intrinsic Fermi energy, Fermi energy, and bandgap with appropriate values for GaAs at room temperature.

Answer

a. The probability is given by the Fermi–Dirac distribution function

$$f(E) = \frac{1}{1 + e^{(E-E_F)/k_BT}}$$

For $E = E_c$: $E_F - E_c = 3\,k_BT$, $E_c - E_F = -3\,k_BT$

$$f(E) = \frac{1}{1 + e^{(E_c-E_F)/k_BT}} = \frac{1}{1 + e^{-3\,k_BT/k_BT}} = \frac{1}{1 + e^{-3}} = 0.999$$

b. The probability is given by the Fermi–Dirac distribution function

$$f(E) = \frac{1}{1 + e^{(E-E_F)/k_BT}}$$

For $E = E_F$, $E_F - E_F = 0$

$$f(E) = \frac{1}{1+e^{0/k_BT}} = \frac{1}{1+e^0} = \frac{1}{1+1} = 0.5$$

c. $$\begin{aligned} n_3 &= N_{c3}\Im_{1/2}(\eta) = 4.4\times10^{17}\,\text{cm}^{-3}\Im_{1/2}(3) \\ &= 4.4\times10^{17}\,\text{cm}^{-3}\times 4.4876 \\ &= 19.7\times10^{17}\,\text{cm}^{-3} \end{aligned}$$

$$\begin{aligned} n_3 &= 19.7\times10^{17}\,\text{cm}^{-3} = 1.97\times10^{18}\,\text{cm}^{-3} \\ &= 1.97\times10^{18}(10^{-2}\,\text{m})^{-3} \\ &= 1.97\times10^{24}\,\text{m}^{-3} \end{aligned}$$

From the table of FDIs $\Im_{1/2}(3) = 4.4876$

d.

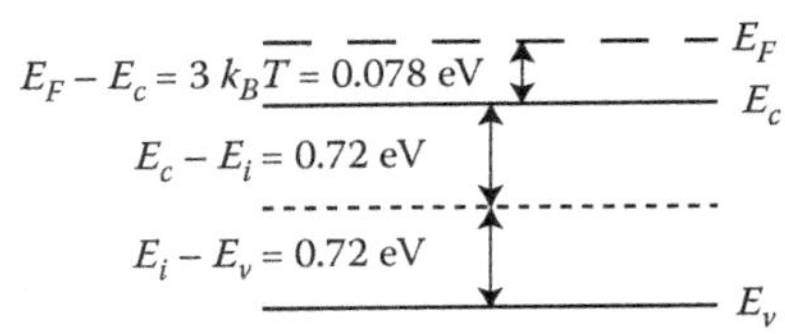

E3.5 Two semiconductors A and B have the same DOS effective masses. The bandgap $E_g = 1$ eV for A and 2 eV for B. Find the ratio n_{iA}/n_{iB} of the intrinsic carrier concentrations.

$$n_i = \sqrt{N_C N_V}\,e^{-E_g/2kT}$$

$$\text{Thus, } \frac{n_{iA}}{n_{iB}} = \frac{\sqrt{N_C N_V}\,e^{-E_{gA}/2kT}}{\sqrt{N_C N_V}\,e^{-E_{gB}/2kT}} = e^{E_{gB}-E_{gA}/2kT} = e^{(2-1)/2(0.026)} = 2.2\times10^8$$

E3.6 GaAs is doped with $N_D = 10^{15}$ cm^{-3} and $N_A = 4\times10^{14}$ cm^{-3}.

a. Find the concentration of electrons and holes.
b. Find the location of the Fermi energy with respect to the conduction band edge ($E_c - E_F$) as well as with respect to the valence band edge ($E_F - E_v$).
c. Draw the energy band diagram clearly indicating the Fermi and intrinsic energy and their numerical values with respect to conduction band and valence band edge.

Answer

a. Thus, we can say that $n_0 = N_D - N_A = 10^{15} - 4\times10^{14} = 6\times10^{14}$ cm^{-3}. If the material is not degenerately doped, then

$$p_0 = \frac{n_i^2}{n_0} = \frac{(2.2\times10^6)^2}{6\times10^{14}} = 8.1\times10^{-3}\,\text{cm}^{-3}$$

b. To locate the Fermi level, we need to find n_0 and p_0. The intrinsic concentration n_i for GaAs is $2.2 \times 10^6\,\text{cm}^{-3}$, so N_A and N_D are both $\gg n_i$.

$$E_C - E_F = -kT\ln\left(\frac{n_0}{N_C}\right) = -0.026\,\text{eV}\ln\left(\frac{6\times 10^{14}}{4.4\times 10^{17}}\right) = 0.17\,\text{eV}$$

$$E_F - E_v = 1.43\,\text{eV} - 0.17\,\text{eV} = 1.26\,\text{eV}$$

c.

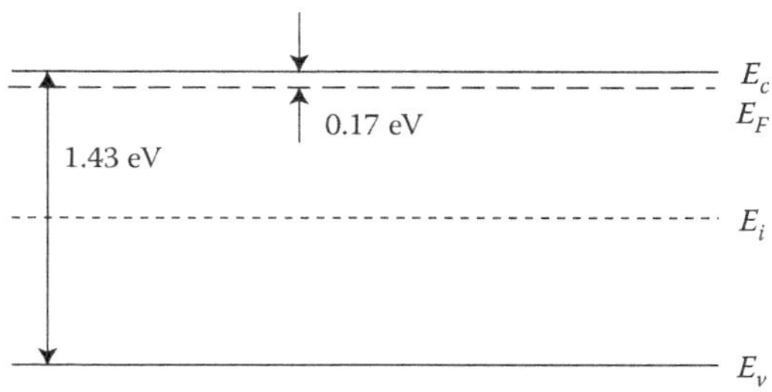

E3.7 A sample of silicon at 300 K is doped with 10^{15} arsenic (Group V) atoms per cm^3 and 10^{14} aluminum (Group III) atoms per cm^3.
a. Is this material n-type or p-type? Explain.
b. Find the equilibrium concentration of electrons and holes.
c. Where is the Fermi level located with respect to the conduction band edge? And, where is the Fermi level located with respect to the intrinsic level?

Answer

a. Arsenic is a donor and aluminum is an acceptor.
b. $N_D = 10^{15}\text{cm}^{-3} > N_A = 10^{14}\text{cm}^{-3}$, n-type

$$n = N_D - N_A = 0.9\times 10^{15}\text{cm}^{-3}$$

$$p = \frac{n_i^2}{N_D - N_A} = 1.3\times 10^5\text{cm}^{-3}$$

c. $E_c - E_F = k_BT\ln\dfrac{N_c}{n} = 0.268\,\text{eV}$

or

$$E_F - E_i = k_BT\ln\frac{n}{n_i} = 0.293\,\text{eV}$$

E3.8 The energy distribution per unit volume per unit energy of electrons $n_E(E)$ in the conduction band is given by

$$n_E(E) = A(E - E_C)^{1/2} \exp\left(-\frac{E - E_F}{k_B T}\right)$$

where A is a constant.

a. Show that the maximum in the energy distribution is at $E - E_c = 1/2\, k_B T$. Hint: n_E is maximum when $dn_E/dE = 0$.
b. Calculate the velocity corresponding to the thermal energy in (a) for GaAs at room temperature.

E3.9 a. $$\frac{dn_E}{dE} = 0$$

$$= A\begin{bmatrix} \frac{1}{2}(E - E_C)^{1/2-1} \exp\left(-\frac{E - E_F}{k_B T}\right) + (E - E_C)^{1/2}\left(-\frac{1}{k_B T}\right) \\ \exp\left(-\frac{E - E_F}{k_B T}\right) \end{bmatrix}$$

$$E - E_c = \frac{1}{2} k_B T$$

b. $$E - E_c = \frac{1}{2} k_B T \Rightarrow \frac{1}{2} m^* v_{th}^2 = \frac{1}{2} k_B T$$

$$v_{th} = \sqrt{\frac{k_B T}{m^*}} = 2.6 \times 10^5 \text{m/s}$$

PROBLEMS

P3.1 A new semiconductor has $N_c = 10^{19}$ cm^{-3}, $N_v = 5 \times 10^{18}$ cm^{-3} at room temperature, and $E_g = 2.0$ eV. It is doped with 10^{17} fully ionized atoms (n-type). Calculate the electron, hole, and intrinsic carrier concentrations at 627°C. Sketch the simplified band diagram showing the position of E_F.

P3.2 (a) Find the equilibrium n_3 and p_3 for Si doped with 10^{17} cm^{-3} boron atoms at 300 K. (b) Find n_3 for Ge doped with 3×10^{13} Sb atoms per cm^3, using the space charge neutrality at 300 K. Intrinsic carrier concentration for Ge is 1.64×10^{13} cm^{-3}.

P3.3 Consider the energy band diagram in the following figure. (a) What is the potential energy of the electron? (b) What is the kinetic energy of the electron? (c) What is the potential energy of the hole? (d) What is the kinetic energy of the hole? (e) The vertical axis is energy, meaning the electron energy. What direction represents increasing hole energy, up or down?

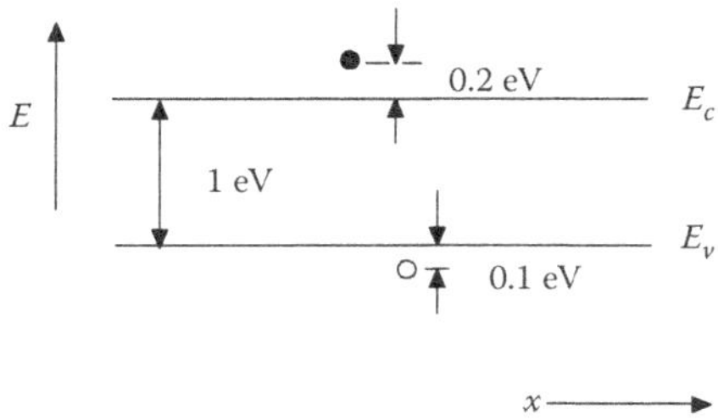

Band energy diagram in equilibrium.

P3.4 What is the probability that a state at the conduction band edge of intrinsic GaAs is occupied at room temperature?

P3.5 The donor ground states for tellurium in GaAs are 5.9 meV below the conduction band edge E_c. (There are two of them because of spin.) At room temperature, what is the probability that a given ground state is occupied if the Fermi level is 0.1 eV below E_c?

P3.6 One possible way to define nondegeneracy is that the Fermi level is $3k_BT$ below the conduction band edge. Find the probability of finding electron at energy $E = E_c$ when at the threshold of nondegeneracy.

P3.7 (a) Silicon is doped with $N_D = 5 \times 10^{15}\text{cm}^{-3}$. Find n_3 and p_3, and locate the Fermi level. Draw the energy band diagram. (b) A new batch of silicon is doped with boron to $N_A = 4 \times 10^{15}\ \text{cm}^{-3}$. Find n_3 and p_3 and locate the Fermi level. Draw the energy band diagram.

P3.8 GaAs is doped with $N_D = 10^{15}\ \text{cm}^{-3}$ and $N_A = 4 \times 10^{14}\ \text{cm}^{-3}$. Draw the energy band diagram and locate the Fermi level.

P3.9 A sample of silicon is doped with $N_D = 4 \times 10^{16}\ \text{cm}^{-3}$ and $N_A = 8 \times 10^{15}\ \text{cm}^{-3}$. Find the equilibrium concentrations of electrons and holes, and locate the Fermi level.

P3.10 In an applied electric field $\mathcal{E} = V/L$, the bands tilt with the decrease in potential energy as one moves from $x = 0$ on one end to $x = L$ on other end. $E_c(x) = E_c(0) - q\mathcal{E}x$ is the fall in the conduction band edge. Consider the energy band diagram in the following figure with V = 1V. (a) Find the electric field, and express the result in V/m. What direction does it point? (b) Find the force on the electron. In what direction is the electron accelerated? (c) Find the force on the hole. In what direction is the hole accelerated? (d) What is the drop in energy (in eV) of the conduction band from $x = 0$ to L?

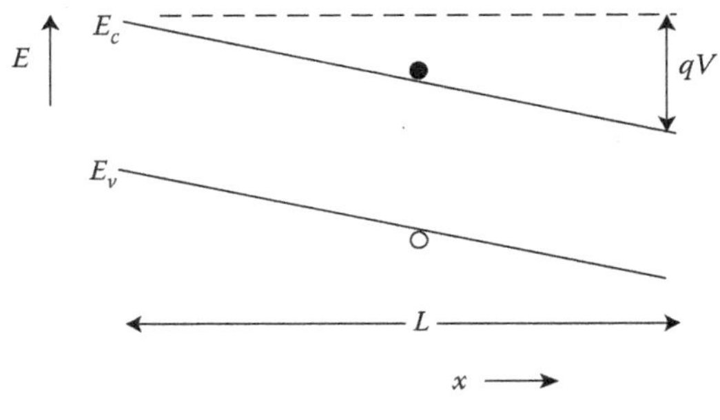

The tilted band diagram in an uniform electric field with dip in energy qV over the sample length L.

P3.11 A strongly degenerate electron gas is characterized by intrinsic velocity $v_{i3Deg} = (3\hbar/4m^*)\left[3\pi^2 n_3\right]^{1/3}$. Show that the de Broglie wavelength for a strongly degenerate gas is given by

$$\lambda_D = \frac{8\pi}{3\left(3n_3\pi^2\right)^{1/3}}$$

Calculate λ_D in nm for $n_3 = 10^{20}, 10^{21}, 10^{22}\ \text{cm}^{-3}$.

P3.12 Find the Fermi energy, Fermi velocity, and Fermi wavelength for copper for which $n_3 = 8.45 \times 10^{22}\ \text{cm}^{-3}$ (strongly degenerate).

P3.13 The Fermi energy of aluminum is $(E_F - E_c) = 11.63$ eV. How many levels are filled in an aluminum layer with thickness 1 nm?

P3.14 Following the same procedure as for the carrier concentration, derive the expression for the intrinsic velocity for $d = 1$, 2, and 3. Discuss its degenerate and ND limits.

P3.15 Estimate the intrinsic velocity of electrons in intrinsic GaAs under ND condition at room temperature when the sample is (i) 3D, (ii) 2D, and (iii) 1D. Discuss any approximation that is made.

P3.16 Estimate the intrinsic velocity of electrons in intrinsic GaAs under degenerate condition when the sample is (i) 3D with $n_3 = 10^{20}\ \text{cm}^{-3}$, (ii) 2D with $n_2 = 10^{13}\ \text{cm}^{-2}$, and (iii) 1D with $n_1 = 10^{7}\ \text{cm}^{-1}$. Discuss any approximation that is made.

P3.17 Obtain expressions for the DOS effective mass for holes in 3D, 2D, and 1D, by including both heavy and light holes.

P3.18 Obtain the DOS effective mass for electrons in germanium.

P3.19 The Fermi level in the channel of GaAs high electron mobility transistor (HEMT) with 2D configuration lies at the lowest quantized level ($E_F - E_c = 0$). What is the areal density n_2? How the Fermi level is affected if the density is increased 100-fold.

P3.20 In silicon, the doping level is 5×10^{19} cm^{-3}, find the Fermi level with respect to conduction band when (i) exact calculations are performed; (ii) Maxwell–Boltzman statistics is used.

CAD/CAE PROJECTS

C3.1 In this project, you will use the function file for the FDI to plot $E_F - E_c$ as a function of carrier concentration n for all dimensionalities. On the same graph, plot the degenerate and ND approximations to assess in what range the ND approximation is good and in what range the degenerate approximation is good. Discuss the results. Once again, no derivations are needed in the report. You need to write relevant equations, define symbols, discuss their significance, and discuss the outcomes from the graph. Note that the inclusion of the computer programs in the report is not necessary and can be easily omitted. However, if you want to include these, please include these in the appendix.

C3.2 Plot $\lambda_D = h/m^* v_{i3}$ as a function of temperature for Si, GaN, and GaAs appropriate for intrinsic velocity v_{i3} for arbitrary degeneracy for three carrier concentrations $n_3 = 10^{17}$, 10^{19}, 10^{21} cm^{-3}. Take $T = 0$–400 K. Discuss the results. Once again, no derivations are needed in the report. You need to write relevant equations, define symbols, discuss their significance, and discuss the outcomes from the graph. Note that the inclusion of the computer programs in the report is not necessary and can be easily omitted. However, if you want to include these, please include these in the appendix.

C3.3 Plot T_{md}/T and T_{id}/T as a function of reduced Fermi energy η_d for $d = 3, 2, 1$. On the same graph, plot approximate values in the ND and degenerate limits. η_d may vary from -5 to $+5$ or any other range you find appropriate. In another graph, plot the same temperatures as a function of n_d/N_{cd}. Discuss the results. Once again, no derivations are needed in the report. You need to write relevant equations, define symbols, discuss their significance, and discuss the outcomes from the graph. Note that the inclusion of the computer programs in the report is not necessary and can be easily omitted. However, if you want to include these, please include these in the appendix.

C3.4 *Carrier statistics*

a. Using the carrier statistics in varying dimensionality ($d = 3$ [bulk], $d = 2$ [nanosheet], $d = 1$ [nanowire]), plot a graph of reduced (normalized) Fermi energy η_d as a function of normalized carrier concentration n_d/N_{cd} for all three dimensionalities on the same graph. Also, plot their ND and degenerate approximations. Discuss the range of validity of these approximations.

b. For parameters appropriate to silicon, plot $(E_F - E_c)$ (in eV) as a function of carrier concentration from $n_3 = 10^{18}$ cm^{-3} to $n_3 = 10^{21}$ cm^{-3}. A simultaneous plot of degenerate and the ND limit will capture the essence of these approximations.

c. Repeat (b) for 2D nanosheet for carrier concentration from $n_2 = 10^{10}$ to 10^{13} cm^{-2}. The appropriate effective mass in this case is $m^* = 0.198\ m_o$ both for conductivity as well as for the DOS. The valley degeneracy is $g_v = 2$.

d. Repeat (b) for 1D nanowire for carrier concentration from $n_1 = 10^5$ to 10^8 cm^{-1}. The appropriate effective mass in this case is $m^* = 0.198m_o$ both for conductivity as well as for the DOS. The valley degeneracy is $g_v = 4$.

C3.5 *Intrinsic velocity*

a. Using the intrinsic velocity in varying dimensionality ($d = 3$ [bulk], $d = 2$ [nanosheet], $d = 1$ [nanowire]), plot a graph of normalized intrinsic velocity as a function of normalized carrier concentration n_d/N_{cd} for all three dimensionalities on the same graph. Also, plot their ND and degenerate approximations. Discuss the range of validity of these approximations.

b. For parameters appropriate to silicon, plot v_{i3} (in 10^5 m/s) as a function of carrier concentration from $n_3 = 10^{18}$ to 10^{21} cm^{-3}. A simultaneous plot of degenerate and the ND limit will capture the essence of these approximations. The conductivity effective mass in this case is $m_c^* = 0.26\ m_o$. The valley degeneracy is $g_v = 6$.

c. Repeat (b) for 2D nanosheet for carrier concentration from $n_2 = 10^{10}$ to 10^{13} cm^{-2}. The appropriate effective mass in this case is $m^* = 0.198m_o$ both for conductivity as well as for the DOS. The valley degeneracy is $g_v = 2$.

d. Repeat (b) for 1D nanowire for carrier concentration from $n_1 = 10^5$ to 10^8 cm^{-1}. The appropriate effective mass in this case is $m^* = 0.198m_o$ both for conductivity as well as for the DOS. The valley degeneracy is $g_v = 4$.

C3.6 *Exploration Project*. Find a company on the Internet that sells quantum dot products for laser, biological, or medical applications. There is a significant debate about the possible adverse health effects of quantum dots in biological bodies. Sometimes, it is attributed to Cd or any other material. Write a report on your exploration of possible uses and effect of quantum dots.

Appendix 3A: Distribution Function

As electrons undergo chaotic motion, there is always a transition from one quantum state to the other. Let us assume only two quantum states, E_1 (lower) and E_2 (upper). Following the exponential distribution law the probability of transition from 1 to 2 (being pumped by thermal energy) will be much lower than from 2 to 1 (spontaneous jump):

$$P_{12} = P_{21}\, e^{-(E_2-E_1)/k_BT} \tag{3A.1}$$

However, the Pauli Exclusion Principle inhibits transition to already occupied quantum state. The transition thus must be modified to take into account the degeneracy level given by the distribution function f. For transition $1 \rightarrow 2$, the 1 should be occupied and 2 should be empty giving degenerate probability

$$P_{12}^{D} = P_{12} f_1(1 - f_2) \tag{3A.2}$$

Similarly, the degenerate probability of $2 \rightarrow 1$ transition is given by

$$P_{21}^{D} = P_{21} f_2(1 - f_1) \tag{3A.3}$$

In equilibrium, there is no net exchange, the probability of upward transition and downward transition must be the same, giving

$$P_{12} f_1(1 - f_2) = P_{21} f_2(1 - f_1) \tag{3A.4}$$

Using Equation 3A.1 gives

$$e^{-(E_2-E_1)/k_BT} f_1(1 - f_2) = f_2(1 - f_1) \tag{3A.5}$$

and rearranging to separate 1 and 2 levels gives

$$e^{E_1/k_BT} \frac{f_1}{1 - f_1} = e^{E_2/k_BT} \frac{f_2}{1 - f_2} \tag{3A.6}$$

This shows that

$$e^{E/k_BT}\frac{f}{1-f} = \text{constant} \tag{3A.7}$$

Taking constant C to be $C = e^{E_F/k_BT}$ transforms Equation 3A.6 to the Fermi–Dirac distribution function given by

$$f(E) = \frac{1}{1+e^{(E-E_F)/k_BT}} \tag{3A.8}$$

This derivation is instructive as it clearly indicates that in moving from one quantum state to the other state, the Fermi level E_F does not change. This is also true even if two different materials or two different types (n- or p-type), as in a diode, are present. The Fermi level in equilibrium is the same throughout the device regions.

Another important feature of this exercise, as learned in the next chapter, is that if 1 and 2 form two ends of a free path ℓ of a carrier, the energy gained or emitted in the free path can become an important consideration (see Figure 3A.1) The bands tilt and so does the Fermi level in an applied electric field. The electron follows the ballistic path between two successive collisions, E_F from one end to the other changes by $q\vec{\mathcal{E}}\cdot\vec{\ell}$. Even if $E_F - E_c$ remains the same throughout the device, the steady-state nonequilibrium requires that an electron moving from one end to the other of a free path will change both E_F and E_c by $-q\vec{\mathcal{E}}\cdot\vec{\ell}$ giving $E_{F2} - E_{F1} = -q\vec{\mathcal{E}}\cdot\vec{\ell}$ and $E_{c2} - E_{c1} = -q\vec{\mathcal{E}}\cdot\vec{\ell}$. The distribution function then becomes

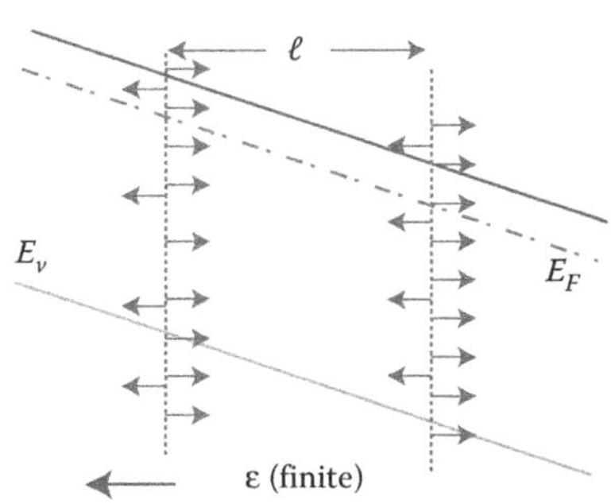

FIGURE 3A.1
The random motion of the electrons transform in an electric field. Only a single mean free fragment is shown.

$$f(E) = \frac{1}{1 + e^{\{E-(E_F - q\vec{\mathcal{E}}\cdot\vec{\ell})\}/k_BT}} \tag{3A.9}$$

This is the origin of the nonequilibrium Arora distribution function (NEADF) discussed in Chapter 4.

Appendix 3B: Electron Concentration for 2D and 1D Nanostructures

3B.1 2D

$$\begin{aligned}
n_2 &= \int_{E_C}^{top\approx\infty} D_2(E) f(E)\, dE \\
&= D_2 \int_{E_C}^{top\approx\infty} \frac{1}{e^{(E-E_F)/k_BT} + 1}\, dE = D_2 \int_{E_C}^{top\approx\infty} \frac{1}{e^{(E-E_c-(E_F-E_c))/k_BT} + 1}\, dE \\
&= D_2 \cdot k_BT \int_0^{\infty} (k_BTx)^0 \cdot \frac{1}{e^{x-\eta} + 1}\, dx \\
&= \frac{m^* k_BT}{\pi\hbar^2} \cdot \Im_0(\eta) = N_{c2} \cdot \Im_0(\eta)
\end{aligned} \tag{3B.1}$$

where

$$x = \frac{E - E_c}{k_BT} \quad \text{and} \quad \eta = \frac{E_F - E_c}{k_BT} \tag{3B.2}$$

The integral $\Im_0(\eta) = \int_0^{\infty} 1/(e^{x-\eta} + 1)dx$ can be evaluated exactly by the following transformation:

$$\begin{aligned}
\Im_0(\eta) &= \int_0^{\infty} \frac{e^{-(x-\eta)}}{1 + e^{-(x-\eta)}}\, dx = \int_{x=0}^{\infty} \frac{-d\left[e^{-(x-\eta)}\right]}{1 + e^{-(x-\eta)}} \\
&= -\ln\left[1 + e^{-(x-\eta)}\right]\Big|_{x=0}^{x=\infty} = \ln\left[1 + e^{\eta}\right]
\end{aligned} \tag{3B.3}$$

3B.2 1D

$$
\begin{aligned}
n_1 &= \int_{E_C}^{top\approx\infty} D_1(E) f(E)\, dE \\
&= \frac{\sqrt{2m^*}}{\pi\hbar} \int_{E_C}^{top\approx\infty} \left(E - E_{cn}\right)^{-1/2} \cdot \frac{1}{e^{(E_k - E_F)/k_B T} + 1} \quad dE \\
&= \frac{\sqrt{2m^*}}{\pi\hbar} \int_{E_C}^{top\approx\infty} \left(k_B T x\right)^{-1/2} \cdot \frac{1}{e^{x-\eta} + 1} \quad k_B T\, dx \\
&= \frac{\sqrt{2m^* k_B T}}{\pi\hbar} \int_{E_C}^{top\approx\infty} \frac{\sqrt{\pi}}{\sqrt{\pi}} \frac{x^{-1/2}}{e^{x-\eta} + 1} \quad dx \\
&= \frac{\sqrt{2m^* k_B T /}}{\hbar} \int_{E_C}^{top\approx\infty} \frac{1}{\sqrt{\pi}} \frac{x^{-1/2}}{e^{x-\eta} + 1} \quad dx \\
&= \frac{\sqrt{2m^* k_B T / \pi}}{\hbar} \int_{E_C}^{top\approx\infty} \frac{1}{\sqrt{\pi}} \frac{x^{-1/2}}{e^{x-\eta} + 1} \quad dx \\
&= N_{c1} \Im_{-1/2}(\eta)
\end{aligned}
\tag{3B.4}
$$

3B.3 Summary of Carrier Statistics

For ND statistics: $n_d = N_{cd} e^{\eta}$ for all dimensionalities. For degenerate statistics, the carrier statistics appear in Table 3B.1.

The Fermi energy for ND statistics is calculated from

$$
\eta = \ln \frac{n_d}{N_{cd}} \Rightarrow (E_F - E_c) = k_B T \ln \frac{n_d}{N_{cd}} \tag{3B.5}
$$

Because $n_d < N_{cd}$, the Fermi energy is below the conduction band edge and hence $(E_F - E_c)$ is negative. Or

$$
(E_c - E_F) = k_B T \ln \frac{N_{cd}}{n_d} \tag{3B.6}
$$

In this case, the Fermi energy is weakly dependent on carrier concentration, but is almost a linear function of temperature.

TABLE 3B.1

Carrier Statistics for 3D, 2D, and 1D

Dimension	Carrier Statistics	EDOS	Degenerate Fermi
	$n_d = N_{cd}\Im_{(d-2)/2}(\eta_d)$	$N_{cd} = 2\left(\frac{m^* k_B T}{2\pi\hbar^2}\right)^{d/2}$	$\Im_j(\eta) = \frac{1}{\Gamma(j+1)}\frac{\eta^{j+1}}{j+1}$
3D	$n_3 = N_{c3}\Im_{1/2}(\eta_3)$	$N_{c3} = 2\left(\frac{m^* k_B T}{2\pi\hbar^2}\right)^{3/2}$	$\Im_{\frac{1}{2}}(\eta) = \frac{4}{3\sqrt{\pi}}\eta^{3/2}$
2D	$n_2 = N_{c2}\Im_0(\eta_2)$	$N_{c2} = \frac{m^* k_B T}{\pi\hbar^2}$	$\Im_0(\eta_2) = \eta$
1D	$n_1 = N_{c1}\Im_{-1/2}(\eta_1)$	$N_{c1} = \left(\frac{2m^* k_B T}{\pi\hbar^2}\right)^{1/2}$	$\Im_{-\frac{1}{2}}(\eta_1) = \frac{2}{\sqrt{\pi}}\eta^{1/2}$

In degenerate statistics, the Fermi energy is a strong function of carrier concentration:

$$\Im_j(\eta) = \frac{1}{\Gamma(j+1)}\frac{\eta^{j+1}}{j+1} = \frac{\eta^{j+1}}{\Gamma(j+2)} \tag{3B.7}$$

$$\frac{n_d}{N_{cd}} = \frac{\eta^{d/2}}{\Gamma(d/2+1)} \Rightarrow \eta = \left[\frac{n_d}{N_{cd}}\Gamma\left(\frac{d}{2}+1\right)\right]^{2/d} \tag{3B.8}$$

For 3D:

$$\eta = \left(\frac{3\sqrt{\pi}}{4}\frac{n_3}{N_{c3}}\right)^{2/3} \text{ or } E_F - E_c = \left[\frac{3\sqrt{\pi}}{4}\frac{n_3}{N_{c3}}\right]^{2/3} k_B T = \frac{\hbar^2\left(3\pi^2 n_3\right)^{2/3}}{2m_n^*} \tag{3B.9}$$

For 2D: In general, $\eta = \ln\left(e^{n_2/N_{c2}} - 1\right)$. For extreme degeneracy, $(n_2/N_{c2}) \gg 1$, $\eta = (n_2/N_{c2})$
or

$$E_F - E_c = k_B T\frac{n_2}{N_{c2}} = k_B T\frac{n_2}{(m^* k_B T)/\pi\hbar^2} = \frac{\hbar^2}{m^*}(\pi n_2) \tag{3B.10}$$

For 1D:

$$\frac{n_1}{N_{c1}} = \frac{2}{\sqrt{\pi}}\eta^{1/2} \Rightarrow \eta = \frac{\pi}{4}\left(\frac{n_1}{N_{c1}}\right)^2$$

$$E_F - E_c = \frac{\pi}{4}k_B T\frac{n_1^2}{(2m^* k_B T)/\pi\hbar^2} = \frac{\pi^2\hbar^2 n_1^2}{8m^*} \tag{3B.11}$$

Appendix 3C: Intrinsic Velocity

3C.1 3D

$$v_{i3} = \frac{1}{n_3}\int_{E_C}^{top\approx\infty} |v| D_3(E) f(E) dE$$
$$= \frac{1}{n_3}\int_{E_C}^{\infty} \sqrt{\frac{2(E-E_c)}{m^*}} \frac{1}{e^{(E-E_F)/k_BT}+1} \frac{1}{2\pi^2}\left(\frac{2m^*}{\hbar^2}\right)^{3/2} (E-E_c)^{1/2} dE \tag{3C.1}$$

By substitution of $E - E_c = k_BT \cdot x$ and $E_F - E_c = k_BT \cdot \eta$, and substituting $v_{th3} = v_{th}\frac{\Gamma(2)}{\Gamma(3/2)} = \frac{2}{\sqrt{\pi}} v_{th}$, the following expression is obtained:

$$v_{i3} = \frac{1}{n_3}\left[v_{th}\frac{\Gamma(2)}{\Gamma(3/2)}\right]\left[\frac{1}{2\pi^2}\left(\frac{2m_n^*}{\hbar^2}\right)^{3/2} (k_BT)^{3/2}\Gamma(3/2)\right]\frac{1}{\Gamma(2)}\int_0^{\infty}\frac{x}{e^{x-\eta_3}+1}dx \tag{3C.2}$$

Simplifying further gives

$$v_{i3} = \frac{1}{n_3} v_{th3}\, N_{c3}\frac{1}{\Gamma(2)}\int_0^{\infty}\frac{x}{e^{x-\eta}+1}\, dx$$
$$= v_{th3}\frac{\Im_1(\eta)}{(n_3/N_{c3})} = v_{th3}\frac{\Im_1(\eta)}{\Im_{1/2}(\eta)} \tag{3C.3}$$

3C.2 2D

The procedure is still the same as above except we use 2D DOS

$$v_{i2} = \frac{1}{n_2}\int_{E_C}^{top\approx\infty} |v| D_2(E) f(E) dE = \frac{1}{n_2}\int_{E_C}^{\infty}\sqrt{\frac{2(E-E_c)}{m^*}} D_2(E) f(E) dE$$
$$= \frac{1}{n_2}\frac{m^*}{\pi\hbar^2}\sqrt{\frac{2}{m^*}}\cdot\int_{E_C}^{\infty}(k_BTx)^{1/2}\cdot\frac{1}{e^{x-\eta}+1}\quad k_BTdx$$
$$= \frac{1}{n_2}\left(\frac{m^*k_BT}{\pi\hbar^2}\right)\sqrt{\frac{2k_BT}{m^*}}\,\Gamma(3/2)\frac{1}{\Gamma(3/2)}\int_{E_C}^{\infty}\frac{x^{1/2}}{e^{x-\eta}+1}\, dx$$
$$= \frac{1}{n_2}N_{c2}\left[v_{th}\frac{\Gamma(3/2)}{\Gamma(1)}\right]\Im_{1/2}(\eta_c)$$
$$= v_{th2}\frac{\Im_{1/2}(\eta)}{n_2/N_{c2}} = v_{th2}\frac{\Im_{1/2}(\eta)}{\Im_0(\eta)} \tag{3C.4}$$

3C.3 1D

Again the same procedure follows with 1D DOS

$$
\begin{aligned}
v_{i1} &= \frac{1}{n_1}\int_{E_C}^{top\approx\infty} |v| D_1(E) f(E)\, dE = \frac{1}{n_1}\int_{E_C}^{\infty}\sqrt{\frac{2(E-E_c)}{m}} D_1(E) f(E)\, dE \\
&= \frac{1}{n_1}\cdot\sqrt{\frac{2k_BT}{m^*}}\cdot\frac{\Gamma(1/2)}{\Gamma(1/2)}\int_{E_C}^{\infty}\frac{\sqrt{2m^*}}{\pi\hbar}\cdot\frac{(k_BT)^{-1/2}}{e^{(E_k-E_F)/k_BT}+1} k_BT\, dx \\
&= \frac{N_{c1}}{n_1}\left(v_{th}\cdot\frac{\Gamma(1)}{\Gamma(1/2)}\right)\left[\frac{1}{\Gamma(1)}\int_0^{\infty}\frac{x^0}{e^{x-\eta}+1} dx\right] = v_{th1}\frac{\Im_0(\eta)}{n_1/N_{c1}} = v_{th1}\frac{\Im_0(\eta)}{\Im_{-1/2}(\eta)}
\end{aligned}
\tag{3C.5}
$$

3C.4 Combined Summary Formula

$$
v_{id} = v_{th}\frac{\Gamma((d+1)/2)}{\Gamma(d/2)}\frac{\Im_{((d-1)/2)}(\eta_d)}{\Im_{((d-2)/2)}(\eta_d)} = v_{thd}\frac{\Im_{((d-1)/2)}(\eta_d)}{\Im_{((d-2)/2)}(\eta_d)} \tag{3C.6}
$$

References

1. B. L. Anderson and R. L. Anderson, *Fundamentals of Semiconductor Devices*. New York, NY: McGraw-Hill, 2005.
2. B. G. Streetman and S. K. Banerjee, *Solid State Electronic Devices*, 6th ed. Upper Saddle River, NJ: Prentice-Hall, 2006.
3. V. K. Arora, Theory of scattering-limited and ballistic mobility and saturation velocity in low-dimensional nanostructures, *Current Nanoscience*, 5, 227–231, May 2009.
4. I. Saad, M. L. P. Tan, I. H. Hii, R. Ismail, and V. K. Arora, Ballistic mobility and saturation velocity in low-dimensional nanostructures, *Microelectronics Journal*, 40, 540–542, Mar. 2009.
5. Y. H. Wu, T. Yu, and Z. X. Shen, Two-dimensional carbon nanostructures: Fundamental properties, synthesis, characterization, and potential applications, *Journal of Applied Physics*, 108, 071301, Oct 1, 2010.
6. A. H. Castro Neto, F. Guinea, N. M. R. Peres, K. S. Novoselov, and A. K. Geim, The electronic properties of graphene, *Reviews of Modern Physics*, 81, 109–162, Jan–Mar. 2009.

7. K. S. Novoselov, S. V. Morozov, T. M. G. Mohinddin, L. A. Ponomarenko, D. C. Elias, R. Yang, I. I. Barbolina et al., Electronic properties of graphene, *Physica Status Solidi B-Basic Solid State Physics*, 244, 4106–4111, Nov. 2007.
8. M. L. P. Tan, V. K. Arora, I. Saad, M. Taghi Ahmadi, and R. Ismail, The drain velocity overshoot in an 80 nm metal–oxide–semiconductor field-effect transistor, *Journal of Applied Physics*, 105, 074503, 2009.
9. I. Saad, M. L. P. Tan, A. C. E. Lee, R. Ismail, and V. K. Arora, Scattering-limited and ballistic transport in a nano-CMOS circuit, *Microelectronics Journal*, 40, 581–583, Mar. 2009.
10. V. K. Arora, M. L. P. Tan, I. Saad, and R. Ismail, Ballistic quantum transport in a nanoscale metal–oxide–semiconductor field effect transistor, *Applied Physics Letters*, 91, 103510, 2007.
11. V. E. Dorgan, M. H. Bae, and E. Pop, Mobility and saturation velocity in graphene on SiO_2, *Applied Physics Letters*, 97, 082112, Aug. 2010.
12. P. H. S. Wong and D. Akinwande, *Carbon Nanotube and Graphene Device Physics*. Cambridge: Cambridge University Press, 2011.
13. I. Gierz, C. Riedl, U. Starke, C. R. Ast, and K. Kern, Atomic hole doping of graphene, *Nano Letters*, 8, 4603–4607, Dec. 2008.
14. V. K. Arora, *Nanoelectronics: Quantum Engineering of Low-Dimensional Nanoensemble*. Wilkes-Barre, PA: Wilkes University, 2013.
15. R. Qindeel, M. A. Riyadi, M. T. Ahmadi, and V. K. Arora, Low-dimensional carrier statistics in nanostructures, *Current Nanoscience*, 7, 235–239, Apr. 2011.
16. M. Abramowitz and I. A. Stegun, *Handbook of Mathematical Functions with Formulas, Graphs, and Mathematical Tables*. Washington, DC: U. S. Government Printing Office, 1972.
17. V. K. Arora, D. C. Y. Chek, M. L. P. Tan, and A. M. Hashim, Transition of equilibrium stochastic to unidirectional velocity vectors in a nanowire subjected to a towering electric field, *Journal of Applied Physics*, 108, 114314–8, 2010.
18. V. K. Arora, M. S. Z. Abidin, M. L. P. Tan, and M. A. Riyadi, Temperature-dependent ballistic transport in a channel with length below the scattering-limited mean free path, *Journal of Applied Physics*, 111, Mar. 1 2012.
19. V. K. Arora, M. S. Z. Abidin, S. Tembhurne, and M. A. Riyadi, Concentration dependence of drift and magnetoresistance ballistic mobility in a scaled-down metal-oxide semiconductor field-effect transistor, *Applied Physics Letters*, 99, 063106, 2011.
20. V. K. Arora, M. L. P. Tan, and C. Gupta, High-field transport in a graphene nanolayer, *Journal of Applied Physics*, 112, 114330, 2012.

4

Nonequilibrium Carrier Statistics and Transport

An ensemble of carriers in equilibrium follows Fermi–Dirac statistics. In equilibrium, randomly oriented velocity vectors or associated mean free paths (mfps) in vector addition give a zero vector sum. The application of an electric field tends to align these tiny electric diploes in the direction of an electric field. In this chapter, nonequilibrium Arora distribution function (NEADF) is discussed [1,2]. NEADF is an anisotropic distribution function that is an outgrowth of the isotropic Fermi–Dirac distribution function. The velocity-field profiles briefly discussed in Chapter 3 are microscopically analyzed as randomly oriented velocity vectors streamline in the direction of an electric field making the velocity unidirectional in an extremely high electric field. This is the source of ultimate saturation of velocity vectors.

4.1 Tilted Band Diagram in an Electric Field

In equilibrium, the bands are flat and velocity vectors are randomly oriented. The bands tilt with drop in potential energy over the length of the sample by $-qV$ as the electric field $\mathcal{E} = V/L$ exists due to the voltage source of potential difference V, as shown in Figure 4.1. The electric field is given by the gradient of the potential $V(x) = U(x)/q$, where $U(x) = qV(x)$ is the potential energy (PE). As the reference level of the potential is arbitrary, it can be taken to be $V(0) = 0$ at the left edge of the resistor of length L in Figure 4.1a. In that case, $U(x) = E_c(x)$ is the potential energy profile. The electric field $\mathcal{E}$ is then given by any of the following forms as each of $E_c(x)$, $E_v(x)$, $E_i(x)$, and $E_{vac}(x)$ differs from the other only by a constant:

$$\mathcal{E} = -\frac{1}{q}\frac{dU(x)}{dx} = -\frac{1}{q}\frac{dE_c}{dx} = -\frac{1}{q}\frac{dE_v}{dx} = -\frac{1}{q}\frac{dE_i}{dx} = -\frac{1}{q}\frac{dE_{vac}}{dx} \tag{4.1}$$

The conduction band edge (as well as the Fermi energy) is the function of the distance along the length of the device

$$E_c(x) = E_{cx=0} - q\mathcal{E}x = E_{cx=0} - q\frac{V}{L}x = E_{cx=0} - qV\frac{x}{L} \tag{4.2}$$

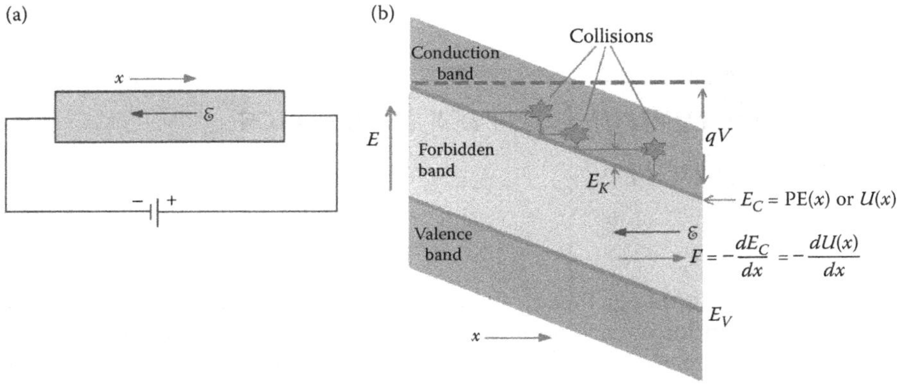

FIGURE 4.1
The tilted band diagram in an electric field when the potential source exists across the sample. (a) Physical sample with voltage applied across its ends with electric field going from right to left. (b) The titled energy band diagram over the length of the sample with dip equal to qV as one moves from left to right.

The potential increases linearly $V(x) = \mathcal{E}x$ as transition takes place from $x = 0$ with $V(0) = 0$ to $x = L$ with $V(L) = V$ with reference voltage level $V = 0$ pegged on the left edge ($x = 0$). However, the electron potential energy (PE) $E_{cxo} - qV(x) = E_{cxo} - q\mathcal{E}x$ decreases as x is increased with reference level E_{cxo} at $x = 0$. PE at $x = L$ is $E_{cxL} = E_{cxo} - qV$, as shown in Figure 4.1b.

4.2 Velocity Response to an Electric Field

The carrier motion in equilibrium is stochastic with net (average) velocity equal to zero, as shown in Figure 4.2. The magnitude of each randomly

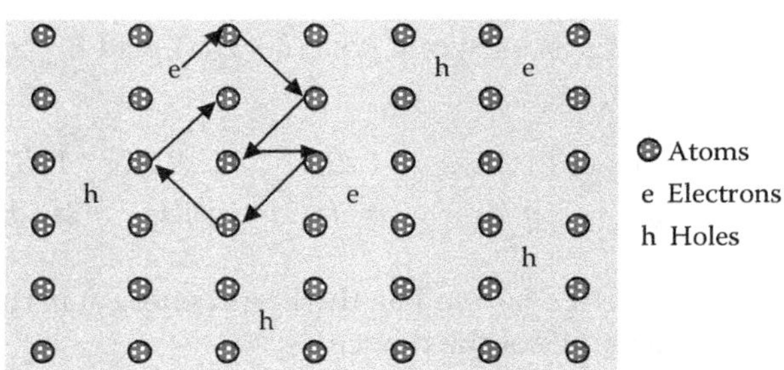

FIGURE 4.2
Example of the path of an electron in a conductor.

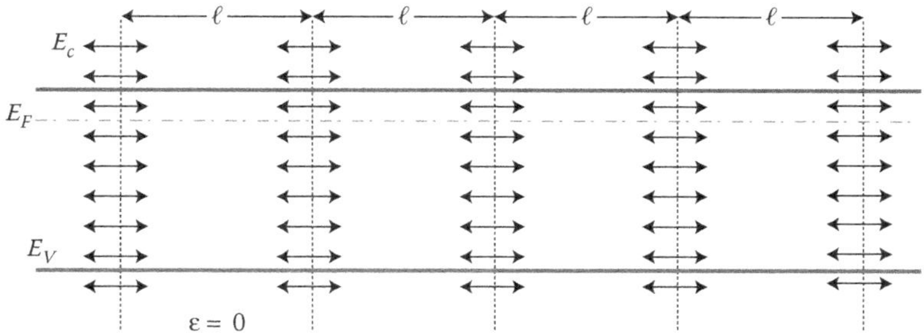

FIGURE 4.3
The series of free paths (with average length ℓ) as electrons moves in equilibrium from one collision site to the other, changing their directions. Only the motion in a single dimension is indicated.

oriented velocity vector is not zero. The average magnitude in an ensemble of carriers or time average of a single carrier is the intrinsic velocity discussed earlier. Figure 4.3 shows a uniformly doped sample in which the bands are flat in the absence of an external stimulation.

The flat bands over the length of the sample are divided into a series of mfps as described by Buttiker [3]. The mfp $\ell_{o\infty}$ in a very large sample ($L \gg \ell_{o\infty}$) is the average distance between two successive collisions. Here, o stands for a low electric field and ∞ for a long sample. The mfp is limited to the length L of the sample when length decreases below the scattering-limited mfp $\ell_{o\infty}$. The associated time interval in traversing $\ell_{o\infty}$ is the mean free time (mft) or collision time $\tau_{co\infty} = \ell_{o\infty}/v_{id}$, where v_{id} is the intrinsic velocity for dimensionality d. The ends of each mfp are Buttiker's thermalizing virtual probes where velocity vectors thermalize and restart their journey for another mfp. In equilibrium, a half of the electrons have velocity vectors directed to the left and the other half directed toward the right during each mfp segment in a homogeneous semiconductor. The net result is a zero drift as both left and right velocity vectors cancel each other out. In a conducting channel with nonzero drift, electrons or holes contribute to the charge conduction. However, the balance of the equally directed velocity vectors is disturbed in the presence of an applied electric field when carriers either accelerate or decelerate, depending on whether the velocity vector is directed in or opposite to the field direction. The motion of the accelerated charge carriers is inhibited by the scattering centers. Scattering centers residing within the sample either emit or absorb energy quanta (phonons) with finite momentum arising from the atomic oscillations. This is an example of inelastic scattering, where, in addition to change in momentum, the energy of the electron also changes. At a low temperature and heavy doping, Coulomb force due to an ionized impurity may play a dominant role in scattering an incoming electron as it changes its direction elastically (without changing energy) on an encounter with an impurity. In either case, carriers change the direction of their path and hence get scattered. In an elastic scattering, there is no change in energy. In inelastic scattering, carriers emit or absorb a quantum of energy.

The typical stochastic motion of an electron in a conductor is a random walk phenomenon as exhibited in Figure 4.2. In a free segment, the carriers are traveling with dimensionality-dependent intrinsic velocity that is randomly oriented in equilibrium. The dependence of this intrinsic velocity v_{id} on temperature, carrier concentration, and dimensionality is discussed in Chapter 3. In the ND approximation, this velocity is independent of carrier concentration and is given by

$$v_{thd} = \frac{\Gamma(d + 1/2)}{\Gamma(d/2)} v_{th} \quad \text{with } v_{th} = \sqrt{\frac{2k_B T}{m^*}} \tag{4.3}$$

The dimensional factor is $\Gamma(d + 1/2)/\Gamma(d/2) = 2/\sqrt{\pi}$ for $d = 3$ (bulk), $\sqrt{\pi}/2$ for $d = 2$ (nanolayer), and $1/\sqrt{\pi}$ for $d = 1$ (nanowire). The magnitude of this intrinsic velocity is approximately 10^5 m/s at room temperature for most semiconductors. Since the thermal velocity does not depend upon the carrier concentration, its use is extensive in the literature even for degenerately doped channels. As strong degeneracy is encountered, an intrinsic velocity is concentration-dependent and temperature-independent. In the strongly degenerate regime, the intrinsic velocity v_{id} is a fraction $(d/[d + 1])$ of a Fermi velocity $v_F = \sqrt{2(E_F - E_c)/m^*}$ corresponding to the Fermi energy $E_F - E_c$. The Fermi energy is above the conduction band for strongly degenerate carrier concentration. Two ends of each mfp acts as virtual thermalizing probes where randomness is re-established and electron starts its journey for another free path. The velocity average is zero, as one expects in equilibrium, because of this stochastic nature of carriers at a virtual thermalizing probe at either end of an mfp.

The application of an electric field tends to stream the motion as velocity vectors change their orientation. This reorientation of velocity vectors changes the left–right balance in the direction of an electric field. This imbalance gives a net nonzero drift velocity in response to an electric field. The stochastic electron network of Figure 4.2 drifts in a direction opposite ($+x$) to that of an applied electric field $\vec{\mathcal{E}}$ directed toward the $-x$-direction, as shown in Figure 4.4. Figure 4.5 shows how the flat bands in equilibrium transform, causing asymmetry in the left–right directions. For holes, this motion will be in the same direction as the electric field. Thus, on the energy band diagram, electrons tend to sink while holes tend to float. The electric field stimulates a carrier by inducing acceleration; a collision or scattering decelerates it by inhibiting its motion.

The rate of change of the velocity vector for an electron of effective mass m_n^* is $d\vec{v}/dt = -q\vec{\mathcal{E}}/m_n^*$ following Newton's second law of motion. The randomizing effect of collisions is akin to friction that tends to dampen the motion. As soon as there is an effort to change the motion, the friction comes into play. The friction analogy makes it convenient to model the collision effect proportional to the change in velocity $\vec{v} - \langle \vec{v}_{id} \rangle$, with $\langle \vec{v}_{id} \rangle = 0$. When the dampening

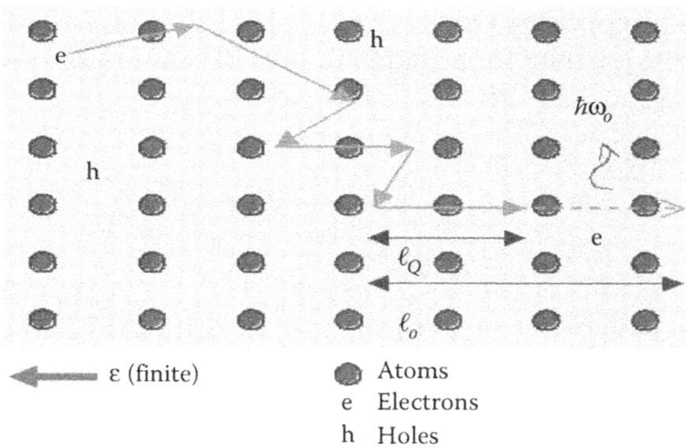

FIGURE 4.4
The drift motion of an electron interrupted by the emission of quantum of energy in the form of a phonon or photon.

effect of collision is also included, the net rate of change of carrier velocity follows the differential equation

$$\frac{d\vec{v}}{dt} = \frac{-q\vec{\mathcal{E}}}{m_n^*} - \frac{\vec{v}}{\tau_{co\infty}} \tag{4.4}$$

with the initial condition

$$v(0) = v_{inj} \tag{4.5}$$

where $q = 1.6 \times 10^{-19}$ C is the electronic charge ($-q$ for an electron and $+q$ for a hole). The injection velocity is important if the carriers are injected, for

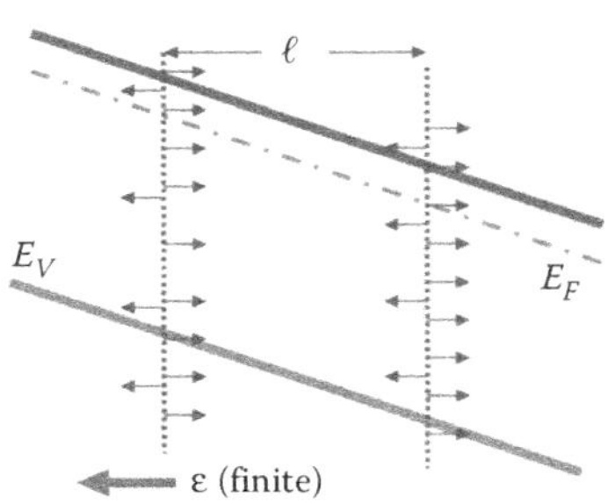

FIGURE 4.5
The random motion of the electrons transforms in an electric field. Only a single mean free fragment is shown.

example, from the contacts. Otherwise, it is the intrinsic velocity v_i of an electron as it starts its journey for a free path until it scatters again. The transient solution of Equation 4.4 is given by [4]

$$\vec{v}(t) = \left(\vec{v}_{inj} + \frac{q\tau_{co\infty}\vec{\mathscr{E}}}{m^*} \right) e^{-t/\tau_{co\infty}} - \frac{q\tau_{co\infty}\vec{\mathscr{E}}}{m^*} \tag{4.6}$$

Equation 4.6 takes a steady-state form if the transit time of an electron spans several free paths (or several collisions), as in a long channel. The transient effects are negligible in a long channel with velocity $\vec{v}(t)$ reaching steady state as $e^{-(t/\tau_{co\infty})} \rightarrow 0$ for $t \gg \tau_{co\infty}$. In fact, a long channel ($L \rightarrow \infty$) can be described with the transient time in at least three collisions ($t \geq 3\tau_{co\infty}$). In this case, the transient factor $e^{-(t/\tau_{co\infty})} \approx 1/20$ is indeed small (5%). In this limit, $\vec{v}(t) \rightarrow \vec{v}_{Dn}$ leads to time-independent drift velocity. The drift response to the electric field is then linear resulting in

$$\vec{v}_{Dn} = -\mu_{o\infty n}\vec{\mathscr{E}} \tag{4.7}$$

with

$$\mu_{o\infty n} = \frac{q\tau_{co\infty n}}{m_n^*} \tag{4.8}$$

where $\mu_{o\infty n}$ is the ohmic mobility of an electron in a long channel for which transit time $t = \tau_t$ is much larger than the collision time ($\tau_t \gg \tau_c$). The long-channel ohmic mobility for intrinsic silicon is $\mu_{o\infty n} = 0.14\ \mathrm{m^2/V \cdot s}$. The symbol "*o*" is used for a low electric field and ∞ for a long sample and n, as used earlier, is for n-type (electrons). The subscript in $\mu_{o\infty n}$ emphasizes the fact that it is low-field mobility in a long channel for electrons. Another subscript "*d*" can be added for its dimensionality. With 3D (bulk) mobility $\mu_{o\infty n} = 0.14\ \mathrm{m^2/V \cdot s}$ for intrinsic silicon and transport (conductivity) effective mass $m_n^* = 0.26\,m_o$, the collision time $\tau_{co\infty}$ is 0.2 ps. An mfp $\ell_{o\infty} = v_{i3n}\tau_{co\infty}$ is in the order of 0.02 μm (20 nm) with 3D intrinsic velocity $v_{i3n} \approx 10^5\ \mathrm{m/s}$. In a long sample ($L \rightarrow \infty$), the transient behavior of injected velocity from the contacts is not important as injection velocity rapidly thermalizes within 3–4 collisions. However, this is not the case if the transit time through a conducting channel is comparable to the collision time. That is the origin of the ballistic mobility and ballistic saturation velocity.

As stated above, in a zero or low electric field, $n_d/2$ electrons travel in each of the $\pm x$ directions with the low electric field applied in $-x$-direction. Therefore, for an electron starting from the left end of the free path, the steady-state velocity acquired on completion of an mfp is

$$v_+ = +v_i + \frac{q\mathcal{E}}{m_n^*}\tau_{co\infty} \tag{4.9}$$

The velocity for an electron starting from the right end of the free path is

$$v_- = -v_i + \frac{q\mathcal{E}}{m_n^*}\tau_{co\infty} \tag{4.10}$$

Here, $v_{inj} = v_i$. The average velocity is the drift velocity $v_D = \langle v \rangle = (v_+ + v_-)/2 = \mu_{o\infty n}\mathcal{E}$ in the +ve direction (opposite to the electric field), as in Equation 4.7. It may be noted that $v_i \approx 10^5$ m/s $\gg \mu_{o\infty n}\mathcal{E}$. The influence of v_i is not noticeable because of the stochastic nature of these velocity vectors, leading to cancelation. However, the drift component does not cancel as it is the same for left- or right-directed velocity vectors giving linear drift in response to an electric field. Implicit is the assumption that in linear domain where ohmic conduction is applicable, the number of electrons n_+ with velocity vector v_+ in the +ve direction are equal to n_- with velocity vector v_- in the negative direction so $n_+ = n_- = n/2$.

The equal distribution of carriers ($n_d/2$ in each of $\pm x$-direction) in an mfp is disturbed as the electric field is increased beyond a critical value $\mathcal{E}_c$ with tendency of velocity vectors to realign opposite to the electric field. The threshold of this imbalance arises when an electron in the $-x$-direction (parallel to $\vec{\mathcal{E}}$) is unable to complete a free path. This is the critical field $\mathcal{E}_c$ at which $v_- = 0$:

$$v_- = -v_i + \frac{q\mathcal{E}_{cn}}{m_n^*}\tau_{co\infty} = 0 \tag{4.11}$$

or

$$\mathcal{E}_{cn} = \frac{v_i}{(q\tau_{co\infty}/m_n^*)} = \frac{v_i}{\mu_{o\infty n}} \tag{4.12}$$

Equation 4.11 indicates the beginning of right–left imbalance in the number of electrons in which the intrinsic velocity plays an important role in determining the electric field beyond which the linear response to an applied electric field is not valid. In a low electric field $\mathcal{E} < \mathcal{E}_c$, the intrinsic velocities from both ends of an mfp cancel. Beyond the critical field $\mathcal{E} > \mathcal{E}_c$, an electron traveling in the $-x$-direction will not be in a position to complete the mfp and reverses its direction at the point where its velocity is zero. The transit time is $\tau_t < \tau_c$ for n_- electrons with the initial velocity vector directed in the $-x$-direction as the electric field rises beyond $\mathcal{E}_c$. In the limits of an extremely large electric field, the transit time will be almost zero for an electron directed in the $-x$-direction:

$$\tau_t = \frac{m_n^* v_i}{q\mathcal{E}} \tag{4.13}$$

Within the limit $\mathscr{E} \to \infty$, the $+x$-directed vector is accelerated to

$$v_+ = +v_i + \frac{q\mathscr{E}}{m_n^*}\tau_t = 2v_i \tag{4.14}$$

on completing an mfp. The average velocity is $v_D = \langle v \rangle = (2v_i + 0)/2 = v_i$ in this limit. This is a familiar situation of a collision where an identical incoming electron with velocity $2v_i$ collides with one at rest of the same mass and transfers its velocity to the resting electron, incoming electron coming to rest. This makes velocity vectors unidirectional directed toward the $+x$-direction. In the extreme limit of a towering electric field ($\mathscr{E} \to \infty$), the velocity vectors instead of being stochastic tend to be unidirectional as shown in Figure 4.6. In the quantum mechanical description, the quantum waves of an electron traveling in the negative direction will reflect back from an infinite potential hill with the reflection coefficient almost unity. In a real semiconductor, a towering electric field is impossible to apply as an electric breakdown will occur long before this condition is reached. The highest measured velocity is normally depicted to be the saturation velocity, as the approach toward saturation is slow. The experimentally reported saturation velocity is always lower than the intrinsic velocity, which is the ultimate velocity an electron can have in a semiconductor.

Another way of depicting this critical field is to utilize $\tau_{co\infty} = \ell_d/v_i$ that gives

$$\mathscr{E}_c = \frac{m_n^* v_i}{q\tau_{co\infty}} = \frac{m_n^* v_i^2}{q\ell_d} \tag{4.15}$$

Here, $\ell_d = \ell_{o\infty}/d$ with $d = 3$ (bulk), 2 (nanolayer), and 1 (nanowire) for dimensionality. Because of a change in orientation of the velocity vectors and an associated change in Fermi energy, an electron temperature related

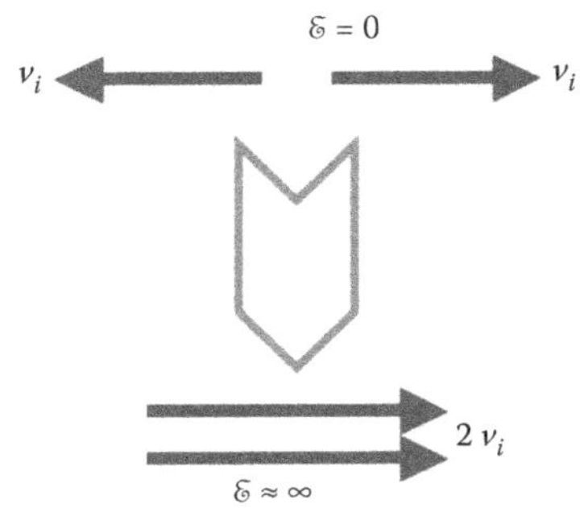

FIGURE 4.6
Transition of randomly oriented vectors to unidirectional velocity vectors in a towering electric field ($\mathscr{E} \to \infty$) with $v_- = 0$ and $v_+ = 2v_i$ giving average, as in equilibrium, $v_D = (0 + 2v_i)/2 = v_i$.

to the intrinsic velocity changes its description. An expression that satisfies both the low-field, equilibrium, and extreme nonequilibrium for all dimensionalities and degeneracy levels is given by

$$\mathcal{E}_{cd} = \mathcal{E}_{co} d \frac{T_{ud}}{T} = d \frac{V_t}{\ell_{o\infty d}} \left(\frac{v_{id}}{v_{thd}} \right)^2 \frac{v_{ud}}{v_{id}} \tag{4.16}$$

This will be discussed in detail as we get to the distribution function. Here, T_{ud} is the unidirectional temperature as discussed in Chapter 3. In the ND regime ($T_{ud} \approx T$), the critical field is approximately equal to $dV_t/\ell_{o\infty}$, where $V_t = k_B T/q$ is the thermal voltage whose value at room temperature is $V_t = 0.0259\,\text{V}$. $\ell_{o\infty}/d = 0.02\,\mu\text{m} = 20\,\text{nm}$ for intrinsic 3D (bulk) silicon with ohmic mobility of $\mu_{o\infty} = 0.14\,\text{m}^2/\text{V}\cdot\text{s}$ giving $\ell_{o\infty} = 60\,\text{nm}$. The critical electric field is $\mathcal{E}_c = 3V_t/\ell_{o\infty}$ that is $\mathcal{E}_c = 0.0259\,\text{V}/0.02\,\mu\text{m} = 1.3\,\text{V}/\mu\text{m}$. The critical electric field lowers with the increase in mfp. For example, for $\ell_{o\infty} = 0.1\,\mu\text{m} = 100\,\text{nm}$, the critical value of $\mathcal{E}_c = 0.0259\,\text{V}/\mu\text{m}$. In degenerately doped samples, the critical electric field $\mathcal{E}_c > \mathcal{E}_{co} = V_t/\ell_{o\infty}$ and hence the onset of nonlinear behavior will arise at much higher values of the electric field. This is the reason that it is difficult to observe high-field effects in metals as intrinsic velocity is extremely high. The above definition of critical electric field is compatible with the tanh function that will be discussed following the description of NEADF.

The electron mfp in a classical model is of the order of the interatomic distance which is a fraction of a nanometer (e.g., in silicon lattice, the lattice constant is 0.543 nm). This is far smaller than the observed mfp of 20 nm as pointed out earlier, indicating that electron can pass through several atoms without a collision. That is possible only when a quantum wave is traversing through the sample. In a perfect crystal, the standing quantum waves give perfect conduction. The scattering or collisions disturb that perfect order. Quantum waves smear out the otherwise point collisions. In a low electric field $\mathcal{E}$, collisions are point collisions as the energy $\Delta E = |q|\,\mathcal{E}\lambda_D$ (field broadening) absorbed by a carrier of charge q ($q = \pm e$ for holes [electrons]) in the de Broglie wavelength λ_D is negligible compared to the collision broadening h/τ_c. This is due to Heisenberg uncertainty principle $\Delta E \Delta t \geq h$. However, as the electric field increases, the field broadening tends to mask the collision broadening and makes its presence negligible in a very high field. The high collision broadening makes carrier transport virtually independent of scattering and hence ballistic. The nonlinear behavior onsets when field broadening is comparable to the collision broadening ($|q|\,\mathcal{E}_c\lambda_D = h/\tau_c$) as electric field is increased. This condition is equivalent to impulse–momentum theorem where the impulse $q\mathcal{E}_c\tau_c$ is comparable to momentum change $m_n^* v_i$ giving $q\tau_c\mathcal{E}_c = m_n^* v_i \Rightarrow \mathcal{E}_c = v_i/\mu_{o\infty}$. In terms of energy, this condition is equivalent to the condition that the energy gained by an electron in a mfp $q\mathcal{E}_c\ell_{o\infty}$ is comparable to the intrinsic energy $(1/2)m_n^* v_i^2$.

4.3 Ballistic Mobility

The concept of ballistic mobility arises as the length of the channel reduces below the scattering-limited mfp. Figure 4.7 is a simplified model of the conducting channel with contacts at its ends.

The electrons are injected from a degenerately doped metal into the highly doped contact region, enter the channel, and exit from the other contact and are collected by the metal. The ballistic transport was initially demonstrated in the experiments of Heiblum et al. [5]. In their experiment, the electrons were injected from the metal with the Fermi velocity traveling either ballistically or thermalizing after undergoing several collisions. They found only 50% of electrons traveled ballistically even though the base length was less than mfp. As $L < \ell_d$, contacts injecting carriers into the channel with high injection velocity v_{inj} play a predominant role. The transit time τ_t is now constrained as it is the fraction of a collision time τ_c ($\tau_t < \tau_c$). Changing the initial condition of Equation 4.5 to $v(0) = \pm v_{inj}$ from left and right contacts with $t = \tau_t$ modifies the transient solution of Equation 4.6 to

$$\vec{v}(\tau_t) = \left(\pm v_{inj} + \frac{q\tau_{co\infty}\vec{\mathcal{E}}}{m^*}\right)e^{-(\tau_t/\tau_{co\infty})} - \frac{q\tau_{co\infty}\vec{\mathcal{E}}}{m^*} \tag{4.17}$$

with

$$\tau_t = \frac{L}{v_{inj}}, \quad \tau_{co\infty} = \frac{\ell_d}{v_{id}}$$

L is the effective length of the channel between contacts. v_{inj} is normally equal to the Fermi velocity of the degenerate contacts. This is the velocity with which the carriers are injected into the contacts through the tunneling process.

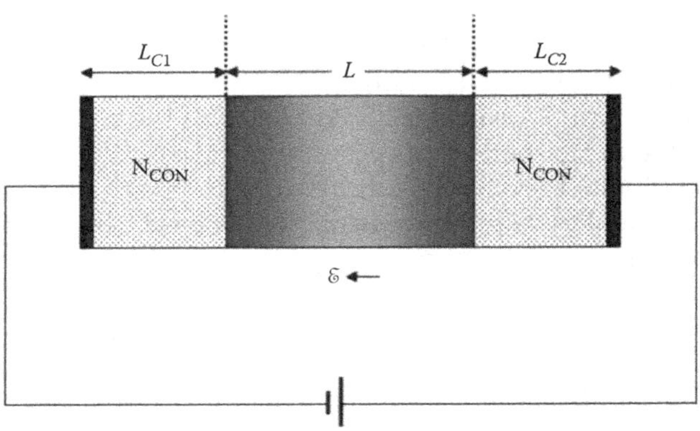

FIGURE 4.7
The simplified view of a ballistic channel with ohmic contacts indicated.

The electron drift velocity is the average of injection from the left and right contacts giving

$$\vec{v}_D = \frac{\vec{v}_+(\tau_t) + \vec{v}_-(\tau_t)}{2} = -\mu_{o\infty}\vec{\mathcal{E}}(1 - e^{-(\tau_t/\tau_{co\infty})}) \tag{4.18}$$

The length-limited mobility $\mu_{oL} = v_D/\mathcal{E}$ in terms of $\mu_{o\infty}(\tau_t \rightarrow \infty \gg \tau_c)$ is given by

$$\mu_{oL} = \mu_{o\infty}(1 - e^{-(\tau_t/\tau_{co\infty})}) = \mu_{o\infty}(1 - e^{-(L/\ell_B)}), \quad \ell_B = \ell_d \frac{v_{inj}}{v_{id}} \tag{4.19}$$

The transient factor, considerable below unity, makes effective collision time and effective mean free path smaller, thereby degrading the mobility. The length-limited collision time τ_{oL} or length-limited mfp ℓ_{oL} is directly obtained from Equation 4.19

$$\tau_{oL} = \tau_{o\infty}(1 - e^{-(\tau_t/\tau_{co\infty})}) = \tau_{o\infty}(1 - e^{-(L/\ell_B)}) \tag{4.20}$$

$$\ell_{oL} = \ell_d(1 - e^{-(\tau_t/\tau_{co\infty})}) = \ell_d(1 - e^{-(L/\ell_B)}) \tag{4.21}$$

with

$$\mu_{o\infty} = \frac{q\ell_{o\infty}}{dm * v_{id}}, \quad \ell_d = \frac{\ell_{ooo}}{d} \tag{4.22}$$

The ballistic mfp $\ell_B > \ell_d$ is much larger than the corresponding long-channel mfp because $v_{inj} > v_i$. The injection velocity from the metal contacts is much larger than the channel intrinsic velocity.

In the long-channel limit ($L \gg \ell_B$), the Ohmic mobility retains its long-length character $\mu_{o\infty}$. The addition of subscript ∞ reinforces the fact that it is long-length mobility unaffected by the sample size. On the other hand, when device length is smaller than the mfp ($L \ll \ell_B$), the Ohmic mobility reaches the ballistic limit

$$\mu_{oB} = \mu_{o\infty}\frac{L}{\ell_B} = \frac{qL}{m_n^* v_{inj}} \tag{4.23}$$

The mobility limited by the length of the device is shown in Figure 4.8. The dotted line is indicative of the fact that the long-channel mfp is unable to explain the decline of mobility in a channel. The solid line with ballistic mfp explains the experimental data very well.

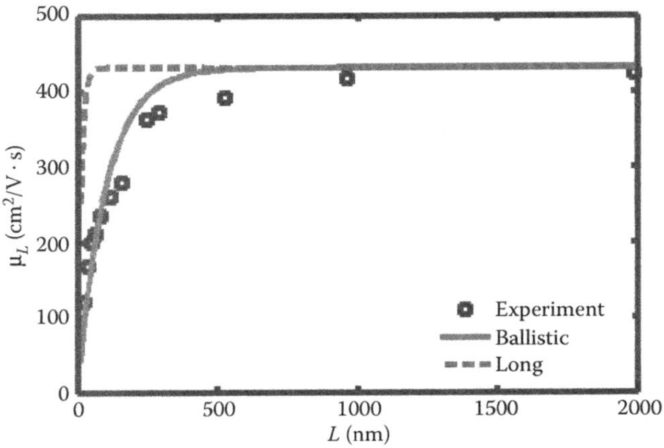

FIGURE 4.8
The length-limited mobility as a function of length. The dotted line shows the mobility degradation when long-channel mfp is used. Solid line uses the ballistic mfp and is in agreement with the experimental data.

4.4 Quantum Emission

The other modification to the classical model arises from the possibility of a quantum emission in a high electric field as shown in Figure 4.4. The quanta of energy in the form of a phonon (acoustic or optical) always reside at a finite temperature. A strong electric field may send an electron into one of the higher quantum states, the possibility of photon emission becomes strong, and the energy of the photon being equal to the difference of the two quantized (digitized) levels. Phonons and photons follow the same Bose–Einstein statistics. The probability of the quantum emission $\hbar\omega_o$ is given by $N_o + 1$ with N_o given by Bose–Einstein distribution

$$N_o = \frac{1}{\exp(\hbar\omega_o/k_BT) - 1} \tag{4.24}$$

The average energy of an emitted quantum is

$$E_Q = (N_o + 1)\hbar\omega_o \tag{4.25}$$

The energy gained $E_Q = q\mathcal{E}\ell_Q$ in the inelastic scattering length ℓ_Q triggers quantum emission of average energy E_Q. Hence,

$$\ell_Q = \frac{E_Q}{q\mathcal{E}} \tag{4.26}$$

Following the transient response model given above, the effective mfp in the high-field limit is given by

$$\ell = \ell_{oL}(1 - e^{-(\ell_Q/\ell_{oL})}) \tag{4.27}$$

Equation 4.26 indicates that the inelastic-scattering quantum length ℓ_Q is infinite in a low electric field. Hence, quantum emission is impossible in a low electric field as mobility will be inhibited by traditional mfp $\ell_{o\infty}$ for a long sample or by ℓ_{oL} for a ballistic channel. Thus, the possibility of quantum emission arises only in a high electric field as exhibited in Figure 4.4. All scattering events change the carrier momentum. However, only some of them, the inelastic scattering events, change the energy of the carriers. Usually, several elastic scatterings are followed by one inelastic scattering by generating a phonon or possibly a photon. However, in a high electric field, inelastic scattering becomes frequent, comparable to the momentum scattering rate. $E_Q = \Delta E$ is the energy spacing between two quantized levels or it can pick up or absorb a phonon. The probability of emitting a photon is considerably enhanced. In the limit of low quantum energy ($\hbar\omega_o \ll k_BT$), the quantum energy is $E_Q \approx k_BT$. In the other extreme of a quantum of large energy $\hbar\omega_o > k_BT$, the quantum $E_Q > \hbar\omega_o$. In fact, when $\hbar\omega_o \gg k_BT$, the effect of quantum emission is negligible as probability of its emission is considerably reduced. With the mfp of Equation 4.27, the mobility degrades with the application of a high electric field. The reduced mobility $\mu(\mathcal{E})$ is given by

$$\mu(\mathcal{E}) = \mu_{oL}(1 - e^{-(E_Q/q\mathcal{E}\ell_{oL})}) \tag{4.28}$$

In this model, the ohmic mobility is $\mu(\mathcal{E}) = \mu_{oL}$ when $E_Q > q\mathcal{E}\ell_{oL}$ as expected in a low electric field. However, as electric field increases toward infinity, the mobility degrades as

$$\mu(\mathcal{E}) \approx \mu_{oL}\frac{E_Q}{q\mathcal{E}\ell_{oL}} = \frac{q\ell_{oL}}{m^*v_{id}}\frac{E_Q}{q\mathcal{E}\ell_{oL}} = \frac{E_Q}{m^*v_{id}\mathcal{E}} \tag{4.29}$$

The saturation velocity is then given by

$$v_{sat} = \mu(\mathcal{E})\mathcal{E} = \frac{E_Q}{m^*v_{id}} \tag{4.30}$$

This simple model does predict the features of a high-field transport, including the saturation velocity limited by the onset of quantum emission. However, it lacks the fact that velocity distribution changes in a high electric field, necessitating the development of a high-field distribution function that evolves from the Fermi–Dirac distribution function.

4.5 High-Field Distribution Function

With the scaling down of a channel to submicron scale, a few nm in the present state of technology, an electric field in a modern device is extremely high. With a logic level of 5 V, a cm-length device of yesteryears had an electric field of 5 V/cm, at which linear velocity–field relation is well obeyed. In a submicron (say 0.5 μm) conducting channel encountered in most transistors, the electric field is 5 V/0.5 μm = 100 kV/cm or 10 V/μm. This is much higher than its critical value ($\mathcal{E} > \mathcal{E}_c$). Under such high electric fields, the linear drift response to the electric field ($v_{dn(p)} = \mp\mu_{o\infty n(p)}\mathcal{E}$) is not obeyed and carrier velocity is saturated with its value comparable to the carrier's intrinsic velocity. Here, $\mu_{o\infty n(p)}$ is used for mobility in a small field. It is appropriate for a long channel for n- or p-type. Normally, subscripts $o\infty$ is left out. These features are retained to make distinct the features of ballistic transport in long/short channels and high electric fields. The high mobility materials are often projected to yield higher saturation velocity. Recent experiments on short-channel transistors tend to indicate that the carrier velocity, not the mobility, controls the transport in these devices. Independently determined decreases in mobility and increases in electron velocity under similar conditions are observed, showing poor correlation between the saturation velocity and ohmic mobility [6].

The understanding of the distribution function is essential in any carrier transport study. The equilibrium distribution function is well known to be the Fermi–Dirac distribution function as discussed in Chapter 3. In equilibrium, the bands are flat unless there is a built-in electric field. The electrons are moving at random with average velocity in each of the three Cartesian directions equal to zero. In an external field, the Fermi energy (electrochemical potential) and bands tilt parallel to each other. The carrier distribution is determined relative to the Fermi level. In zero bias, the distribution is independent of position and the Fermi level (chemical potential) is constant. In an external field, however, the Fermi level and bands tilt as shown in Figure 4.5. The carrier distribution is now a function of spatial coordinates. When electrons are accelerated in the field, they move from a region of higher Fermi energy $n(E - EF)$ to that of lower Fermi energy. These electrons can dissipate their net additional energy to the lattice by emitting phonons, thereby causing lattice heating. The carriers drop close to the conduction band edge and the motion is repeated. From an intuitive argument, a limiting velocity comparable to the random thermal velocity at lattice temperature is expected. On the other hand, the electrons in the direction of the electric field rise to the higher Fermi energy $n(E - [E_F + q\mathcal{E}\ell])$. The carrier motion in an external field is, therefore, not random. It has a finite component in the field direction. This makes distribution of carriers asymmetric in the direction of the applied electric field. Electrons roll down the energy hill and holes bubble up the energy hill. The electric field thus tries to organize the otherwise completely random motion. Understandably, the Fermi level will be affected by the applied electric field. All these features are contained in the nonequilibrium Arora's distribution function (NEADF) [2] that shows the asymmetric distribution of electrons in and opposite to the direction of an applied electric field. The distribution is modification of the Fermi–Dirac distribution by adding the energy gained in a mfp $q\vec{\mathcal{E}} \cdot \vec{\ell} = q\mathcal{E}\ell\cos(\theta)$, where $0 \leq \theta \leq \pi$ is the polar angle that a mfp (or a randomly oriented velocity vector) makes with the polar direction in which electric field is applied. The distribution function is given by

$$f(E,\mathcal{E},\theta) = \frac{1}{e^{(E-(E_F-q\vec{\mathcal{E}}\cdot\vec{\ell})/k_BT)}+1} = \frac{1}{e^{x-H(\theta)}+1} \tag{4.31}$$

with

$$H(\theta) = \eta - \delta_o \cos\theta \tag{4.32}$$

$$x = \frac{E-E_c}{k_BT}, \quad \eta = \frac{E_F-E_c}{k_BT}, \quad \delta_o = \frac{\mathcal{E}}{\mathcal{E}_{co}}, \quad \mathcal{E}_{co} = \frac{k_BT}{q\ell_{oL}} = \frac{V_t}{\ell_{oL}}, \quad V_t = \frac{k_BT}{q} \tag{4.33}$$

$q\vec{\mathcal{E}} \cdot \vec{\ell} = q\mathcal{E}\ell\cos(\theta)$ is positive from $0 \leq \theta \leq \pi/2$ with extreme value $q\vec{\mathcal{E}} \cdot \vec{\ell} = +q\mathcal{E}\ell$ at $\theta = 0^o$. In this range, electrons are moving in the positive polar direction with the electric field applied in the negative polar direction. $q\vec{\mathcal{E}} \cdot \vec{\ell} = q\mathcal{E}\ell\cos(\theta)$ is negative from $\pi/2 \leq \theta \leq \pi$ with extreme value $q\vec{\mathcal{E}} \cdot \vec{\ell} = -q\mathcal{E}\ell$ at $\theta = \pi$. In general, $\ell(E) = \tau(E)v$ is the energy-dependent mfp. An averaged constant value ℓ cuts down the numerical work and brings out the salient features of the extreme nonequilibrium distribution function of Equation 4.31. NEADF has a very simple interpretation. The electrochemical potential E_F during the free flight of a carrier changes by $q\mathcal{E}\ell$ as electrons tend to sink and holes tend to float on the tilted energy band diagram. This observation may suggest that an applied electric field tends to organize the otherwise complete random motion. An electric dipole $q\ell$ due to the quasi-free motion of the carriers tends to organize in the direction of the electric field for holes and in the opposite direction for electrons. The collisions tend to bring the electrons closer to the conduction band edge. If the electric field is strong, this unidirectional motion gives carrier drift comparable to the intrinsic velocity, which is the average of the magnitude of electron velocity v, as discussed in the last chapter. A quasi-ballistic behavior of carriers thus follows in a strong electric field. The normalized reduced Fermi energy $H(\theta)$ is now directional as compared to isotropic zero-field reduced Fermi energy $\eta_d = (E_{Fd} - E_c)/k_BT$.

The Fermi energy η_d for dimensionality d is now necessarily a function of the electric field $\mathcal{E}$ (or δ). It is re-evaluated (see Appendix 4A at the end of Chapter 4) through renormalization. The integral over the product of

NEADF and the density of states gives the number of filled quantum states and hence, the carrier concentration n_d ($d = 3, 2, 1$) is given by

$$n_3 = \left(\frac{N_{c3}}{2}\right)\int_0^{\pi} d\theta\, \Im_{1/2}(H(\theta))\sin(\theta)$$

$$= \left(\frac{N_{c3}}{2}\right)\int_{-1}^{+1} d[\cos(\theta)]\,\Im_{1/2}(H(\theta) \quad (4.34)$$

$$n_2 = \left(\frac{N_{c2}}{2\pi}\right)\int_0^{2\pi} d\theta\, \Im_o(H(\theta)) = (N_{c2}/2\pi)\int_0^{2\pi} d\theta \ln(1 + e^{H(\theta)}) \quad (4.35)$$

$$n_1 = \left(\frac{N_{c1}}{2}\right)[\Im_{-1/2}(\eta_1 + \delta) + \Im_{-1/2}(\eta_1 - \delta)] \quad (4.36)$$

$\Im_j(H(\theta)) = \exp(H(\theta))$ in the ND regime. Equations 4.34 through 4.36 are tractable in ND approximation. The simplified expressions are given by

$$n_3 = \left(\frac{N_{c3}}{2}\right)\left[2e^{\eta_3}\frac{\sinh(\delta)}{\delta}\right] = N_{c3}\, e^{\eta_3}\frac{\sinh(\delta)}{\delta} \quad (4.37)$$

$$n_2 = N_{c2}\, e^{\eta_2}\, 2\pi I_o(\delta) \quad (4.38)$$

$$n_1 = N_{c1}\, e^{\eta_1}\cosh(\delta) \quad (4.39)$$

$$I_o(\delta) = \frac{1}{\pi}\int_0^{\pi} d\theta\, e^{\delta\cos\theta} \quad (4.40)$$

$I_o(\delta)$ is the modified Bessel function of order 0.

The normalized electrochemical potential $\eta_d(\delta)$ is related to the normalized chemical potential η_{do} by

$$\eta_3 = \eta_{3o} - \ln\left(\frac{\sinh\delta}{\delta}\right) \quad (4.41)$$

$$\eta_2 = \eta_{2o} - \ln[I_o(\delta)] \quad (4.42)$$

$$\eta_1 = \eta_{1o} - \ln(\cosh\delta) \quad (4.43)$$

The Fermi energy decreases with the electric field almost linearly. This may indicate that in the high electric field ($\delta > 1$), the ND approximation of the nonequilibrium distribution function may be justified and simplifies the framework considerably. However, there is a transformation in the nature of Fermi energy as electrons are transferred from those going in the direction of the electric field in the degenerate regime. This makes the Fermi energy appropriate to $2n$ electrons going in the $+x$-direction. In fact, when the velocity vectors become unilateral in the direction of an electric field, the Fermi energy changes for that appropriate for $2n$ electrons as only quantum states that are occupied in one direction (opposite to the electric field for degenerate electrons). In this extreme, $\eta_d \approx \eta_{ud} - (q\mathcal{E}\ell_{oL}/k_{BT})$. η_{ud} is the unidirectional reduced Fermi energy appropriate for $2n$ electrons. In the ND regime, the Fermi energy is weakly dependent on the carrier concentration, so it is not affected by redistribution of electrons. However, in the degenerate regime where carrier concentration is high, it can make a tremendous difference. As stated earlier, the Fermi level η_{ud} is that appropriate to $2n$ electrons as compared to η_{do} in equilibrium. In extreme nonequilibrium, only half the phase space is occupied due to the unidirectional motion of carriers.

4.6 ND Drift Response

It is fairly straightforward now to evaluate the drift velocity as the average of $|v|$ with the distribution function of Equation 4.31 assuming it preserves its ND character especially when $\mathcal{E} > \mathcal{E}_c$. The results are derived in Appendix 4A at the end of the chapter. The result for the drift velocity is

$$v_D = v_{thd}M(\delta) = v_{thd}M\left(\frac{\mathcal{E}}{\mathcal{E}_c}\right), \quad d = 1, 2, 3 \tag{4.44}$$

The thermal velocity v_{thd} is the approximation to the intrinsic velocity v_{id} in the nondegenerate regime as discussed in Chapter 3. The modulating function $M(\delta)$ that defines the probability of conversion of the random motion to the streamlined one as the electric field rises from its zero value is given by

$$M_3(\delta) = \mathcal{L}(\delta) = \coth(\delta) - \frac{1}{\delta} \quad \text{(3D)} \tag{4.45}$$

$$M_2(\delta) = \left(\frac{I_1(\delta)}{I_o(\delta)}\right) \quad \text{(2D)} \tag{4.46}$$

$$M_1(\delta) = \tanh(\delta) \quad \text{(1D)} \tag{4.47}$$

where $\mathcal{L}(\delta)$ is the Langevin function that was originally invented by Langevin to define the alignment of magnetic moments in the direction of a magnetic field. Here, it plays a similar role for realignment of electric dipoles $q\ell$ in the direction of an applied electric field. The conversion of randomly oriented dipoles $q\ell$ in equilibrium to streamlined ones in an applied electric field follows the same general pattern as observed for magnetic dipoles by Langevin, except drift triggers in the presence of an electric field and electric dipole moments $q\mathcal{E}\ell_{oL}$ are aligned. $I_j(\delta)$ (with $j = 0$ or 1) is the modified Bessel function of order j. In a low-field limit ($\delta \to 0$), as is expected, Equation 4.44 gives linear response to the electric field:

$$M_3(\delta) = \mathcal{L}(\delta) \approx \frac{\delta}{3} \quad (\delta \to 0) \tag{4.48}$$

$$M_2(\delta) = \frac{I_1(\delta)}{I_o(\delta)} \approx \frac{\delta}{2} \quad (\delta \to 0) \tag{4.49}$$

$$M_2(\delta) = \tanh(\delta) \approx \delta \quad (\delta \to 0) \tag{4.50}$$

Figure 4.9 shows the relative drift velocity as compared to the intrinsic velocity in each case. Also shown is the empirical rise that is derived from experimental data [7]. The empirical equation that fits a number of experimental data is given by [7]

$$v_D = v_{sat} \frac{\mathcal{E}/d\mathcal{E}_c}{\left[1 + (\mathcal{E}/d\mathcal{E}_c)^\gamma\right]^{1/\gamma}} \tag{4.51}$$

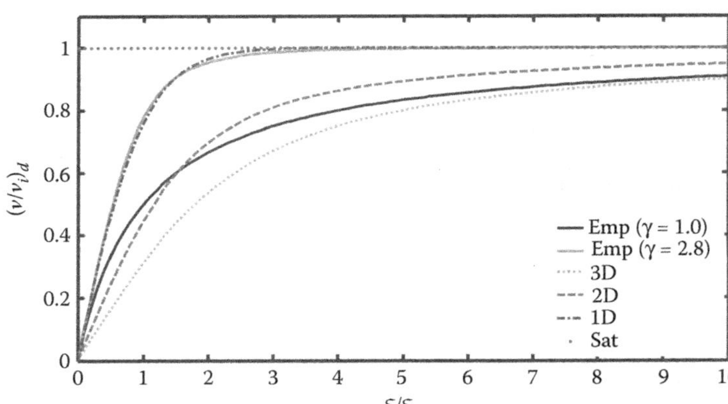

FIGURE 4.9
The normalized drift velocity with ratio v_{Dd}/v_{id} versus the normalized electric field $\mathcal{E}/\mathcal{E}_c$. In the high-field limit all curves reach 1, approach being slower or faster depending on the dimensionality.

where γ is a fitting parameter, v_{sat} is the experimental saturation velocity, and $\mathcal{E}_c = v_{sat}/\mu_{oL}$ is the critical electric field at the onset of transition toward saturated behavior.

At the expense of marginal error that may occur at intermediate fields, $M_d(\delta)$ can be approximated as $\tanh(\delta/d)$:

$$M_d(\delta) \approx \tanh(\delta/d) \tag{4.52}$$

This approximate curve fits very well the nanowire ($d = 1$ case). The general expression of Equation 4.51 correctly predicts the linear drift response to the low electric field as given by Equations 4.48 through 4.50, which can be summed in a single equation for a long sample

$$v = v_{thd}\frac{\mathcal{E}}{d\mathcal{E}_{co}} = \frac{v_{thd}}{d\mathcal{E}_{co}}\mathcal{E} = \mu_{o\infty d}\mathcal{E} \tag{4.53}$$

It may be noted that the quantum emission does not play any active role in low-electric-field domain as $\ell_Q \to \infty$ and hence $\mathcal{E}_c \approx \mathcal{E}_{co} = V_t/\ell_{o\infty}$. An expression for low-field long-channel mobility $\mu_{o\infty d}$ emerges as

$$\mu_{o\infty d} = \frac{1}{d}\frac{v_{thd}}{V_t}\ell_{o\infty} = \frac{2qv_{thd}}{m_n^* v_{th}^2}\ell_{o\infty} = \frac{1}{d}\frac{2q\ell_{o\infty}}{m_n^* v_{thd}}\left(\frac{v_{thd}}{v_{th}}\right)^2 \tag{4.54}$$

where effective v_{thd} differs from v_{th} by a dimensionality-dependent numerical factor as given by

$$\frac{v_{thd}}{v_{th}} = \frac{\Gamma((d+1)/2)}{\Gamma(d/2)} \tag{4.55}$$

The generalization to include degenerate drift response is discussed next.

4.7 Degenerate Drift Response

The degenerate limit of the distribution function is difficult to implement as integral over angular coordinates and energy are required to be solved. In this case, it is impossible to give a general paradigm. However, a modification in general paradigm is made by the replacement of intrinsic velocity v_{id} to v_{ud} as a coefficient of modulating function $\tanh(\delta_d/d)$ of Equation 4.52. The critical electric field is renormalized by replacement of the lattice temperature to the electron temperature T_{ud} as defined in Chapter 3.

As indicated in Appendix 2A, the differential element in 3D is $dk_x\,dk_y\,dk_z = k^2\,dk\sin\theta\,d\theta\,d\varphi$ where k is the wave vector related to the energy $E = \hbar^2k^2/2m^*$. In a nonequilibrium situation, we seek drifts of carriers with respect to the electric field applied in a polar direction, bringing out the importance of polar coordinates for which the differential element is $k^2\,dk\sin\theta\,d\theta\,d\varphi$. In equilibrium statistics, the angular component $\sin\theta\;d\theta\;d\varphi$ is integrated to give a factor of 4π, 2π coming from the integration of φ from 0 to 2π and 2 coming from integration of $\sin\theta\,d\theta = -d(\cos\theta)$ with θ from 0 to π (or $\cos\theta$ from –1 to +1). When an electric field is applied, the term $q\vec{\mathcal{E}}\cdot\vec{\ell} = q\mathcal{E}\ell\cos\theta$ breaks that assumed isotropy. The presence of $\cos\theta$ in $q\vec{\mathcal{E}}\cdot\vec{\ell} = q\mathcal{E}l\cos\theta$ not only makes distribution anisotropic, but it makes it also difficult to integrate to get a simplified expression. To make it tractable, it is desirable to have an average value of $\pm|\cos\theta|$ in a hemisphere containing parallel-to-the-electric-field direction and the other hemisphere containing antiparallel-to-the-electric-field direction, assuming the electric field is in a polar direction. The component of a randomly oriented velocity vector with component $v_z = v\cos\theta$ in the direction of the electric field requires an average of $\cos^2\theta$ in angular differential element $\sin\theta\;d\theta$. However, integration of $\cos^2\theta\sin\theta\,d\theta$ is to be broken into two parts: 0 to $\pi/2$ (with a finite component parallel-to-the-electric-field direction) and $\pi/2$ to π (hemisphere in antiparallel-to-the-electric-field direction). The average $\langle\cos\theta\rangle$ in the direction of the electric field gives

$$\langle\cos\theta\rangle = -\int_{\theta=0}^{\pi/2}\cos^2\theta\,d(\cos\theta) = \frac{1}{3} \tag{4.56}$$

The similar value when integrated from $\pi/2$ to π gives $-1/3$. Hence, $\langle\cos\theta\rangle = \pm 1/3 = \pm 1/d$, where $d = 3$ is the dimensionality in the bulk (3D) case. Similarly, in circular coordinates in 2D, $\langle\cos^2\theta\rangle$ gives $|\cos\theta| = \pm 1/2$ and in 1D it is obvious that for $\theta = 0$, $\cos\theta = 1$ and for $\theta = \pi$, $\cos\theta = -1$, so $|\cos\theta| = \pm 1$. To conclude, the average of $\cos\theta$ depends on dimensionality, $|\cos\theta| = \pm 1/d$, that will be used in defining degeneracy temperature for electrons and holes. This explains why the dimensionality-dependent mfp $\ell_d = \ell_{o\infty}/d$ was used earlier. With this transformation, Equations 4.34 through 4.36 transform to $n_\pm$ with electrons in and opposite to the direction of the electric field given by

$$n_{3\pm} = \frac{N_{C3}}{2}\Im_{1/2}\left(\eta \pm \frac{\delta}{3}\right) \tag{4.57}$$

$$n_{2\pm} = \frac{N_{C2}}{2}\Im_{0}\left(\eta \pm \frac{\delta}{2}\right) \tag{4.58}$$

$$n_{1\pm} = \frac{N_{C1}}{2}\Im_{-1/2}\left(\eta \pm \frac{\delta}{1}\right) \tag{4.59}$$

Collectively, these equations can be condensed to a single equation where $d = 1, 2, 3$ of bulk (3D), nanolayer (2D), and nanowire (1D). The generalized equation is

$$n_{d\pm} = \frac{N_{Cd}}{2}\Im_{(d-2/2)}\left(\eta \pm \frac{\delta}{d}\right) \tag{4.60}$$

Our goal is to measure anisotropy of the electrons with velocity component in and opposite to the direction of the electric filed. This imbalance $\Delta n_{d+} = (n_{d+} - n_{d-})$ in carrier direction is given by

$$\Delta n_{d+} = \left(n_{d+} - n_{d-}\right) = \frac{N_{cd}}{2}\left[\Im_{(d-2/2)}\left(\eta + \frac{\delta}{d}\right) - \Im_{(d-2/2)}\left(\eta - \frac{\delta}{d}\right)\right] \tag{4.61}$$

In the linear domain, this expression can be linearized by expansion of $\Im_j\left(\eta \pm \delta/d\right)$ resulting in

$$\Im_j\left(\eta \pm \frac{\delta}{d}\right) = \Im_j\left(\eta\right) + \frac{\delta}{d}\frac{d\Im_j}{d\eta} \tag{4.62}$$

where $j = (d - 2/2) = (d/2) - 1$. The FDI of order j follows the recursive property

$$\frac{d\Im_j}{d\eta} = \Im_{j-1}\left(\eta\right) \tag{4.63}$$

The fraction of electrons going in the positive direction (opposite to the electric field) is now given by

$$\frac{\Delta n_d}{n_d} = \frac{\delta}{d}\frac{\Im_{j-1}\left(\eta\right)}{\Im_j\left(\eta\right)} = \frac{q\mathcal{E}\ell}{dk_BT}\frac{\Im_{j-1}\left(\eta\right)}{\Im_j\left(\eta\right)} \tag{4.64}$$

This way the drift velocity in the ohmic domain is given by

$$v_D = \frac{\Delta n_d}{n_d}v_{id} \tag{4.65}$$

where v_{id} is the intrinsic velocity as discussed in Chapter 4.

In the ND statistics $\Im_j(\eta) = \exp(\eta)$ regardless of order j. This gives us a simplified expression for ND statistics as

$$v_D = \frac{q\mathcal{E}\ell}{dk_BT}v_{id} \tag{4.66}$$

where $v_{id} = v_{thd}$ in ND domain. The comparison between Equations 4.64 and 4.66 gives a recipe for defining the ohmic degeneracy temperature T_{od} in the ohmic domain given by

$$T_{od} = T\frac{\Im_j(\eta)}{\Im_{j-1}(\eta)} \tag{4.67}$$

In the ND domain, $T_{od} = T$ as expected. In a strong degenerate (Deg) domain, T_{od} of Equation 4.67 simplifies to

$$T_{od}(\text{Deg}) = 2T\frac{\eta}{d} \tag{4.68}$$

The mobility expression is the coefficient of $\mathcal{E}$ in (4.66). With the use of T_{od} in place of T, the mobility expression takes a natural form

$$\mu_{od} = \frac{q\ell}{dk_B T_{od}} v_{id} \tag{4.69}$$

As explained in Chapter 3, the transformation of electrons antiparallel to the electric field increases the degeneracy in that direction and decreases in the other direction. In the strong-field limit, the v_{id} transforms to v_{ud} as appropriate for $2n$ electrons in one direction and none in the other, keeping the average n, but depleting the k-space in the parallel direction to the electric field. When this correction is initiated, the electron degeneracy temperature T_{ed} can be obtained as

$$T_{ed} = T\frac{\Im_j(\eta)}{\Im_{j-1}(\eta)}\frac{v_{ud}}{v_{id}} \tag{4.70}$$

A modification in the general paradigm is made by the replacement of intrinsic velocity v_{id} to v_{ud} as the coefficient of modulating function $\tanh(\delta_d/d)$ of Equation 4.64. The critical electric field is renormalized by replacement of the lattice temperature to the electron temperature T_{ed}. The drift velocity response then becomes

$$v_{Dd} = v_{ud}\tanh\left(\frac{\delta_d}{d}\right) \tag{4.71}$$

with

$$\delta_d = \frac{\mathcal{E}}{\mathcal{E}_{cd}}, \quad \mathcal{E}_{cd} = \mathcal{E}_{co}d\frac{T_{ud}}{T} \tag{4.72}$$

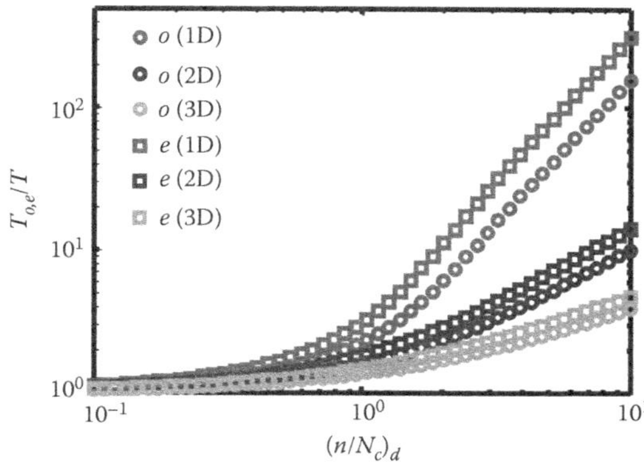

FIGURE 4.10
Ohmic and electron degeneracy temperature for all dimensionalities, the top for $d = 1$ and bottom for $d = 3$. Ohmic temperature is depicted by circles and electron temperature with unidirectional velocity correction for high fields is indicated by squares.

Ohmic and electron degeneracy temperatures are shown in Figure 4.10. With replacement of $\ell_{o\infty}$ with ℓ_{oL}, $\mathscr{E}_{co}$ the critical electric field changes to $\mathscr{E}_{co} = k_BT/q\ell_{oL}$; all other features remaining the same. v_{ud} replaces intrinsic velocity v_{id} in the strong degenerate limit when only half the total quantum states are available for filling, as shown in Figures 4.11 and 4.12.

As shown in Figures 4.11 and 4.12, in the effective mass approximation, the energy E is a parabolic function of $k_{x,y,z}$ as given by

$$E_{k3} = \frac{\hbar^2(k_x^2 + k_y^2 + k_z^2)}{2m*}, \quad -\infty \le k_{x,y,z} \le +\infty \tag{4.73}$$

In the strong degenerate limit, all quantum states are filled up to the Fermi level E_{Fo}, as shown in Figure 4.11. Figure 4.12 shows the other extreme when

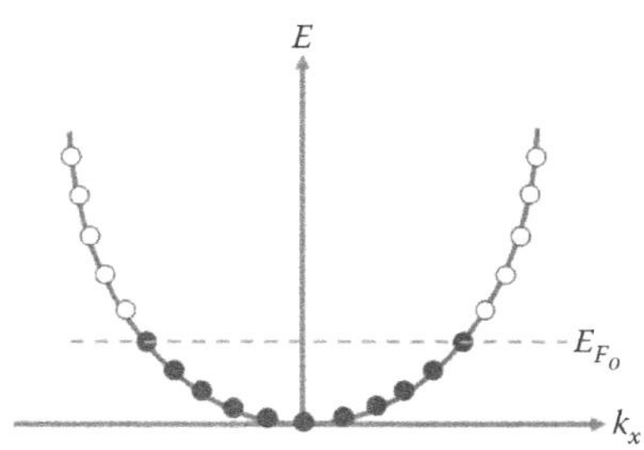

FIGURE 4.11
Stochastic velocity vectors in equilibrium with equal number of electrons in each of the $\pm k_x$ directions filling the states up to the Fermi level E_{F_o} in equilibrium.

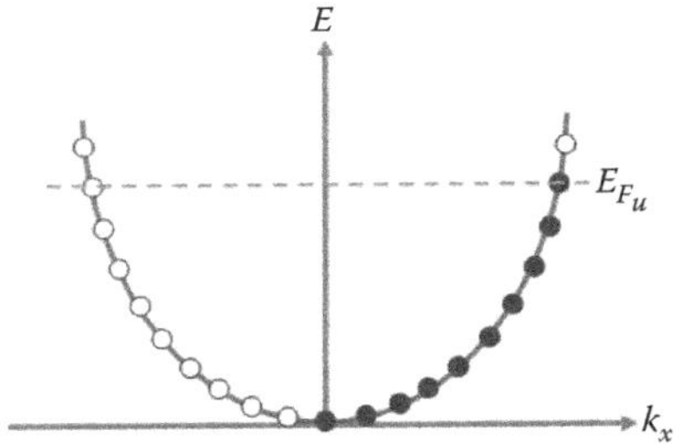

FIGURE 4.12
The extreme nonequilibrium when all velocity vectors are streamlined in the +*x*-direction, with the Fermi level E_{Fu} in the ultimate position.

the velocity vectors for the electrons have been directed (or realigned) in the antiparallel direction to an electric field. The Pauli Exclusion Principle requires that if a state is filled in the degenerate limit, an electron must go to the next higher state. That is why those electrons that transform from $-k_x$ to $+k_x$ have no quantum state available at the level of $+k_x$, so these electrons must occupy the higher quantum states and hence, the Fermi level moves to a higher level in the strong degenerate limit. So, only half the phase space is populated in extreme nonequilibrium as velocity vectors become unidirectional. Equivalently, it is the Fermi level appropriate to $2n$ electrons populating the momentum k-space. v_{ud} and v_{id} have same value ($v_{ud} \approx v_{id}$) in the ND regime, but differ substantially only in the strongly degenerate regime, where v_{id} is appropriate for concentration n and v_{ud} is appropriate for concentration $2n$. In a strong degenerate limit $v_{ud}/v_{id} = 2^{1/d}$. v_{ud}/v_{id} is $2^{1/3}$ in the 3D (bulk) semiconductor, $2^{1/2}$ in 2D nanostructure, and 2 in 1D nanostructure. Quantum filling is more significant as dimensionality is reduced. Both v_{ud} and v_{id} are functions of carrier concentration, temperature, and dimensionality in the absence of any quantum emission.

The saturation velocity can be lower than the intrinsic velocity if the quantum emission is present. In that case, an mfp is limited to the inelastic quantum length ℓ_Q during which an electron gains enough energy to emit a quantum of energy E_Q, that is, $q\mathcal{E}\ell_Q = E_Q$. The nature of the emitted quantum depends on the experimental setup. It may be the energy of an optical phonon or photon as many envision. This vision can be extended to embrace any quantum emission that may also include transitions among quantized energy levels in active and neighboring layers of a nanostructure. In this situation, the velocity saturates to a value

$$v_{satd} \approx v_{ud} \tanh\left(\frac{E_Q}{dk_BT}\right) \tag{4.74}$$

where v_{satd} is limited by the quantum emission and is smaller than the ultimate unidirectional velocity v_{ud}. When the emitted quantum has energy larger than the thermal energy $(E_Q \gg k_BT)$, the saturation velocity is the same as the ultimate velocity. For low-energy quanta, $E_Q = k_BT$ saturation velocity will become $v_{satd} \approx v_{ud}\tanh(1/d)$. For high-energy quanta, $\tanh(E_Q/k_BT) \approx 1$, and the saturation velocity will be the unidirectional velocity v_{ud}.

The distinction between v_{ud} and v_{id} is necessary only when high level of degeneracy is encountered. That is the case when the carrier concentration n_d is high compared to the effective density of quantum states N_{cd} ($n_d \geq N_{cd}$). This is a serious issue in semiconductors with low effective mass, high carrier concentration, and low ambient temperature.

4.8 Direct and Differential Mobility

As Figure 4.9 shows, the slope of the curve changes as we move from a low to high electric field. The mobility as defined by the linear response relation $v_D = \mu\mathcal{E}$ does not hold good. The direct mobility as a function of electric field is defined as $\mu(\mathcal{E}) = v_D(\mathcal{E})/\mathcal{E}$ whose expression is obtained as follows:

$$\mu(\mathcal{E}) = \mu_o \frac{\tanh(\mathcal{E}/d\mathcal{E}_{cd})}{\mathcal{E}/d\mathcal{E}_{cd}} \tag{4.75}$$

The differential or incremental mobility is defined as

$$\mu_i(\mathcal{E}) = \frac{dv_D(\mathcal{E})}{d\mathcal{E}} = \mu_o \text{sech}^2\left(\frac{\mathcal{E}}{d\mathcal{E}_c}\right) \tag{4.76}$$

The relative mobility $\mu(\mathcal{E})/\mu_o$ and $\mu_i(\mathcal{E})/\mu_o$ as a function of normalized electric field $\mathcal{E}/\mathcal{E}_{co}$ are shown in Figure 4.13 [8].

Signal (incremental or ac) mobility degrades faster than the incremental mobility. Also shown is the experimental data and the dashed curves obtained from the empirical relation

$$\frac{\mu}{\mu_o} = \frac{1}{\left[1 + (\mathcal{E}/\mathcal{E}_{cd})^{2.0}\right]^{(1/2.0)}}, \quad \frac{\mu_i}{\mu_o} = \frac{1}{\left[1 + (\mathcal{E}/\mathcal{E}_{cd})^{2.0}\right]^{1.5}} \tag{4.77}$$

4.9 Bandgap Narrowing and Carrier Multiplication

The hot electron concept is used in the literature to demonstrate when the carrier becomes hot. This concept has been disputed in Reference 9. In this

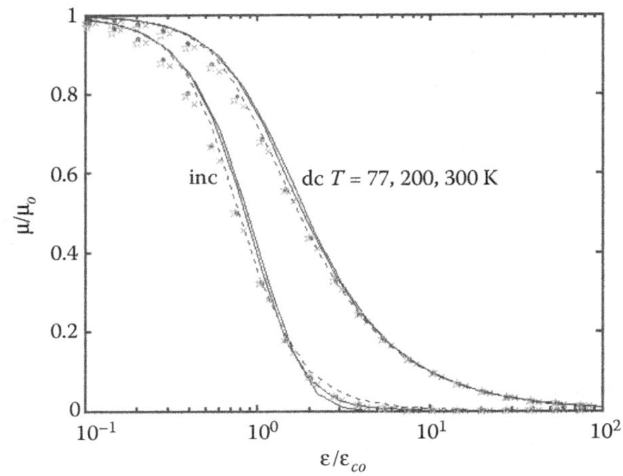

FIGURE 4.13
The normalized mobility as a function of normalized electric field along with the experimental data with markers. Incremental mobility ratio μ_{inc}/μ_o degrades faster than direct mobility ratio μ_{dc}/μ_o.

reference, contexts permitting mythical hot-electron temperatures are discussed. Temperature, as we know from thermodynamics, is a measure of randomness or entropy. Higher temperature leads to higher randomness and higher entropy. However, we have seen above that the presence of high-electric field streams the electrons unidirectionally opposite to an extremely large electric field. NEADF does not support hot-electron theories. However, it does support the circumstances in which apparent effects of hot electrons may become visible. Once such concept is a decrease in the apparent bandgap in a high electric field as a two-band mechanism is considered. It is clear from the tilted band diagram discussed that electrons tend to sink in the antidirection and holes tend to float in the direction of an applied electric field. The sinking energy $q\mathcal{E}\ell_e$ in an mfp of an electron and floating energy is $q\mathcal{E}\ell_h$ in an mfp of a hole. The electrons and holes then find themselves closer on a bandgap, with apparent effective bandgap E_{geff} becoming narrow $E_{geff} = E_g - q\mathcal{E}(\ell_e + \ell_h)$ resulting in a multiplication of intrinsic carriers.

To appreciate this bandgap narrowing for a bandgap semiconductor, we take the example of a bandgap semiconductor in 3D. Of course, similar formalism is valid for 2D and 1D configurations where bandgap narrowing due to an electric field and bandgap enhancement because of quantized zero-point energy will compete. Equation 4.37 for electrons and holes can be written as

$$n_3 = N_{c3}\, e^{\eta_{c3}} \frac{\sinh(\delta_e)}{\delta_e}, \quad p_3 = N_{v3}\, e^{\eta_{v3}} \frac{\sinh(\delta_e)}{\delta_e} \tag{4.78}$$

with

$$\eta_{c3} = \frac{E_F - E_c}{k_B T}, \quad \eta_{v3} = \frac{E_v - E_F}{k_B T} \tag{4.79}$$

$$\delta_e = \frac{q\mathcal{E}\ell_e}{k_B T}, \quad \delta_h = \frac{q\mathcal{E}\ell_h}{k_B T} \tag{4.80}$$

In an intrinsic semiconductor $n_3 = p_3 = n_{i3}$ giving

$$n_{i3}(\mathcal{E}) = n_{i3o}\left[\frac{\sinh(\delta_e)}{\delta_e}\frac{\sinh(\delta_h)}{\delta_h}\right]^{1/2}, \quad n_{i3o} = \sqrt{N_{c3}N_{v3}}\;\; e^{-E_g/2k_BT} \tag{4.81}$$

Here, n_{i3o} is the equilibrium intrinsic carrier density in a zero electric field. In fact, the effective bandgap E_{geff} in terms of true bandgap can be written as

$$E_{geff} = E_g - k_B T \ln\left[\frac{\sinh(\delta_e)}{\delta_e}\frac{\sinh(\delta_h)}{\delta_h}\right] \tag{4.82}$$

which in a high electric field approximates to $E_{geff} \approx E_g - q\mathcal{E}(\ell_e + \ell_h)$.

The mfps can be obtained from the mobility as below

$$\ell_e = \frac{\mu_e m_e v_{th3e}}{q}, \quad \ell_h = \frac{\mu_h m_h v_{th3h}}{q} \tag{4.83}$$

As one can expect, the intrinsic carrier concentration rises dramatically as a high electric field is encountered. The rise is exponential as $n_{i3}(\mathcal{E})/n_{i3o} \approx \exp[(\delta_e + \delta_h)]/(\delta_e\delta_h)$. Figure 4.14 shows the carrier multiplication $n_i(\mathcal{E})/n_{io}$ as the electric field is increased for doping levels 10^{16} (top curve), 10^{17} (intermediate curve), and 10^{18}. Figure 4.15 shows the bandgap narrowing effect. Enhanced impurity ions reduce the mobility of the carriers and hence the mfp.

An alternative description for carrier multiplication and bandgap narrowing can be given in terms of bandgap temperature T_b that gives the same effect as if the temperature of the sample is increased, giving enhanced carrier concentration. This bandgap temperature T_b is defined as

$$\frac{n_i(\mathcal{E})}{n_{io}} = \left(\frac{T_b}{T}\right)^{3/2} \exp\left[-\frac{E_g}{2k_B}\left(\frac{1}{T_b} - \frac{1}{T}\right)\right] \tag{4.84}$$

which compares $n_i(\mathcal{E})$ at an enhanced temperature T_b to n_{io} with lattice temperature T, but with the same bandgap. This equation is transcendental

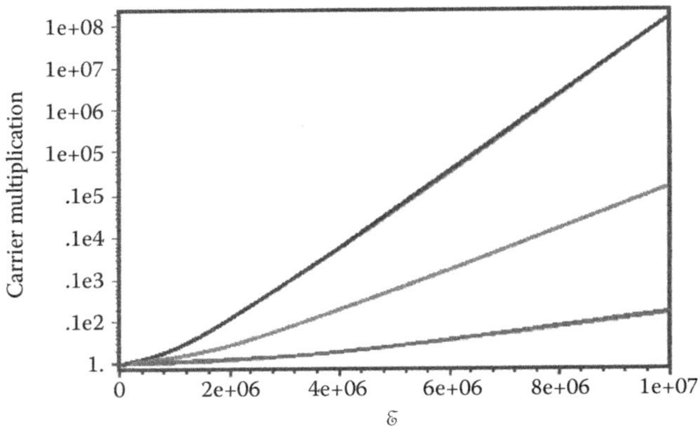

FIGURE 4.14
Intrinsic carrier concentration (ions/cm^3) versus electric field (V/m) for doping levels of 10^{16} (top curve), 10^{17} (intermediate curve), and 10^{18} ions/cm^3 (bottom curve).

in nature and cannot be solved directly. However, an iterative solution is obtained by using lattice temperature T as the seed value and convergence is sought. An approximate form can be assumed by the predominance of an exponential term that can be easily inverted for T_b/T. This can be taken as a seed value to calculate the exact value by iterative functional routine. Figure 4.16 shows the bandgap temperature as a function of the electric filed. Higher mfp leads to a higher temperature. This can be construed as one of the ways to define hot-electron temperature.

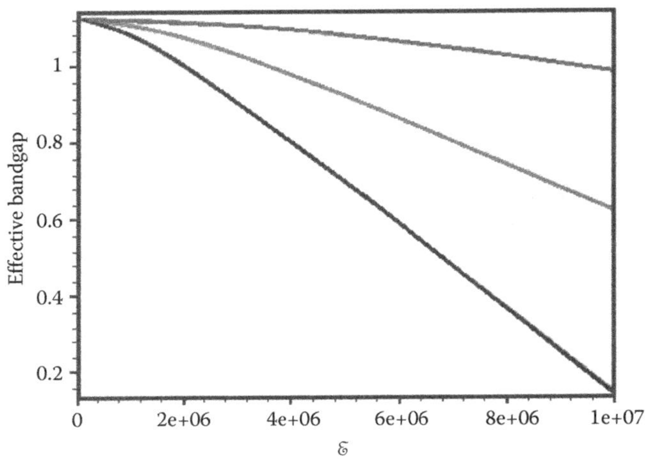

FIGURE 4.15
Effective bandgap (eV) versus electric field (V/m) for doping levels of 10^{16} (top curve), 10^{17} (intermediate curve), and 10^{18} ions/cm^3 (bottom curve).

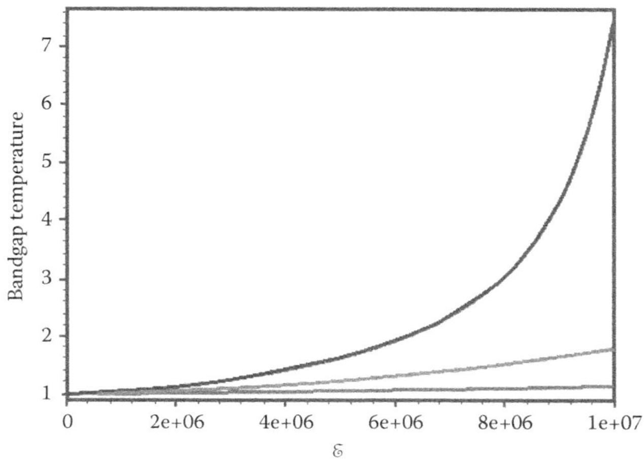

FIGURE 4.16
Normalized bandgap temperature T_b/T versus electric field (V/m) for the doping levels as in Figures 4.14 and 4.15.

EXAMPLES

E4.1 a. GaAs is known to have room-temperature long-channel bulk electron mobility $\mu_\infty = 8000\,(\text{cm}^2/\text{V}\cdot\text{s})$. What is the mfp between collisions for itinerant (drifting) electrons?

b. Calculate the room-temperature thermal velocity of electrons in GaAs.

c. Calculate the room-temperature electron mfp.

d. At approximately what value of the electric field will the room-temperature electron drift velocity not follow the linear relationship between the drift velocity and electric field?

e. At approximately what value of the applied voltage, Ohm's law will not be valid for a microresistor of 2-μm length.

Answer

a. $\mu_{o\infty} = 8000\,(\text{cm}^2/\text{V}\cdot\text{s}) = 0.8\,(\text{m}^2/\text{V}\cdot\text{s})$

$$\mu_{o\infty} = \frac{q\tau_c}{m^*} \Rightarrow \tau_c = \frac{m^*\,\mu_{o\infty}}{q}$$

$$= \frac{0.067 \times 9.11 \times 10^{-31}\,\text{kg} \times 0.8\left(\text{m}^2/\text{V}\cdot\text{s}\right)}{1.6 \times 10^{-19}\,\text{C}}$$

$$= 3.0 \times 10^{-13} = 0.3\ \text{ps}$$

Note: $C \cdot V = J = \text{kg m}^2/\text{s}^2$

b. $$\frac{1}{2}m^* v_{th}^2 = \frac{3}{2}k_B T \Rightarrow v_{th} = \sqrt{\frac{3k_B T}{m^*}}$$
$$= \sqrt{\frac{3 \times 1.38 \times 10^{-23}\,\text{J/K} \times 300\,\text{K}}{0.067 \times 9.11 \times 10^{-31}\,\text{kg}}}$$
$$= 4.51 \times 10^5\ \text{m/s}$$

c. $$\ell_{0\infty} = v_{th}\tau_c = 4.51 \times 10^5\,\text{m/s} \times 3.0 \times 10^{-13}$$
$$s = 1.35 \times 10^{-7}\,\text{m} = 0.1\ \mu\text{m}$$

d. $$\mathcal{E}_c = \frac{V_t}{\ell_{0\infty}} = \frac{0.0259\ \text{V}}{1.35 \times 10^{-7}\ \text{m}} = 1.91 \times 10^5\ \text{V/m} = 1.91\ \text{kV/cm}$$

e. $$V_c = \mathcal{E}_c L = 1.91 \times 10^5\ \text{V/m} \times 2 \times 10^{-6}\ \text{m} = 0.38\ \text{V}$$

E4.2 Consider the energy band diagram of silicon as shown below:

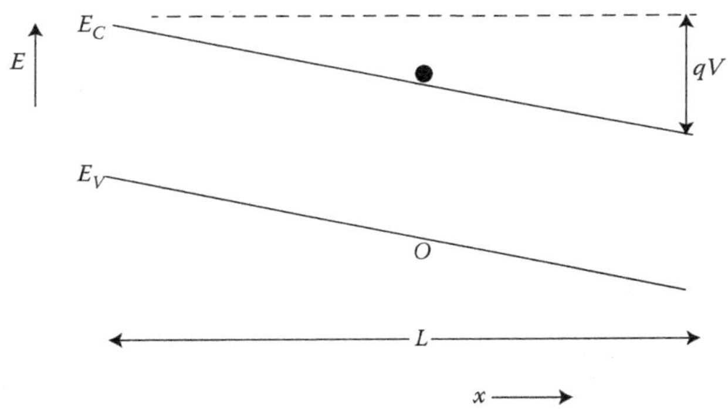

a. Find the electric field, and express the result in V/m. What direction does it point?
b. Find the force on the electron. In what direction is the electron accelerated? Find the acceleration. Find electron's drift velocity (ohmic).
c. Find the force on the hole. In what direction is the hole accelerated? Find the acceleration. Find hole's drift velocity (ohmic).
d. Given that the semiconductor is doped with $N_D = 5 \times 10^{15}\,\text{cm}^{-3}$ donors, find the current density in the sample.

Answer

a. The electric field is voltage/distance. The potential energy changes by 1 eV over a distance of 2 m. Since the voltage $\Delta V = \Delta U/(-q)$, the electric potential change is

$$\Delta V = \Delta U/(-q) = \left(\frac{-1\ \text{eV}}{-\text{e}}\right) = 1\ \text{V}$$

The electric field is $\mathcal{E} = -\dfrac{dV}{dx} = \dfrac{-1\,\text{V}}{2\times 10^{-6}\,\text{m}} = -5\times 10^{5}\,\text{V/m}$

The field points to the left (uphill).

b. From $F = -q\mathcal{E} = -(1.6\times 10^{-19}\,\text{C})(-5\times 10^{5}\,\text{V/m}) = 8.0\times 10^{-14}\,\text{N}$. The force is to the right.

$$a = \frac{F}{m^*_{ce}} = \frac{8.0\times 10^{-14}\,\text{N}}{0.26\times 9.1\times 10^{-31}\,\text{kg}} = 3.38\times 10^{17}\,\text{m/s}^2$$

$$v_d = \mu_n\mathcal{E} = 0.14\,\text{m}^2/\text{V}\cdot\text{s}\times 5\times 10^{5}\,\text{V/m} = 0.7\times 10^{5}\,\text{m/s}$$

c. The force on the hole is $F = +q\mathcal{E} = (+1.6\times 10^{-19}\,\text{C})(5\times 10^{5}\,\text{V/m}) = -8.0\times 10^{-14}\,\text{N}$, to the left.

$$a = \frac{F}{m^*_{ch}} = \frac{8.0\times 10^{-14}\,\text{N}}{0.36\times 9.1\times 10^{-31}\,\text{kg}} = 2.44\times 10^{17}\,\text{m/s}^2$$

$$v_d = \mu_p\mathcal{E} = 0.04\,\text{m}^2/\text{V}\cdot s\times 5\times 10^{5}\,\text{V/m} = 0.2\times 10^{5}\,\text{m/s}$$

d. $$J \approx J_n = nqv_d \approx 5\times 10^{21}\,\text{m}^{-3}\times 1.6\times 10^{-19}\,\text{C} \times 0.7\times 10^{5}\,\text{m/s} = 5.6\times 10^{7}\,\text{A/m}^2$$

E4.3 A "motivated" conduction electron in a silicon sample responds to an electric field of 500 V/cm by accelerating in the direction opposite to that of an electric field. If it was inert with zero velocity while "unmotivated," how much velocity does it gain while under the influence of the applied electric field for a short duration of 0.2 ps.

$$F = -q\mathcal{E} = -1.6\times 10^{-19}\,\text{C}\times 500\frac{\text{V}}{10^{-2}\,\text{m}} = -8.0\times 10^{-15}\,\text{N}$$

$$a = \frac{F}{m*} = -\frac{q\mathcal{E}}{m*} = \frac{-8.0\times 10^{-15}\,\text{N}}{0.26\times 9.1\times 10^{-31}\,\text{kg}} = -3.38\times 10^{16}\,\text{m/s}^2$$

$$v = v_o + a\tau_c = 0 + -3.38\times 10^{16}\,\text{m/s}^2\times 0.2\times 10^{-12}\,\text{s}$$
$$= -6.7\times 10^{3}\,\text{m/s}^2$$

E4.4 Metals have a very high conductivity (low resistivity). The collision time τ_c can be calculated from the conductivity $\sigma = n_3 q^2\tau_c/m_o$. Appendix 4A lists the conductivity of metals.

a. Estimate the mean collision time for copper.
b. Calculate the Fermi velocity of electrons in copper.
c. Estimate the mfp assuming the electrons are traveling with the Fermi velocity.

In metals, as one may surmise, there can be frequent collisions with the densely packed electrons. The mfp $\ell = v_F\tau \approx 3.9\times 10^{-8}\,\text{m} = 39\,\text{nm}$ with τ calculated from the conductivity and $v_F = 1.57\times 10^{6}\,\text{m/s}$ for copper as an example.

The lattice constant in copper is 0.361 nm. This shows that mfp is of the order of 100 lattice constants. The mfp is only $\ell \approx 2.8 \times 10^{-9}\,\text{m} = 2.8\,\text{nm}$ if the thermal velocity is used in place of the Fermi velocity which is about 8 lattice constants, indicating that the scattering is infrequent in strongly degenerate metals as compared to sparse population in a semiconductor. Mobility and hence collision time τ_c is a strong function of temperature. As temperature decreases, mobility increases due to suppression of phonon scattering. At low temperatures, the scattering due to ionized impurities is shown to increase in 3D semiconductors. In pure copper at $T = 4.2\,\text{K}$ (liquid helium temperature), it is possible to obtain $\tau_c \approx 1\,\text{ns}$ with a resulting mfp $\ell = v_F\tau = \hbar(3\pi^2 n_3)^{1/3}\tau_c/m_o \approx 3\,\text{mm} = 3000\,\mu\text{m}$ with $n_3 = 8.45 \times 10^{28}\,\text{m}^{-3}$. The possibility of ballistic transport is very high in metals at low temperatures.

PROBLEMS

P4.1 Calculate the ND intrinsic velocity v_{i3n} and rms thermal velocity v_{rms3n} of electrons in bulk silicon at room temperature.

P4.2 Calculate the drift velocity of an electron in intrinsic silicon in a sample of length $L = 2.0\,\mu\text{m}$ across which 5 V is applied.

P4.3 The electron mobility of intrinsic silicon is $\mu_{o\infty n} = 1400\,\text{cm}^2/\text{V}\cdot\text{s}$. Calculate the collision time τ_c and mfp $\ell_{o\infty}$.

P4.4 The critical electric field is defined as the field at which the electron going in the direction of the applied electric field will have zero velocity within the duration of the collision. Calculate the critical electric field for silicon.

P4.5 Electrons are injected from the aluminum contacts where they have the Fermi energy $E_F - E_c = 13\,\text{eV}$. What is the highest injection velocity v_{inj} of the electrons? What is the intrinsic velocity v_{i3} of electron in the channel of silicon. Assuming that the length of a silicon sample is $L = 100\,\text{nm}$, calculate the length-limited mfp ℓ_L and length-limited mobility μ_L, given that the long-channel mobility of silicon is $\mu_{o\infty} = 1400\,\text{cm}^2/\text{V}\cdot\text{s}$.

P4.6 The mfp is limited by the onset of a quantum emission in a very high electric field. Assuming $\mathcal{E} = 10\,\text{kV/cm}$, and energy of the quantum $E_Q = 36\,\text{meV}$, calculate the inelastic scattering length $\ell_Q = E_Q/q\mathcal{E}$. If the length-limited mfp $\ell_L = 20\,\text{nm}$, what is the effective mfp ℓ in a field of $\mathcal{E} = 10\,\text{kV/cm}$? What is ℓ if the field is $\mathcal{E} = 100\,\text{kV/cm}$?

P4.7 Using the same procedure as used for finding carrier density in terms of FDI, show how Equation 4.34 is obtained. Show that it can be simplified to Equation 4.37 in ND approximation.

P4.8 In a simplified model, the drift velocity v_D as a function of the electric field is given by $v_D = v_u \tanh(\mathcal{E}/\mathcal{E}_c)$ in the absence of a quantum emission. Show that for a low electric field ($\mathcal{E} \ll \mathcal{E}_c$), the drift velocity is a linear function of an electric field $v_D = \mu\mathcal{E}$. Obtain expression for μ.

P4.9 For a silicon sample, estimate $\mathcal{E}_c$, drift velocity at an electric field $\mathcal{E} = 0.1\mathcal{E}_c$, $\mathcal{E} = \mathcal{E}_c$, $\mathcal{E} = 2\mathcal{E}_c$, and $\mathcal{E} = 3\mathcal{E}_c$ from $v_D = v_i \tanh(\mathcal{E}/\mathcal{E}_c)$. Calculate the drift velocity from linear formula $v_D = \mu\mathcal{E}$. What can you say about the correctness of the linear model that is extensively used?

P4.10 For a heavily doped (strongly degenerate) silicon with electron concentration $n_3 = 10^{20}\,\text{cm}^{-3}$, calculate the intrinsic velocity, mobility velocity v_{m3}, and rms velocity v_{rms3}.

P4.11 In a ballistic GaAs channel, the channel length is $L = 100\,\text{nm}$ (including contacts). The low-field-long-channel mfp is $\ell_{o\infty} = 40\,\text{nm}$. Ohmic contacts have strongly degenerate doping concentration of $10^{20}\,\text{cm}^{-3}$ at room temperature.

a. Find the injection velocity from the contacts assuming they have a 3D configuration.

b. The channel between the Ohmic contacts has 2D configuration and is lightly doped. Find the approximate mobility velocity $v_{i2} = (\Gamma(1)/\Gamma(1/2))v_{th}$ of the channel.

c. Find the low-field-long-channel mobility $\mu_{o\infty} = (q\ell_{o\infty}/m * v_{i2})$.

d. Find the value of the ballistic mfp $\ell_B = \ell_{o\infty}(v_{inj}/v_{i2})$.

e. Calculate the length-limited ballistic mobility $\mu_L = \mu_{o\infty}$ $(1 - e^{-(L/\ell_B)})$ of the channel.

P4.12 a. Find the saturation (intrinsic) velocity $v_{i2Deg} = (\hbar/m*)\left[8\pi n_2/9\right]^{1/2}$ of a degenerately doped ($n_2 = 10^{12}\,\text{cm}^{-2}$) n-$In_{0.15}Ga_{0.85}As$ 2D channel with electron effective mass $m* = 0.057\,m_o$.

b. The critical electric field for this channel is experimentally found to be $\mathcal{E}_c = 3.8\,\text{kV/cm}$ at room temperature. Knowing that $\mathcal{E}_c = V_t/\ell_{o\infty}$, determine the mfp $\ell_{o\infty}$.

c. With the voltage $V = 2.0\,\text{V}$ applied across the length $L = 0.1\,\mu\text{m}$, what is the drift velocity of an electron? $v_D = v_{i2Deg}\tanh(\mathcal{E}/\mathcal{E}_c)$.

d. Estimate the low-field long-channel mobility of the channel.

e. Estimate the collision time τ_c.

CAD/CAE PROJECTS

C4.1 *CAD/CAE Design Project.* Plot velocity-field curves for bulk (3D) silicon at room temperature. Estimate any parameters needed and state any approximation needed. Also, show on the same graph, the linear behavior and saturation behavior. Discuss your findings.

C4.2 *Intrinsic Degeneracy Temperature.* When the Maxwell–Boltzmann approximation (ND statistics) is valid, the average (intrinsic) energy of a carrier is given by $(d/2)k_BT$, where d is the dimensionality ($d = 3$ [bulk], $d = 2$ [nanosheet], $d = 1$ [nanowire]) and T is the crystal temperature. However, for degenerately doped semiconductors, as intrinsic energy is elevated over this ND limit, so does the apparent intrinsic temperature T_{id} for a given dimensionality.

a. Obtain an expression of $T_{id}/T = E_{id}/E_{idND}$ where ND stands for nondegenerate.

b. Plot T_{id}/T versus normalized carrier concentration n_d/N_{cd} for all three dimensionalities on the same graph. Also, plot their ND and degenerate approximations. Discuss the range of validity of these approximations.

c. For parameters appropriate to silicon, plot T_{i3} as a function of carrier concentration from $n_3 = 10^{18}$ to $10^{21}\,\text{cm}^{-3}$. A simultaneous plot of degenerate and ND limit will capture the essence of these approximations. The conductivity effective mass in this case is $m_c^* = 0.26\,m_o$. The valley degeneracy is $g_v = 6$.

d. Repeat (c) for a 2D nanosheet for carrier concentration from $n_2 = 10^{10}$ to $10^{13}\,\text{cm}^{-2}$. The appropriate effective mass in this case is $m^* = 0.198\,m_o$ both for conductivity as well as for the density of states. The valley degeneracy is $g_v = 2$.

e. Repeat (c) for 1D nanowire for carrier concentration from $n_1 = 10^5$ to $10^8\,\text{cm}^{-1}$. The appropriate effective mass in this case is $m^* = 0.198\,m_o$ both for conductivity as well as for the density of states. The valley degeneracy is $g_v = 4$.

C4.3 *Ohmic and Electron Degeneracy Temperature.* Repeat C4.2 to plot on the same plot T_{od}/T and T_{ed}/T.

C4.4 *2D Bandgap Narrowing, Carrier Multiplication, and Bandgap Temperature.* Repeat Section 4.9 for 2D electron gas say in GaAs.

C4.5 *1D Bandgap Narrowing, Carrier Multiplication, and Bandgap Temperature.* Repeat Section 4.9 for 1D electron gas. You can choose silicon nanowire. An alternative will be bulk semiconductors in a strong magnetic field, which is also 1D character. Or, you can do a comparative study.

Appendix 4A: Derivation of Velocity-Field Characteristics

4A.1 3D Bulk

The average energy is evaluated from

$$v_D = \langle v_z \rangle = \frac{\sum_{k_x,k_y,k_z,s} v_z f(E,\mathcal{E})}{\sum_{k_x,k_y,k_z,s} f(E,\mathcal{E})} \tag{4A.1}$$

where s stands for spin. The denominator is normalization of the Fermi energy:

$$N_3 = \sum_{k_x,k_y,k_z,s} f(E,\mathcal{E}) \tag{4A.2}$$

Assuming that the electric field is applied in the polar direction ($\vec{\mathcal{E}} \parallel \hat{z}$), the summation can be expanded to

$$N_3 = \frac{L_x L_y L_z}{(2\pi)^3} 2\iiint dk_x\, dk_y\, dk_z\, f(E,\mathcal{E} \tag{4A.3}$$

$$n_3 = \frac{N_3}{L_x L_y L_z} = \frac{2}{(2\pi)^3} \int_{k=0}^{\infty} dk \int_{\theta=0}^{\pi} d\theta \int_{\phi=0}^{2\pi} d\phi\, k^2 \sin\theta f(E,\mathcal{E}) \tag{4A.4}$$

where $n_3 = N_3/L_x L_y L_z$ is the carrier concentration per unit volume. The density of states in momentum space is given by $L/2\pi$ in each direction, as stated earlier. That density allows us to change a summation to integral for an analog variable where spacing between quantum levels is diminutive. Here, $dk_x\, dk_y\, dk_z = k^2\, dk \sin\theta\, d\phi$ is converted to the polar coordinates. The energy in the effective mass approximation is given by

$$E = E_c + \frac{\hbar^2 k^2}{2m^*} = E_c + \frac{\hbar^2 (k_x^2 + k_y^2 + k_z^2)}{2m^*} \tag{4A.5}$$

The energy depends only on the magnitude of $|k|$.
Therefore

$$k = \left(\frac{2m^*}{\hbar^2}\right)^{1/2} (E - E_c)^{1/2}\, dk = \left(\frac{2m^*}{\hbar^2}\right)^{1/2} \frac{1}{2}(E - E_c)^{-1/2}\, dE \tag{4A.6}$$

Substituting this in Equation 4A.4 with

$$f(E) = \frac{1}{e^{(E-q\mathcal{E}\,\ell\cos\theta-E_F/k_BT)}+1} \tag{4A.7}$$

gives

$$n_3 = \frac{2}{(2\pi)^3}\frac{1}{2}\left(\frac{2m^*}{\hbar^2}\right)^{3/2}\int_{E=0}^{\infty} dE(E-E_c)^{1/2}\int_0^{\theta} d\theta\,\sin\theta\frac{1}{e^{x-H(\theta)}+1}2\pi \tag{4A.8}$$

where

$$x = \frac{E-E_c}{k_BT}\,\eta = \frac{E_F-E_c}{k_BT}\quad \mathrm{H}(\theta) = \eta+\delta\cos\theta\,\delta = \frac{q\mathcal{E}\,\ell}{k_BT} = \frac{\mathcal{E}}{\mathcal{E}_c}$$

$$n_3 = \frac{1}{(2\pi)^3}\left(\frac{2m^*k_BT}{\hbar^2}\right)^{3/2}\int_{E=0}^{\infty} x^{1/2}\,dx\int_{y=-1}^{y=+1} dy\frac{1}{e^{x-\eta-\delta y}+1} \tag{4A.9}$$

Here, $y = \cos\theta$ which means $dy = -\sin d\theta$ and $\theta = 0$, $y = 1$ and $\theta = \pi$, $y = -1$. This integral can be written as

$$n_3 = \frac{1}{(2\pi)^3}\left(\frac{2m^*k_BT}{\hbar^2}\right)^{3/2}\Gamma\left(\frac{3}{2}\right)\int_{y=-1}^{y=+1} dy\Im_{1/2}(\eta+\delta\,y) \tag{4A.10}$$

The integral in Equation 4A.10 is difficult to evaluate for arbitrary degeneracy. However, for the Maxwell–Boltzmann approximation 1 in the denominator of $f(x,y) = 1/(1+e^{x-\eta-\delta y})$, it can be neglected giving a simpler expression. In an ND approximation, $\Im_j(\eta) \approx e^{\eta}$ for all orders. Hence,

$$\begin{aligned}
n_3 &= \frac{1}{(2\pi)^3}\left(\frac{2m^*k_BT}{\hbar^2}\right)^{3/2}\frac{\sqrt{\pi}}{2}\int_{y=-1}^{y=+1} e^{\eta+\delta y}\,dy \\
&= \frac{N_{c3}}{2}e^{\eta}\int_{y=-1}^{y=+1} e^{\delta y}\,dy = \frac{N_{c3}}{2}e^{\eta}\,e^{\delta y}\Big|_{-1}^{+1} \\
&= \frac{N_{c3}}{2}e^{\eta}\frac{e^{\delta}-e^{-\delta}}{\delta} \\
&= N_{c3}e^{\eta}\frac{\sinh(\delta)}{\delta}
\end{aligned} \tag{4A.11}$$

where

$$N_{c3} = 2\left(\frac{m^* k_B T}{2\pi\hbar^2}\right)^{3/2}$$

Therefore

$$n_3 = N_{c3}e^{\eta}\left[\frac{\sinh(\delta)}{\delta}\right] \tag{4A.12}$$

or

$$\eta = \ln\left[\frac{n_3}{N_{c3}}\frac{\delta}{\sinh\delta}\right] = \eta_o - \ln\left(\frac{\delta}{\sinh\delta}\right) \tag{4A.13}$$

The numerator of Equation 4A.1 can be similarly evaluated with $v_z = v\cos\theta = \hbar k/m^* y$ with k given by Equation 4A.6. The numerator transforms to

$$\sum_{k_x,k_y,k_z,s} v_z f(E,\mathcal{E}) = \frac{N_{c3}}{2}\frac{2}{\sqrt{\pi}}v_{th}\,e^{\eta}\int_{-1}^{+1} y\,e^{\delta y}dy \tag{4A.14}$$

$$n_3 v_{D=}N_{c3}\frac{2}{\sqrt{\pi}}v_{th}\,e^{\eta}\frac{1}{\delta}\left[\cosh\delta - \frac{1}{\delta}\sinh\delta\right] \tag{4A.15}$$

$$n_3 v_{D=}N_{c3}\frac{2}{\sqrt{\pi}}v_{th}\,e^{\eta}\left[\frac{\cosh\delta}{\delta} - \frac{1}{\delta^2}\sinh\delta\right] \tag{4A.16}$$

$$v_D = \frac{2}{\sqrt{\pi}}v_{th}\left[\frac{(\cosh\delta/\delta) - (1/\delta^2)\sinh\delta}{(\sinh\delta/\delta)}\right] = \frac{2}{\sqrt{\pi}}v_{th}\mathcal{L}(\delta) = v_{th3}\mathcal{L}(\delta) \tag{4A.17}$$

where $\mathcal{L}(\delta) = \coth\delta - (1/\delta)$ is a Langevin function.

4A.2 2D Sheet

Following the same method as in the 3D, 2D drift velocity is evaluated as

$$v_D = \frac{\sum_{k_x,k_y,s} v_x f(E,\mathcal{E},\theta)}{\sum_{k_x,k_y,s} f(E,\mathcal{E},\theta)} \tag{4A.18}$$

The denominator of Equation 4A.18 is

$$N = \frac{L_x L_y}{(2\pi)^2} 2 \int_{-\infty}^{+\infty}\int dk_x\, dk_y \frac{1}{e^{x-\delta\cos\theta-\eta}+1} \tag{4A.19}$$

$$E = E_{c2} + \frac{\hbar^2 k^2}{2m^*} \tag{4A.20}$$

$$k = \left(\frac{2m^*}{\hbar^2}\right)^{1/2} (E - E_{c2})^{1/2} \quad dk = \left(\frac{2m^*}{\hbar^2}\right)^{1/2} \frac{1}{2}(E - E_{c2})^{-1/2} \tag{4A.21}$$

$$dk_x\, dk_y = k\, dk\, d\theta \tag{4A.22}$$

The transformation gives

$$\begin{aligned} n_2 &= \frac{N}{L_x L_y} = \frac{1}{(2\pi)^2} 2 \int_0^{+\infty}\int_0^{2\pi} dk\, d\theta \left(\frac{2m^*}{\hbar^2}\right)^{1/2} (E - E_{c2})^{1/2} \frac{1}{e^{x-\delta\cos\theta-\eta}+1} \\ &= \frac{2}{(2\pi)^2} \int_0^{2\pi} d\theta \int_0^{+\infty} \left(\frac{2m^*}{\hbar^2}\right)^{1/2} \frac{1}{2}(E - E_{c2})^{1/2} \left(\frac{2m^*}{\hbar^2}\right)^{1/2} (E - E_{c2})^{1/2} \frac{1}{e^{x-\delta\cos\theta-\eta}+1} \\ &= \frac{N_{c2}}{2} \int_0^{2\pi} d\theta\, \Im_0(\delta\cos\theta + \eta) \end{aligned} \tag{4A.23}$$

with $N_{c2} = (m^* k_B T/\pi\hbar^2)$ and $\Im_0(\delta\cos\theta + \eta) = \ln\left[e^{\delta\cos\theta+\eta} + 1\right]$.

In an ND approximation,

$$\begin{aligned} n_2 &= \frac{N_{c2}}{2} \int_0^{2\pi} d\theta\, e^{\delta\cos\theta+\eta} \\ &= \frac{N_{c2}}{2} e^{\eta} 2 \int_0^{\pi} d\theta\, e^{\delta\cos\theta+\eta} \\ &= N_{c2}\, e^{\eta} I_o(\delta) \end{aligned} \tag{4A.24}$$

$$e^{\eta_2} = \left(\frac{n_2}{N_{c2}}\right)\left(\frac{1}{I_o(\delta)}\right) \tag{4A.25}$$

$$v_D = \frac{(N_{c2}/2)e^{\eta}v_{th2}\int_0^{2\pi}\cos\theta\, e^{\delta\cos\theta}}{N_{c2}\, e^{\eta} I_o(\delta)}$$

$$= v_{th2}\left[\frac{I_1(\delta)}{I_o(\delta)}\right] \tag{4A.26}$$

where

$$I_1(\delta) = I_1(-\delta) = \int_0^{\pi}\cos\theta\, e^{\delta\cos\theta} \quad \text{and} \quad v_{th2} = \frac{\sqrt{\pi}}{2}\left[\frac{I_1(\delta)}{I_o(\delta)}\right]$$

4A.3 1D Nanowire

$$v_D = \frac{\sum_{k_x} v_x f(E,\mathcal{E},I)}{\sum_{k_x} f(E,\mathcal{E},I)} \tag{4A.27}$$

$$N = \frac{L_x}{2\pi} 2\int_{-\infty}^{+\infty} dk_x\, f(E,\mathcal{E},I)$$

$$= \frac{L_x}{2\pi} 2\int_0^{+\infty} dk_x\, f(E,\mathcal{E},+) + \int_0^{+\infty} dk_x\, f(E,\mathcal{E},-) \tag{4A.28}$$

$$E = E_{c1} + \frac{\hbar^2 k^2}{2m^*} \tag{4A.29}$$

$$dk = \left(\frac{2m^*}{\hbar^2}\right)^{1/2} \frac{1}{2}(E - E_{c1})^{1/2} \tag{4A.30}$$

$$n_1 = \frac{N}{L_x} = \frac{N_{c1}}{2}\left[\Im_{-1/2}(\eta+\delta) + \Im_{-1/2}(\eta-\delta)\right] \tag{4A.31}$$

with $N_{c1} = 2(2m^* k_B T/\pi\hbar^2)^{1/2}$

$$v_D = v_{th1}\frac{\Im_0(\eta+\delta) - \Im_0(\eta-\delta)}{\Im_{-1/2}(\eta+\delta) + \Im_{-1/2}(\eta-\delta)} \tag{4A.32}$$

where $v_{th1} = (1/\sqrt{\pi})v_{th}$.

ND approximation gives

$$v_D = v_{th1}\frac{e^{\eta+\delta} - e^{\eta-\delta}}{e^{\eta+\delta} + e^{\eta-\delta}} = v_{th1}\tanh(\delta) \tag{4A.33}$$

$$n_1 = \frac{N_{c1}}{2}\left[e^{\eta+\delta} + e^{\eta-\delta}\right] = N_{c1}\, e^{\eta}\cosh\delta \tag{4A.34}$$

$$e^{\eta} = \frac{n_1}{N_{c1}\cosh\delta} \tag{4A.35}$$

References

1. V. K. Arora, High-field electron mobility and temperature in bulk semiconductors, *Physical Review B*, 30, 7297–7298, 1984.
2. V. K. Arora, High-field distribution and mobility in semiconductors, *Japanese Journal of Applied Physics*, Part 1: Regular Papers & Short Notes, 24, 537–545, 1985.
3. M. Büttiker, Role of quantum coherence in series resistors, *Physical Review B*, 33, 3020, 1986.
4. V. K. Arora, Quantum engineering of nanoelectronic devices: The role of quantum emission in limiting drift velocity and diffusion coefficient, *Microelectronics*, 31, 853–859, 2000.
5. M. Heiblum and M. Fischetti, Ballistic hot-electron transistors, *IBM Journal of Research and Development*, 34, 530, 1990.
6. K. K. Thornber, Relation of drift velocity to low-field mobility and high-field saturation velocity, *Journal of Applied Physics*, 51, 2127–2136, 1980.
7. D. R. Greenberg and J. A. d. Alamo, Velocity saturation in the extrinsic device: A fundamental limit in HFETs, *IEEE Transactions on Electron Devices*, 41, 1334–1339, 1994.
8. V. K. Arora, Drift diffusion and Einstein relation for electrons in silicon subjected to a high electric field, *Applied Physics Letters*, 80, 3763–3765, May 2002.
9. V. K. Arora, Hot electrons: A myth or reality? in *SPIE Proceedings Series*, 563–569, 2002.

5

Charge Transport

Information processing by transporting carriers in nanoelectronic devices and circuits is severely affected by the breakdown of Ohm's law when the applied voltage V in a nanoscale ($L < 1000$ nm) device exceeds the critical voltage $V_c = V_t L/\ell_{o\infty}$ ($V > V_c$). Here V_t is the thermal voltage with a value 0.0259 V at room temperature and $\ell_{o\infty}$ (typically 100 nm) is the mfp. This nonohmic behavior is the cause of the current saturation leading to the resistance surge. The saturation arises due to realignment of randomly oriented velocity vectors to the unidirectional streamlined ones in a high electric field when voltage applied across a resistor exceeds the critical value. The surge accelerates for signal propagation as dc voltage is increased. Both the digital and analog signal processing will be affected by the presence of nonohmic nonlinear behavior. This surge changes the RC time constants, power consumption, and voltage and current division laws. Ballistic processes overpower the scattering-limited transport in a high electric field and in channels where length is smaller than the scattering-limited mfp. The transient switching delay in a micro-/nanoscale circuit containing resistive and reactive elements is sternly affected by the surge in the resistance arising out of sublinear current–voltage (I–V) characteristics limited to the drift velocity leading to current saturation. The frequency response $f = 1/2\pi\tau_t$, where τ_t is the transit time through the conducting channel and is lower than that predicted from the application of Ohm's law. The resistance surge boosts the RC time constant and attenuates the $\mathcal{L}/R$ time constant dramatically. These results are necessary for extraction of transport parameters and assessing the limitations of parasitic elements in a micro-/nanoscale circuit.

5.1 Primer

The ever-decreasing length of a conducting channel has now entered the decananometer regime with a typical length in the order in 10–100 nm. Many unexpected effects that are normally negligible in the long channels demonstrate their prowess in nanoscale channels. One such effect is the surge in the resistance of a signal due to saturated current [1]. This increase in resistance affects all timing delays that are normally based on Ohm's law where resistance is constant due to current rising linearly with the applied voltage.

The failure of Ohm's law [1] naturally transforms all time constants affecting the overall switching delay, both for digital and analog signals. Greenberg and del Alamo [2] carried out an experimental study revealing that velocity and current saturation occurring in the extrinsic source and drain contacts results in resistance surge. The enhanced contact resistance sets a fundamental limit on the maximum intrinsic channel current and gate swing in a heterojunction field-effect transistor (HFET), for example. The cause of the parasitic-resistance surge is due to velocity-approaching saturation in the extrinsic device that in turn degrades the transconductance and transistor frequency of an HFET. The contact regions, therefore, play an important role in the switching delay and hence the high-frequency performance.

The high-frequency behavior depends on the carrier response as it transits through the channel in response to the driving stimulus. The information carriers are affected by the switching delay as a signal propagates through the circuit with the simultaneous presence of reactive and resistive elements. The transit time delay arises due to the breakdown of linear velocity response to the electric field resulting in velocity saturation that is limited by material parameters. Saad et al. [3] discuss ballistic effects limiting the mobility and saturation velocity. The saturation velocity is limited by the temperature and doping density. The cause of this saturation is the intrinsic velocity due to the realignment of otherwise randomly oriented velocity vectors in equilibrium to the unidirectional streamlined ones in a towering electric field [4]. The resistance surge effect in microcircuit engineering has been demonstrated [5] by Saxena et al. The reactive elements control the circuit speed through $\tau_{R_oC} = R_oC$ and $\tau_{L/R_o} = \mathcal{L}/R_o$ time constants, where R_o is the ohmic resistance, C is the series capacitance in an R–C circuit, and $\mathcal{L}$ is the series inductance in an R–L circuit. τ_{RC} is boosted and $\tau_{L/R} = \mathcal{L}/R$ is attenuated as blown-up resistance R replaces R_o in a micro/nanocircuit.

5.2 Ohmic (Linear) Transport

A bulk resistor in the form of a sheet resistor with dimensions $L_{x,y,z} = L, W,$ and T that are much larger than the de Broglie wavelength λ_D is a medium for propagating quantum waves in any direction of the applied electric field. The ohmic contacts have their own resistive effect that can be neglected in the primary stage. As the voltage is applied to a homogeneous sample, the volume carrier concentration $n_3 = N/V$ of the itinerant electrons per unit volume drift in response to the applied voltage establishing an electric field $\mathcal{E} = V/L$. Naturally, this drift leads to the current in the conducting (or resistive) channel.

Figure 5.1 shows the schematic of a planar resistor with length L, width W, and thickness T. Carriers entering the resistor from one end leave the other end, while emptying out the entire resistor during the transit time delay t in traversing the channel. The current $I = Q/t$ as the rate of charge flow in a homogeneous sample with no concentration gradients is

$$I_n = \frac{Q_n}{t} = \frac{n_3(LWT)(-q)}{t} = -n_3\frac{L}{t}qA_c = -n_3 v_{Dn} q A_c \tag{5.1}$$

where $A_c = WT$ is the area of cross-section that is perpendicular to the direction of charge flow (current). $v_{Dn} = L/T$ is the drift response to the electric field $\mathcal{E}$ that is normally modeled by the linear velocity-field relation:

$$v_{Dn} = -\mu_{o\infty n}\mathcal{E} \tag{5.2}$$

where $\mu_{o\infty n}$ is the ohmic mobility. The linear response is the origin of Ohm's law, which is obtained by substitution of Equation 5.2 into Equation 5.1 and using $\mathcal{E} = V/L$ is given by

$$I_n = \frac{V}{R_{on}} \quad \text{with} \quad R_{on} = \frac{1}{qn_3\mu_{on}}\frac{L}{A_c} = \rho_{3n}\frac{L}{A_c} \tag{5.3}$$

where R_{on} is the ohmic resistance that depends on material properties of the sample given by $\rho_3 = (1/qn_3\mu_{o\infty n})$ and geometry ratio (L/A_c). Designers use the sample properties and geometry ratio to design, characterize, and assess the electric response to the applied stimulation due to an applied electric field.

A semiconducting sample contains both electrons and holes; the hole current with each hole charge $+q$ and drift response to the applied electric field as $v_p = +\mu_{o\infty p}\mathcal{E}$ is similarly evaluated to give current in the same direction of the electric field

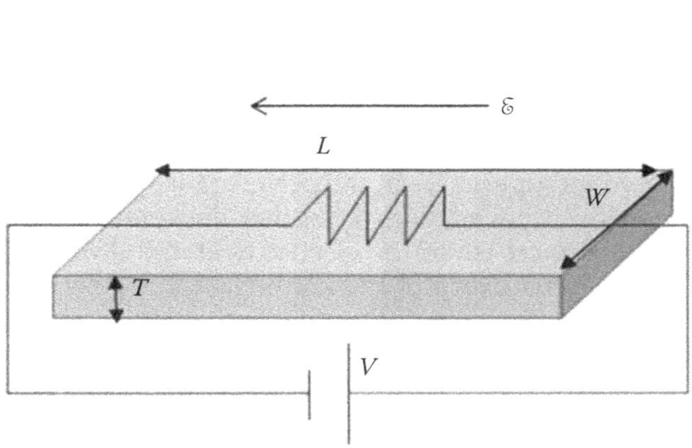

FIGURE 5.1
A sheet resistor with all dimensions $L_x = L$, $L_y = W$, and $L_z = T$.

$$I_p = \frac{Q_p}{t} = \frac{p_3(LWT)(+q)}{t} = +p_3\frac{L}{t}qA_c = +p_3 v_{Dp} q A_c \tag{5.4}$$

The total current $I = I_n + I_p$ is unidirectional in the direction of the applied electric field and hence electron and hole currents always add. The total current (I) in response to the applied voltage V is thus given by

$$I = I_n + I_p = \frac{V}{R_o}, \quad R_o = \rho_3\frac{L}{A_c}, \quad \rho_3 = \frac{1}{\sigma_3} = \frac{1}{(n_3\mu_{o\infty n} + p_3\mu_{o\infty p})q} \tag{5.5}$$

Here $\sigma = \sigma_n + \sigma_p = n_3\mu_{o\infty n}q + p_3\mu_{o\infty p}q$ is the sample conductivity that is the sum of electron and hole conductivities. Ohmic resistance R_o (in Ω) depends on 3D sample resistivity ρ_3 (in $\Omega \cdot m$) and carrier concentrations $n(p)_3$ (in m^{-3}) and $\mu_{o\infty n(p)}$(in $m^2/V \cdot s$).

In metals, the conductivity can present a paradox if a 3D model is used for the nanometer-scale cross-section. The conductivity is $\sigma = 5.9 \times 10^7$ S/m for copper. The resistance per unit length of copper is $R_o/L = \rho/A_c = 4\rho/\pi D^2 = 5.4 \times 10^{-5}$ Ω/m for diameter $D = 20$ mm, which is negligible for a finite sample of a few cms. As the diameter is reduced to $D = 20$ μm, 1000 times smaller, $R_o/L = 54$ Ω/m. And, for a nanowire of $D = 20$ nm, staggering $R_o/L = 5.4 \times 10^7$ Ω/m is obtained. The mobility decreases by a factor of A_c/λ^2 for a nanowire. This will give further enhancement in resistance if the diameter is reduced below the Fermi wavelength.

Equation 5.1 indicates that when velocity saturates ($v_D = -v_{sat}$), the current is proportional to the area of a cross-section A_c, independent of the channel length. Similarly, hole current is $I_p = p_3 v_{satp} q A_c$.

5.3 Discovery of Sat Law

Ohm's law enjoyed its superiority in the performance assessment of all conducting materials until it was discovered that the velocity cannot increase indefinitely with the increase in the electric field and eventually saturates to a value $v_{satn(p)}$. The discovery of the nonlinear drift velocity response to an electric field stunned many in the twentieth century. Even in the current century, performance evaluation and characterization of conducting channels is evaluated based on Ohm's law. A number of relations to relate drift response to the electric field were put on trial to see which one fits the experimental data best. The most prominent (tentatively called Sat Law) of these relations is

$$v_D = \frac{-\mu_o \mathcal{E}}{\left[1 + (\mathcal{E}/\mathcal{E}_{cd})^{\gamma_{n(p)}}\right]^{1/\gamma_{n(p)}}} \quad \text{with} \quad \mathcal{E}_{cd} = d\frac{v_{satn(p)}}{\mu_o} \tag{5.6}$$

where the critical electric field is modified to include dimensionality d so that $\mathcal{E}_{cd} = d\mathcal{E}_c$. Wide combinations of $\mathcal{E}_{cd}$ and $\gamma_n = 1$–2.8 have been utilized to give a fit to the experimental data that can be changed at will in most simulation programs.

The normalized drift response to the normalized electric field is shown in Figure 5.2 for $\gamma = 1$ (normally considered for holes), $\gamma = 2$ (normally considered valid for electrons), and the most recently obtained value of $\gamma = 2.8$ for electrons in InGaAs microchannel. Also, shown in Figure 5.2 are the extreme linear and saturation limits of the empirical curves. As one can see, a value of γ does not affect the linear or saturation limits. It only affects the rise toward saturation. A larger value of γ brings velocity closer to its saturation faster; as the electric field is increased, the mobility and saturation value are unaffected by γ. The theoretical solid curve is the outcome of the theory discussed in Chapter 4. According to this theoretical formalism, the velocity response to an applied electric field for all dimensionalities ($d = 3$, 2, and 1) is given by

$$v_D = v_{sat} \tanh\left(\frac{\mathcal{E}}{d\mathcal{E}_{co}}\right) \tag{5.7}$$

The critical electric field is $\mathcal{E}_{cd} = d\mathcal{E}_{co} = dV_t/\ell_{oL}$, where $V_t = k_B T/q$ is the thermal voltage or its degenerate-doping counterpart with an extreme value of Fermi voltage $V_F = k_B T_o/q = E_F/q$. $\mathcal{E}_{cd}$ or $V_{cd} = \mathcal{E}_{cd} L$ defines the boundary between Ohm's law with linear rise to the saturation regime with velocity saturation.

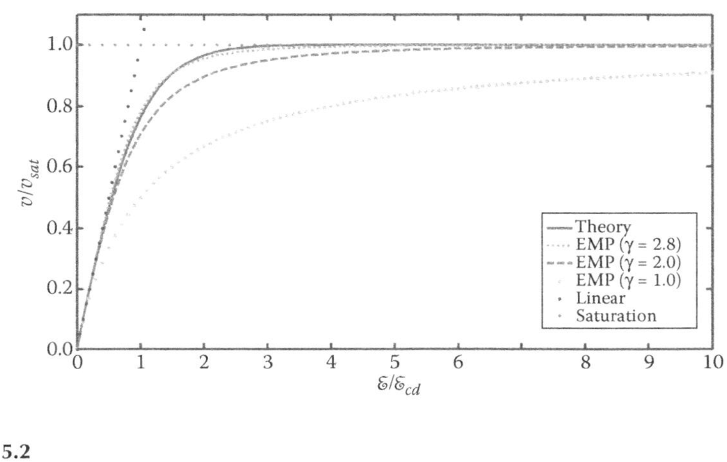

FIGURE 5.2
Drift velocity response to the electric field.

The current saturation is a direct result of this velocity saturation. ℓ_{oL} is the length-limited mfp that depends on the length of the sample. For long samples, it can be replaced with $\ell_{o\infty}$. Typical mobility in a semiconductor is 0.1 $m^2/V \cdot s$. The critical electric field, considering these facts, is $\mathcal{E}_c = 10^6$ V/m = 1.0 MV/m = 1.0 V/μm. The applied electric field depends on the length of the device. With a typical voltage of 5 V applied to a cm-length color-coded laboratory resistor $\mathcal{E} = V/L = 5\ V/0.01$ m = 500 V/m is much smaller than $\mathcal{E}_c = 10^6$ V/m. As our microprocessor embraces 1-μm transistors or even lower in a nanometer (nm) regime, the electric field even with $L = 1$ μm is $\mathcal{E} = V/L = 5\ V/1.0$ μm = 5 V/μm; surpassing its critical value of $\mathcal{E}_c = 1.0$ V/μm, the drift velocity gets closer to its saturation value. The critical voltage $V_c = \mathcal{E}_c L$ at the onset of nonlinear behavior is equal to 10 kV for a cm-long resistor and merely 1.0 V for the μm-long resistor. So, any reasonable voltage applied to a microresistor is sure to trigger the breakdown of Ohm's law.

It is easier to discover Sat Law for a microresistor as Equation 5.6 or 5.7 is substituted in Equation 5.1. The result with the empirical Equation 5.6 is

$$I = \frac{V}{R_o} \frac{1}{[1 + (V/V_{cd})^{\gamma}]^{1/\gamma}} \quad \text{with} \quad V_{cd} = \mathcal{E}_{cd} L \tag{5.8}$$

I–V characteristics with the use of Equation 5.7 are obtained as

$$I = I_{sat} \tanh\left(\frac{V}{V_{cd}}\right) = \frac{V}{R_o} \frac{\tanh(V/V_{cd})}{(V/V_{cd})} \tag{5.9}$$

where $I_{sat} = V_{cd}/R_o = n_3 q v_{satn} A_c$ is the saturation current. It is worth noting that the saturation current depends only on the area of the cross-section, but not on the length of the sample. The ohmic resistance depends on both the area of cross-section and on the length of the sample. This is an important distinction between two regimes as resistors are designed for complementary channels. Both Equations 5.8 and 5.9 in the extreme regimes can be approximated as

$$I = \frac{V}{R_o} \frac{1}{[1 + (V/V_{cd})^{\gamma}]^{1/\gamma}} = \begin{cases} \dfrac{V}{R_o}, & V < V_{cd} \\ I_{sat} = n_3 q v_{sat} A_c, & V \gg V_{cd} \end{cases} \tag{5.10a}$$

$$I = I_{sat} \tanh\left(\frac{V}{V_{cd}}\right) = \begin{cases} \dfrac{V}{R_o}, & V < V_{cd} \\ I_{sat} = n_3 q v_{sat} A_c, & V \gg V_{cd} \end{cases} \tag{5.10b}$$

The critical voltage V_{cd} thus marks a transition point between the ohmic and sat region. In the following, simplified V_c is used in place of V_{cd}.

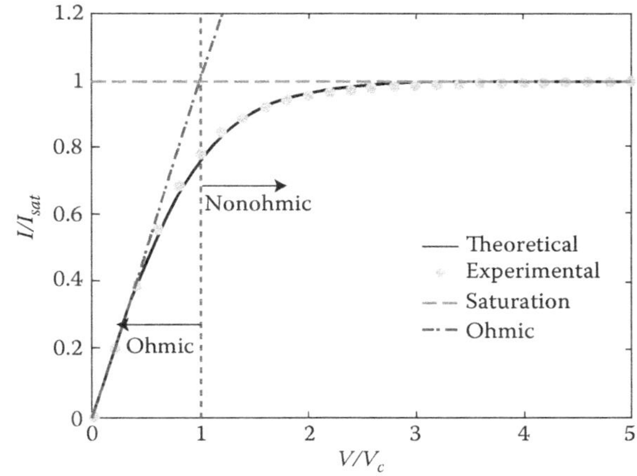

FIGURE 5.3
Theoretical and experimental I–V characteristics of a micro-/nanoresistor.

Normalized I–V characteristics with I/I_{sat} plotted as a function of normalized voltage V/V_c follow the same pattern as in Figure 5.2 with the axis appropriately labeled with I/I_{sat} to replace v/v_{sat} and V/V_c to replace $\mathcal{E}/\mathcal{E}_c$, as shown in Figure 5.3.

As I–V characteristics become nonlinear, the resistance $R = V/I$ and signal resistance $r = dV/dI$ is given by

$$R = \frac{V}{I} = R_o\left[1 + \left(\frac{V}{V_c}\right)^{\gamma}\right]^{1/\gamma} = R_o\frac{1}{\left[1-(I/I_{sat})^{\gamma}\right]^{1/\gamma}} \tag{5.11}$$

$$\begin{aligned} r &= \frac{dV}{dI} = R_o\left[1 + \left(\frac{V}{V_c}\right)^{\gamma}\right]^{1+(1/\gamma)} \\ &= R_o\frac{1}{\left[1-(I/I_{sat})^{\gamma}\right]^{1+(1/\gamma)}} = \frac{R^{(\gamma+1)}}{R_o^{\gamma}} \end{aligned} \tag{5.12}$$

Figures 5.4 and 5.5 show this resistance surge effect as a function of voltage and current for $\gamma = 2.8$. The resistance is close to its ohmic value for $V < V_c$, but rises dramatically beyond V_c, incremental (signal) resistance rising much faster than the direct resistance. Similarly, since the resistance remains closer to its ohmic value so far, the current is about 50% of its saturation value and then rises toward infinity as the current-saturation approaches, the signal resistance rising much sharply. In a prototype uniform sheet resistor, when a voltage V is applied across its length L, the resulting electric field is $\mathcal{E} = V/L$.

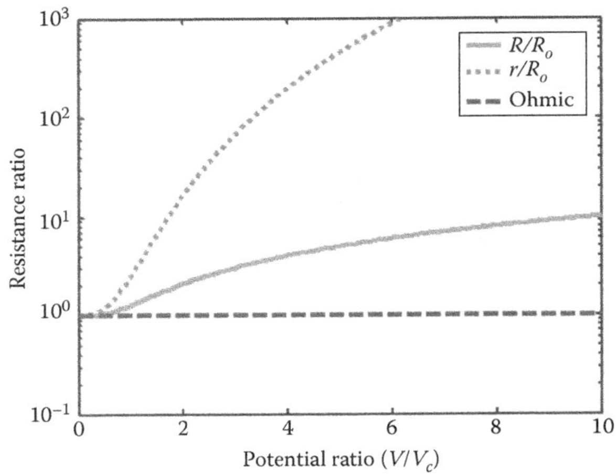

FIGURE 5.4
Resistance blow-up ratio R/R_o and r/R_o as a function of the applied voltage ratio V/V_c with $\gamma = 2.8$.

The trend agrees with Arora's formalism [6], consistent with Equation 5.7. In this form, R and r are given by the applied voltage V or the ensuing electric field $\mathcal{E}$ is given by

$$R = \frac{R_o(V/V_c)}{\tanh(V/V_c)} \qquad r = R_o \cosh^2\left(\frac{V}{V_c}\right) \tag{5.13}$$

The current I in response to the applied voltage V for low dimensions follows naturally [6] from Equation 5.13 by using $I = n_3 qvA_c = n_2 qvW = n_1 qv$ ($n_2 = n_3 T$ is the surface concentration per unit area and $n_1 = n_3 WT$ is the resistance per unit length) for a bulk resistor of cross-sectional area A_c or sheet resistor of cross-sectional width W perpendicular to the current [7]. It is given by

$$I = \frac{V}{R_o}\frac{\tanh(V/d\,V_c)}{V/d\,V_c} \approx \begin{cases} \dfrac{V}{R_o}, & V < V_{cd} \\ I_{sat}, & V \gg V_{cd} \end{cases} \tag{5.14}$$

with

$$I_{sat} = n_3 qv_{sat} A_c = n_2 qv_{sat} W = n_1 qv_{sat} = \frac{V_{cd}}{R_o} \tag{5.15}$$

where I_{sat} is the saturation current for a given resistor that depends on the volume carrier concentration n_3 (or surface concentration n_2 or line concentration n_1). Figure 5.3 shows the current response to the applied voltage obtained from Equation 5.14. The agreement with the experimental data

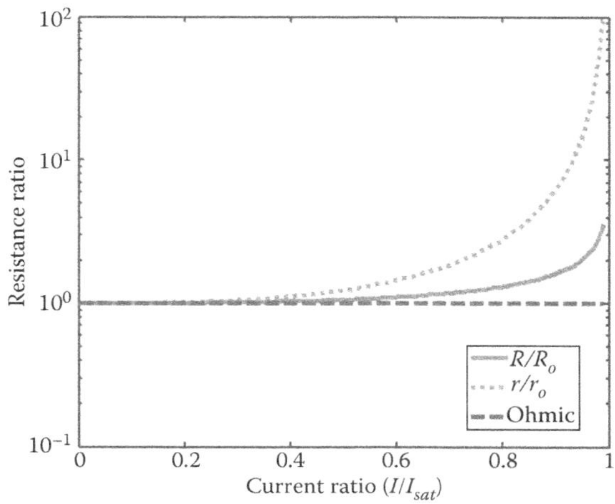

FIGURE 5.5
Resistance blow-up ratio R/R_o and r/R_o as a function of current ratio I/I_{sat} with $\gamma = 2.8$.

(dots) of Greenberg and del Alamo [2] is extremely good. Extreme ohmic and saturation current curves intersect at $V = V_c$. Figures 5.4 and 5.5 show the resistance blow-up effect where resistance rises dramatically as voltage V goes beyond its critical value V_c or approaches saturation current I_{sat}.

The dc resistance $R = V/I$ is distinct from signal (incremental) resistance $r = dV/dI$ at a dc bias point (I, V) when the current response is sublinear. In the ohmic regime $(V < V_c)$, $R = r = R_o$ as current I is a linear function of the applied voltage. The ratio R/R_o as a function of the applied voltage is given by

$$\frac{R}{R_o} = \frac{V/V_c}{\tanh(V/V_c)} = \begin{cases} 1, & V < V_c \\ \dfrac{V}{V_c}, & V \gg V_c \end{cases} \tag{5.16}$$

The incremental (differential) resistance ratio r/R_o for a signal floating at a dc bias point V is given by

$$\frac{r}{R_o} = \cosh^2\left(\frac{V}{V_c}\right) = \begin{cases} 1, & V < V_c \\ \dfrac{\exp(2V/V_c)}{4}, & V \gg V_c \end{cases} \tag{5.17}$$

Equations 5.16 and 5.17, through the use of Equation 5.14 for the current ratio I/I_{sat}, transform to

$$\frac{R}{R_o} = \frac{\tanh^{-1}(I/I_{sat})}{I/I_{sat}} \tag{5.18}$$

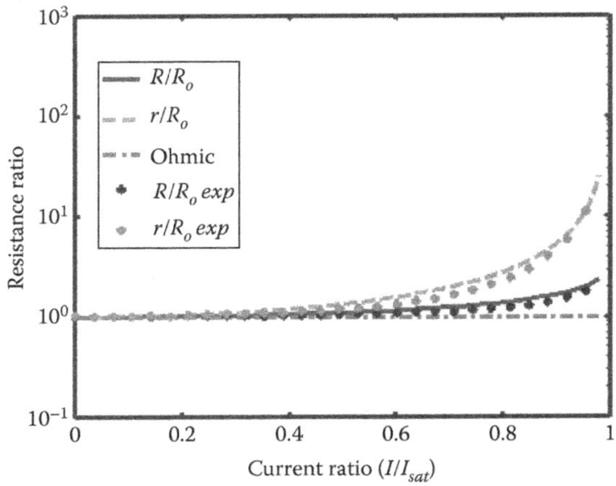

FIGURE 5.6
Comparison of theoretical and experimental resistance ratio versus normalized current.

$$\frac{r}{R_o} = \frac{1}{1-(I/I_{sat})^2} \tag{5.19}$$

The exponential rise of the signal resistance r with V can have deleterious effects for analog and digital signal propagation in a VLSI circuit. Once understood, a circuit designer will be able to customize the signal resistance that can be varied with the dc bias and may have useful applications in a multiplexer, for example. Figure 5.6 shows the comparison of resistance from tanh relation with experimental values as obtained from the empirical formula.

5.4 Charge Transport in 2D and 1D Resistors

The charge transport in the 2D sheet resistor follows the same pattern as in the 3D sample except the surface carrier concentration $n_2 = N/LW$ (in m^{-2}) becomes important; so the current is given by

$$I = \frac{Q}{t} = \frac{n_2(LW)(q)}{t} = n_2\frac{L}{t}qW = n_2 v_D q W \tag{5.20}$$

In the ohmic domain $v = \mp\mu_o\mathcal{E}$, Equation 5.20 reduces to the familiar Ohm's law $I = V/R_o$, but the resistance R_o is now given by

$$R_o = \rho_2\frac{L}{W} \quad \text{with} \quad \rho_2 = \frac{1}{(n_2\mu_{on} + p_2\mu_{op})q} \tag{5.21}$$

where ρ_2 is the sheet resistivity (in $\Omega/\square$). In the saturation regime, $I_{sat} = n_2 q v_{sat} W = (V_{c2}/R_o)$.

Similarly, for 1D nanowires, the current and resistance are given by

$$I = \frac{Q}{t} = \frac{n_1(L)(q)}{t} = n_1 \frac{L}{t} q = n_1 v_D q \tag{5.22}$$

$$R_o = \rho_1 L \quad \rho_1 = \frac{1}{(n_1 \mu_{on} + p_1 \mu_{op})q} \tag{5.23}$$

where n_1 is the carrier concentration per unit length (in m^{-1}) and ρ_1 is the line resistivity (Ω/m). The saturation current $I_{sat} = n_1 q v_{sat} = (V_{c1}/R_o)$.

5.5 Charge Transport in a CNT

Graphene and CNTs follow the same pattern except that the mobility expression is different because of a linear E–k relationship. A rudimentary analysis of the mobility in terms of mfp is to change the mobility expression $\mu_{o\infty} = q\tau/m^* = q\ell_{o\infty}/m^* v$ by replacing $m^* v \rightarrow \hbar k = (E_F - E_{Fo})/v_F$. This gives a simple mobility expression that has been utilized in Reference 8 in extracting mfp. The expression obtained from this analogy is

$$\mu_{o\infty} = \frac{q \ell_{o\infty} v_F}{E_F - E_{Fo}} \tag{5.24}$$

The resistance can then be obtained from Equation 5.21 with $n_2 \rightarrow n_g$ replacement for areal density for graphene and $n_1 \rightarrow n_{CNT}$ for CNT in Equation 5.23. The velocity response to the high electric field is discussed in Reference 8.

NEADF's transformation of equilibrium stochastic velocity vectors into a streamlined mode in extreme nonequilibrium leads to velocity saturation in a towering electric field. In a metallic CNT, the randomly oriented velocity vectors in equilibrium are of uniform Fermi velocity $v_{F0} = 1.0 \times 10^6$ m/s as elucidated in detail in References 6 and 9. The saturation current $I_{sat} = n_{CNT} q v_{Fo}$ arises naturally from this saturation, where $n_{CNT} = 1.53 \times 10^8$ m^{-1} is the linear carrier concentration along the length of the tube consistent with the experimentally observed $I_{sat} = 21$ μA [10]. q is the electronic charge. The carrier statistics [9] gives $E_F = 67.5$ meV which is larger than the thermal energy for all temperatures considered ($T = 4$, 100, and 200 K), making applicable statistics to strongly degenerate. The transition from ohmic to nonohmic saturated behavior initiates at the critical voltage $V_c = (k_B T/q\ell)L$ for nondegenerate statistics with energy $k_B T$ and $V_c = (E_F/q\ell)L$ for degenerate statistics with energy E_F. The mfp ℓ extracted from $R_o = 40$ kΩ is $\ell = 70$ nm that gives mobility [8] $\mu_o = q\ell v_{Fo}/E_F \approx 10{,}000$ cm^2/V · s. The possibility of ballistic transport is miniscule given $\ell \ll L = 1\ \mu$m. The ballistic transport in 2D systems is extensively discussed by Arora et al. [11,12], where it is shown that the ballistic conduction substantially degrades the mobility in a 2D ballistic conductor with length smaller than the ballistic mfp. It may be tempting to apply the same formalism to 1D nanowire or nano-CNT. However, the surge in resistance in a 1D resistor contradicts the expected vanishing resistance for a ballistic conductor. A high-field resistance model [13] that employs the onset of phonon emission consistent with phonon-emission-limited mfp ℓ_Q of Tan et al. [14] explains very well the saturation in 2D GaAs/AlGaAs quantum well. Phonon-emission-limited mfp is generalized to any energy quantum by Arora et al. [8]. ℓ_Q is the distance that a carrier travels before gaining enough energy $q\mathcal{E}\ell_Q = \Delta_Q$ to emit a quantum of energy Δ_Q with the probability of emission given by the Bose–Einstein statistics. $\ell_Q = \Delta_Q/q\mathcal{E}$ is infinite in equilibrium, very large in a low electric field, and a limiting factor only in an extremely high electric field becoming comparable or smaller than the low-field mfp. That is why in the published literature on CNT, it is considered a high-field mfp, distinct from low-field scattering-limited mfp. It is ℓ_Q that was used by Yao et al. [10] to interpret the linear rise $R/R_o = 1 + (V/V_c)$ in resistance with the applied voltage. Here, $R = V/I$ is the direct resistance. This direct resistance R cannot replicate the incremental signal resistance $r = dV/dI$. Therefore, the description of Yao et al. [10] is deficient in not employing the distribution function and hence, does not correctly attribute the source of current saturation, the transition point to current saturation, and the paradigm leading to the rise of direct and incremental resistance.

NEADF has a recipe for nonohmic transport leading to current saturation consistent with velocity saturation. An equal number of electrons has directed velocity moments in and opposite to the electric field ($n_+ = n_- = n/2$) in equilibrium, where n is the total concentration and $n_\pm/n$ is the fraction antiparallel (+) and parallel (–) to the applied electric field. However, in the presence of an electric field $n_+ \gg n_-$ as $n_\pm/n = \exp(\pm q\mathcal{E}\ell/k_BT)$. The fraction of electrons going in the opposite direction to an applied electric field $E = V/L$ is then $\Delta n_+/n = \tanh(q\mathcal{E}\ell/k_BT)$. The current–voltage relation with tanh function ($I = \Delta n_+ qv_{Fo}$) is a derivative of rigorous degenerate statistics [15] with $V_c = (E_F/q\ell)L$ and magnitude of velocity vector equal to the Fermi velocity v_{Fo}. The current–voltage characteristics in a CNT are given by

$$I = I_{sat}\tanh\left(\frac{V}{V_c}\right) \qquad (5.25)$$

Figure 5.7 is a plot of Equation 5.25 along with the experimental data of Yao et al. [10]. Also shown are the lines at temperature $T = 4$, 100, and 200 K following the rigorous degenerate statistics [15].

The distinction between direct $R = V/I$ and differential $r = dV/dI$ mode of resistance is crucial when I–V relation is nonlinear. As discussed earlier, R and r are given by

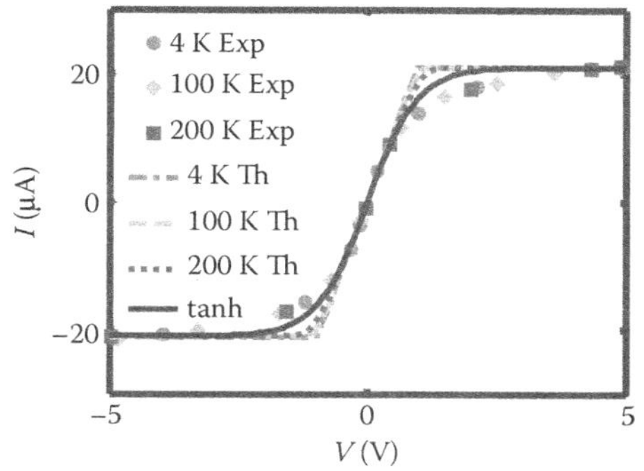

FIGURE 5.7
I–V characteristics of a CNT of length 1 μm. Th stands for theoretical curves derived from degenerate statistics. tanh curves are the display of Equation 5.25.

$$\frac{R}{R_o} = \frac{(V/V_c)}{\tanh(V/V_c)} \tag{5.26}$$

$$\frac{r}{R_o} = \cosh^2\left(\frac{V}{V_c}\right) \tag{5.27}$$

This relationship is in direct contrast to $R/R_o = 1 + (V/V_c)$ with $V_c = I_{sat}R_o$ used by Yao et al. [10], which can be obtained from Equation 5.26 by using approximation $tanh\,(x) \approx x/1 + x$. I_o of Yao et al. is the same as I_{sat}.

As shown in Figure 5.8, the rise in r/R_o is exponential compared to the linear rise in R/R_o. The potential divider rule between the channel and contacts will make the lower-length resistor more resistive [1,5]. Hence, great care is needed to ascertain the critical voltage V_c of the contact and channel regions.

Figure 5.7 makes it clear that both direct slope I/V giving inverse resistance R^{-1} and incremental slope dI/dV giving differential (incremental) resistance r^{-1} decrease as voltage is increased ultimately reaching zero in the regime of saturation. However, in the work cited [8], r is shown to decrease, while conductance dV/dI increases with applied voltage. It may be noted that incremental resistance increases almost exponentially as indicated in Equation 5.9 for $V \gg V_c$ and hence the curves are limited to a mV range to indicate the superlinear surge of incremental resistance. Direct resistance does follow the linear rise with applied voltage. It may be noted that Equations 5.7 and 5.8 are strictly applicable for nondegenerate statistics. These are adaptable to degenerate statistics when the thermal voltage V_t is replaced by Fermi voltage $V_F = (E_F - E_C)/q$, where $E_F - E_C$ is the Fermi energy above the conduction band edge. The curvature effect in Figure 5.7 is the result of ignoring the effect of degeneracy temperature that is much more challenging to define for CNT and graphene as compared to that for a parabolic semiconductor with an effective mass.

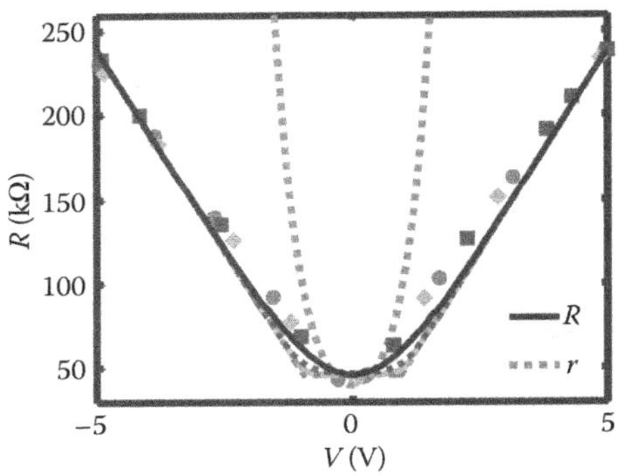

FIGURE 5.8
R–*V* characteristics of a CNT of length 1 μm. Markers and lines have the same legend as in Figure 5.7. The differential resistance *r* (Equation 5.27) rises sharply than the direct resistance *R* (Equation 5.26).

The following observations are made consistent with the experimental data:

1. Ohmic transport is valid when the applied voltage across the length of the channel is below its critical value ($V < V_c$).
2. The transition to the nonlinear regime at the onset of the critical electric field corresponding to energy gained in an mfp is comparable to the thermal energy for nondegenerate statistics and Fermi energy for degenerate statistics [6,16,17].
3. Resistance surge effect in ballistic channels corroborates well with that observed by Yao et al. [10] preceded by what was pointed out by Greenberg and del Alamo [2] in 1994. The surge in the contact region will change the distribution of voltage between contacts and the channel. In this light, Yao et al. [10] correctly conjectured that the measured resistance is a combination of the resistance due to the contacts and the scattering-limited resistance of the CNT channel. The application of NEADF in CNT [9] gives not only the comprehensive overview of the metallic and semiconducting band structure of CNT, but also elucidates the rise of resistance due to the limit imposed on the drift velocity by the Fermi velocity.
4. The onset of quantum emission lowers the saturation velocity. However, if quantum emission is larger than the thermal energy, its effect on transport is negligible [6]. It is important to employ Bose–Einstein statistics [8] to phase in the possible presence of acoustic phonon emissions in addition to optical phonons or, for that matter, photons as transitions are induced by transfer to a higher quantum level induced by an electric field. The phonon emission, generalized to quantum emission with Bose–Einstein statistics, is effective in

lowering the saturation velocity only if the energy of the quantum emission is comparable to the thermal energy. Quantum emission does not affect the ohmic mobility or for that matter ohmic resistance.

5.6 Power Consumption

Velocity saturation limits the current in any resistor. It is clear that the outcomes of Equation 5.7 agree very well with the empirical Equation 5.6. As discussed earlier, the current–voltage (I–V) characteristic based on Equation 5.14 is given by

$$I_{n(p)} = I_{satn(p)} \tanh\left(\frac{V}{V_{cn(p)}}\right) \tag{5.28}$$

where $I_{satn(p)} = (V_{cn(p)}/R_{on(p)})$ is the saturation current. Naturally, the question arises: What is power consumption? Let us take $R_{on} = 33.6\ \Omega$ for an InGaAs resistor with $W/L = 20$. The resistance of two different geometries is the same if W/L ratio is the same. Two resistors are taken with identical W/L and hence R_o, one with $W_1 = 100.0\ \mu m$ and $L_1 = 5.0\ \mu m$ and the other with the same W/L ratio of 20 with $W_1 = 200.0\ \mu m$ and $L_1 = 10.0\ \mu m$. The critical electric field is 3.8 kV/cm if both resistors are fabricated from the same diffused sheet giving $V_{c1} = 1.9$ V and $V_{c2} = 3.8$ V. The power consumed by each of these resistors is given by

$$P = VI_n = VI_{satn} \tanh\left(\frac{V}{V_{cn}}\right) = \frac{V^2}{R_o} \frac{\tanh(V/V_{cn})}{(V/V_{cn})} \tag{5.29}$$

In the ohmic limit ($V < V_c$), the usual ohmic expression is obtained

$$P_o = \frac{V^2}{R_o} \tag{5.30}$$

In the saturation limit ($V \gg V_c$), $P_{sat} = I_{sat}V$ is the linear function of the voltage. Thus, the voltage law changes and the transition occurs at $V = V_c$. This transition is clearly visible in Figure 5.9. The solid line shows quadratic behavior for both resistors, consistent with Equation 5.30. However, when current saturation is considered, the one with a lower length (or lower V_c) consumes less power than the one with a higher length (or higher V_c).

This is important information for circuit designers as the power consumption by the devices on the chip and heat removal is an important feature in the VLSI design.

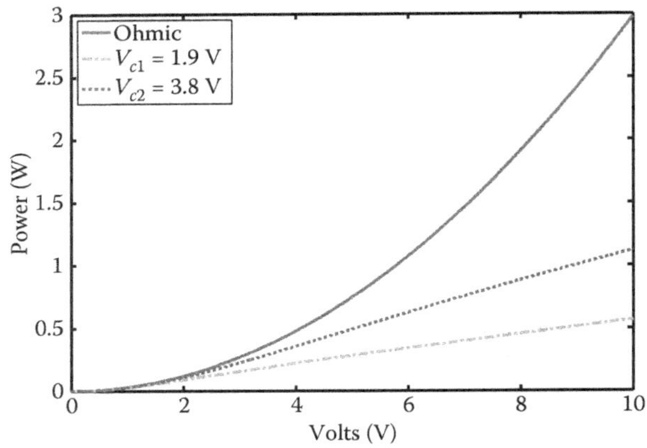

FIGURE 5.9
Power as a function of voltage for two resistors with differing V_c but same ohmic resistance R_o.

5.7 Transit Time Delay

The transit time delay τ_t can be calculated from the conducting channel length L divided by the drift velocity

$$\tau_t = \frac{L}{v_{sat}\tanh(V/V_c)} \approx \begin{cases} \tau_{to} = \dfrac{L^2}{\mu_o V}, & V < V_c \\ \tau_{t\infty} = \dfrac{L}{v_{sat}}, & V \gg V_c \end{cases} \tag{5.31}$$

where v_{sat} is the intrinsic velocity for a sheet resistor that in the degenerate limit [16,17] is estimated to be

$$v_{sat} = \frac{2}{3}\left(\frac{2\pi\hbar^2}{m^*}n_s\right)^{1/2} \tag{5.32}$$

The saturation velocity is $v_{sat} \approx 3.41 \times 10^5$ m/s for an interpolated effective mass $m^* = 0.057\ m_o$ for an $In_{0.15}Ga_{0.85}As$ channel and surface carrier density of $n_s = 10^{12}$ cm^{-2}. The transit time delay as compared to its ohmic counterpart, τ_t/τ_{to}, is given by

$$\frac{\tau_t}{\tau_{to}} = \frac{V/V_c}{\tanh(V/V_c)} \tag{5.33}$$

The transit time delay approaches $\tau_{t\infty} = L_c/v_{sat}$ when $V \gg V_c$. The rise of the relative transit time delay is attributed to decreasing τ_{to} as voltage is increased giving the impression that the transit time delay is reduced. The underestimated ohmic transit time delay in $V > V_c$ regime does not significantly contribute to the overall delay in a nanoscale channel.

5.8 RC Time Delay

In digital signal processing, the transit time delay (τ_t) and RC time constants (τ_{RC}) compete in limiting the speed of a signal. Considerable progress has been made in reducing the transit time delay due to scaling down of the size of the devices that is now in the nanoregime. Efforts are underway to utilize low-resistivity materials and low-*k* dielectrics to shorten the RC time delay. However, there are intrinsic factors that enhance the RC timing delay due to resistance blow up when the step voltage V exceeds the critical voltage V_c for the onset of nonohmic behavior. A prototype RC circuit with a sheet resistor and a capacitive load is shown in Figure 5.10. Figure 5.11 shows the equivalent circuit where a digital voltage signal rises from low to high ($V = 0$ to V at time $t = 0$). The capacitor will not respond to this change in voltage instantaneously. The voltage rises slowly. The current through the resistor responds to change in the voltage as given by

$$i_R(t) = I_{sat} \tanh\left(\frac{v_R(t)}{V_c}\right) \tag{5.34}$$

with

$$i_R(0) = I_{sat} \tanh\left(\frac{V}{V_c}\right) \tag{5.35}$$

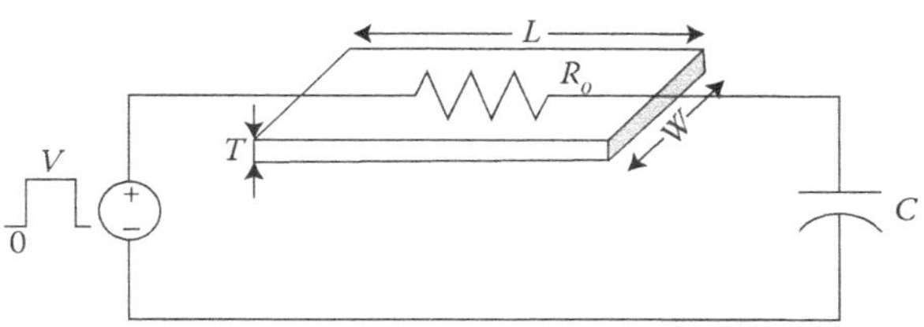

FIGURE 5.10
A prototype *RC* circuits with the resistor of a few nanometers in length.

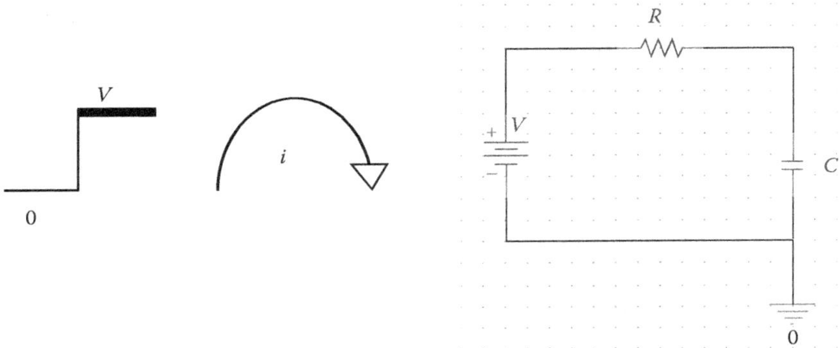

FIGURE 5.11
Equivalent circuit diagram for the *RC* transient circuit as the input is excited from 0 *V* to *V* volts.

The potential *V* is divided across the resistor and the capacitor as given by

$$V = v_R(t) + v_C(t) = V_c \tanh^{-1}\left(\frac{i(t)}{I_{sat}}\right) + \frac{q(t)}{C} \tag{5.36}$$

Differentiating Equation 5.36 with respect to *t* gives

$$0 = V_c \frac{1}{1-(i/I_{sat})^2} \cdot \frac{1}{I_{sat}} \frac{di}{dt} + \frac{i}{C} \tag{5.37}$$

Considering that $R_o = V_c/I_{sat}$, the equation can be written in the differential format

$$\frac{1}{1-(i/I_{sat})^2} \frac{di}{dt} + \frac{i}{R_o C} = 0 \tag{5.38}$$

Separating the variables *i*(*t*) and *t* on each side of the equation and substituting $\tau_o = R_o C$ yields

$$\frac{1}{i} \frac{di}{1-(i/I_{sat})^2} = -\frac{1}{\tau_o} dt \tag{5.39}$$

The denominator in Equation 5.39 can be split into partial fractions. An integration after splitting into partial fractions with the constant of integration ln *K* gives

$$\int di\left[\frac{1}{i}+\frac{1}{2}\frac{1}{I_{sat}-i}-\frac{1}{2}\frac{1}{I_{sat}+i}\right]=-\frac{t}{\tau_o}+\ln K \tag{5.40}$$

Integration yields

$$\ln i-\ln(I_{sat}-i)^{1/2}-\ln(I_{sat}+i)^{1/2}=-\frac{t}{\tau_o}+\ln K \tag{5.41}$$

Equation 5.41 can be further condensed to

$$\ln\left[\frac{i}{K(I_{sat}-i)^{1/2}(I_{sat}+i)^{1/2}}\right]=-\frac{t}{\tau_o} \tag{5.42}$$

or

$$\frac{i}{(I_{sat}^2-i^2)^{1/2}}=Ke^{-t/\tau_o} \tag{5.43}$$

The constant K can be evaluated by invoking the initial condition (5.35). This gives

$$K=\frac{\tanh(V/V_c)}{(1-\tanh^2(V/V_c))^{1/2}} \tag{5.44}$$

Utilizing $\text{sech}^2(x)=1-\tanh^2(x)$ identity yields

$$K=\sinh\left(\frac{V}{V_c}\right) \tag{5.45}$$

Therefore

$$\frac{i}{(I_{sat}^2-i^2)^{1/2}}=\sinh\left(\frac{V}{V_c}\right)e^{-t/\tau_o} \tag{5.46}$$

In the ohmic domain ($i \ll I_{sat}$ and $V \ll V_c$), Equation 5.46 gives

$$\frac{i}{I_{sat}}=\frac{V}{V_c}e^{-t/\tau_o}\Rightarrow i(t)=\frac{V}{R_o}e^{-t/\tau_o} \tag{5.47}$$

Equation 5.46 is organized to solve for $i(t)$

$$i^2=(I_{sat}^2-i^2)\sinh^2\frac{V}{V_c}e^{-2t/\tau_o} \tag{5.48}$$

or

$$i^2\left(1+\sinh^2\frac{V}{V_c}e^{-2t/\tau_o}\right) = I_{sat}^2\sinh^2\frac{V}{V_c}e^{-2t/\tau_o} \tag{5.49}$$

$$i^2 = \frac{I_{sat}^2\sinh^2(V/V_c)e^{-2t/\tau_o}}{1+\sinh^2(V/V_c)e^{-2t/\tau_o}} \Rightarrow i(t) = I_{sat}\frac{\sinh(V/V_c)e^{-t/\tau_o}}{\left(1+\sinh^2(V/V_c)e^{-2t/\tau_o}\right)^{1/2}} \tag{5.50}$$

Equation 5.50 is checked for initial current $i(0)$ to correctly give

$$i(0) = I_{sat}\frac{\sinh(V/V_c)}{(\cosh^2(V/V_c))^{1/2}} = I_{sat}\tanh\frac{V}{V_c} \tag{5.51}$$

Equation 5.50 also gives $i(\infty) = 0$ as expected. In the limit $L \to \infty$, Equation 5.50 reduces to the ohmic expression

$$i(t) = \frac{V}{R_o}e^{-t/\tau_o} \tag{5.52}$$

Hence, the solution of Equation 5.50 satisfies all boundary conditions correctly (Figure 5.12). Figure 5.12 indicates that the initial current $i_{NO}(0)$ is considerably lower than $i_O(0)$ expected from ohm's law.

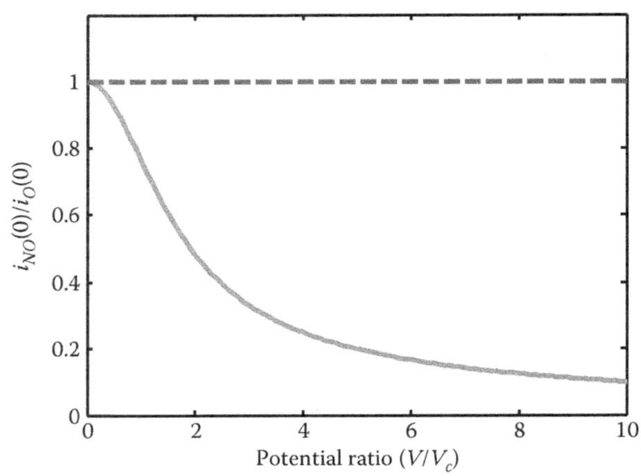

FIGURE 5.12
The ratio of the initial charging current in the nonohmic model to that in the ohmic model. The dashed line represents ohmic current. The relative power ratio follows the same pattern.

Equations 5.50 and 5.52 show that the initial current will be substantially higher in the ohmic model when compared to that obtained from the nonohmic current-saturation-limited model. In the ohmic model, the initial current $i_O(0) = V/R_o$. In the nonohmic model, it will be $i_{NO}(0) = (V_c/R_o)\tanh(V/V_c)$. The ratio of the nonohmic to ohmic initial current and the related power consumption ratio P/P_o as shown in Figure 5.12 is given by

$$\frac{i_{NO}(0)}{i_O(0)} = \frac{P}{P_o} = \frac{\tanh(V/V_c)}{(V/V_c)} \tag{5.53}$$

This ratio approaches 1 in the ohmic regime $V < V_c$ as expected. However, in the regime $V \gg V_c$, the ratio decreases as $(V/V_c)^{-1}$. This drop in the initial current as the capacitor starts charging is shown in Figure 5.13. The power consumption $P = VV_c/R_o$, in the nonohmic regime, is not only smaller but is also a linear function of the applied step voltage when compared to the quadratic behavior in the ohmic regime ($V < V_c$). This transformed behavior affects the figure of merit with trade-off between frequency and power.

Figure 5.13 indicates the charging response of the capacitor $C = 1$ pF connected in series with a resistor of $W = 100$ μm and $R_o = 16.8\Omega$. The approach toward full potential of the capacitor is very slow, especially for the shortest resistor when compared to the ohmic response indicative of a considerable enhancement of the RC timing delay. This observation is consistent with the linear resistance rise with the applied voltage when

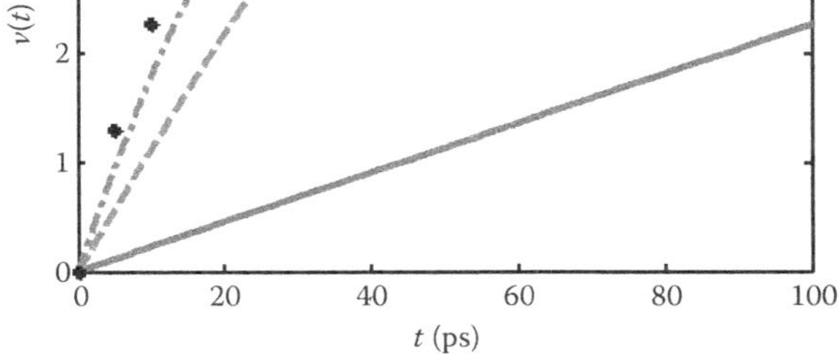

FIGURE 5.13
The response of the capacitor in a nano-RC circuit as voltage is increased from low to high.

$V > V_c$. For an alternate current (ac) signal, the differential resistance rise is even larger.

For comparison, $t = \tau_{RC}$ is defined as the time at which the capacitor potential is $(1 - e^{-1})$ of the higher-logic potential V. The ratio of this transformed time delay to its ohmic value is obtained as

$$\frac{\tau_{RC}}{\tau_{R_oC}} = \ln\left[\frac{\sinh(V/V_c)}{\sinh(V/eV_c)}\right] \tag{5.54}$$

In the regime where $V < V_c$ ($L \to \infty$), the ratio reaches unity. In the other extreme ($L \to 0$), the ratio is

$$\frac{\tau_{RC}}{\tau_{R_oC}}(L \to 0) \approx (1 - e^{-1})\frac{V}{V_c} \tag{5.55}$$

This ratio linearly rises with potential in the nonohmic regime.

Figure 5.14 shows the comparison of two time delays for channels of length $L = 1$, 5, and 20 μm. In the long-channel limit ($L \to \infty$), the transit time delay is $L^2/\mu_o V$. Here, μ_o is the ohmic mobility and $E = V/L$ is the applied electric field. In the short-channel limit ($L \to 0$), it is L/v_{sat} independent of the applied voltage.

The response of a resistive circuit to a digital signal is slow due to the enhancement of the RC time delay. However, power consumption is smaller. A trade-off in the circuit design therefore exists. These results are useful in the extraction of parasitics in the contact regions.

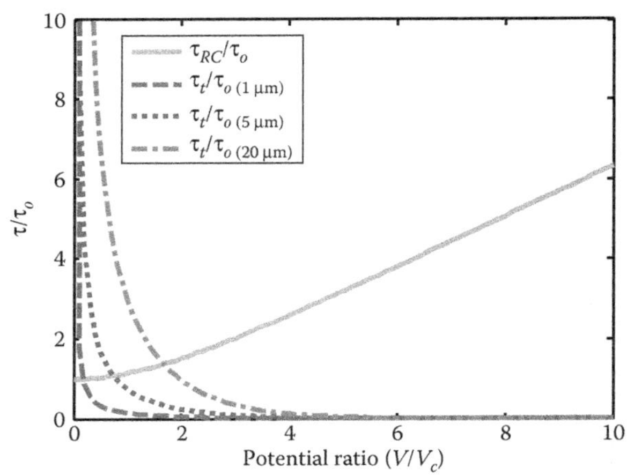

FIGURE 5.14
The normalized RC time delay and transit time delay as a function of normalized applied voltage.

5.9 $\mathcal{L}/R$ Transient Delay

An R–L series circuit excited by a step voltage V at $t = 0$ is shown in Figure 5.15.

The transient response is expressed as

$$V = v_R(t) + v_L(t) = V_c \tanh^{-1}\left(\frac{i(t)}{I_{sat}}\right) + \mathcal{L}\frac{di}{dt} \tag{5.56}$$

In terms of time constant $\tau_{L/R_o} = \mathcal{L}/R_o$, the differential equation can be written as

$$\tau_{L/R_o}\frac{di}{dt} + I_{sat}\tanh^{-1}\left(\frac{i}{I_{sat}}\right) = \frac{V}{R_o} \tag{5.57}$$

with the initial condition

$$i(0) = 0, \quad i(\infty) = I_{sat}\tanh\left(\frac{V}{V_c}\right) \tag{5.58}$$

This differential equation does not have an analytical solution and is solved numerically. The extracted $\tau_{L/R}$ from the numerical solution when compared to its ohmic counterpart $\tau_{L/R_o} = \mathcal{L}/R_o$ is given by the solid line in Figure 5.16.

An analytical solution of Equation 5.57 is tractable if $\tanh(x) \approx x/(1 + x)$ is considered consistent with the empirical model with $\gamma = 1$ [2]. The complete solution is presented in Appendix 5A. With this approximation, $\tau_{L/R}/\tau_{L/R_o}$ is extracted as follows:

$$\frac{\tau_{L/R}}{\tau_{L/R_o}} = \frac{1 + (1 - e^{-1})(V/V_c)}{(1 + V/V_c)^2} \tag{5.59}$$

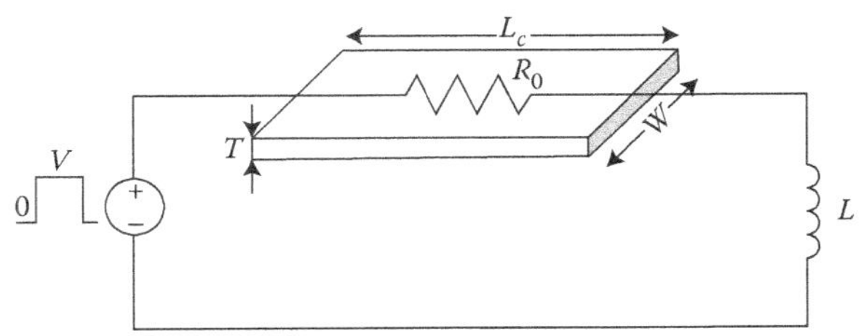

FIGURE 5.15
A schematic RL circuit with the resistor of a few nanometers in length.

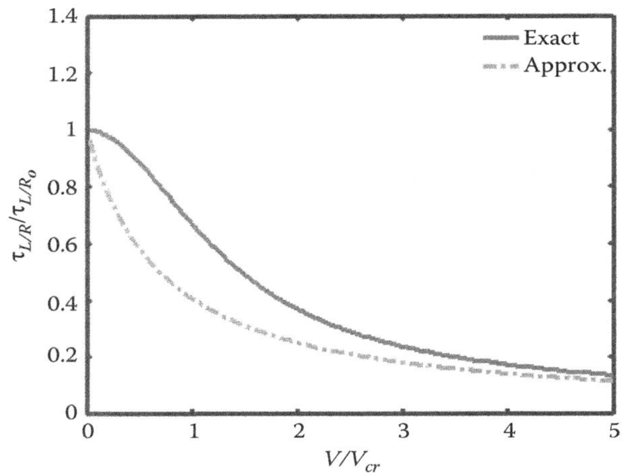

FIGURE 5.16
The normalized *L*/*R* time delay as a function of the normalized applied voltage.

In the ohmic domain ($V < V_c$), the solution of Equation 5.57 agrees well with the well-known result

$$i_o(t) = \frac{V}{R_o}(1 - e^{-t/\tau_{L/R_o}}) \tag{5.60}$$

The steady-state current in the ohmic domain when $t \to \infty$ is $i_o(\infty) = V/R_o$. $\tau_{L/R_o} = \mathcal{L}/R_o$ is the time required for an inductor to attain the inductor current $i(\tau_{L/R_o}) = (1 - e^{-1})i_o(\infty)$, starting from zero at $t = 0$. $\tau_{L/R}$ decreases considerably due to the resistance surge effect. The steady-state current in the nonohmic case is

$$i_{NO}(\infty) = \frac{V}{R_o}\frac{\tanh(V/V_c)}{V/V_c} = \frac{V_c}{R_o}\tanh(V/V_c) \tag{5.61}$$

The new *L*/*R* time constant $t = \tau_{L/R}$ can be analogously defined at which

$$i(\tau_{L/R}) = (1 - e^{-1})i_{NO}(\infty) \tag{5.62}$$

The approximate analytical ratio (lowest dash-dotted curve of Figure 5.17) $\tau_{L/R}/\tau_{L/R_o}$ of Equation 5.59 is consistent with the exact numerical solution especially when $V \gg V_c$. The resistance surge does not offer any serious limitation for an *L*–*R* circuit in the nonohmic domain. Perhaps, that is the reason that inductive coupling is insignificant in VLSI circuits. Efforts are needed to limit the *RC* time delay that appears to be predominant.

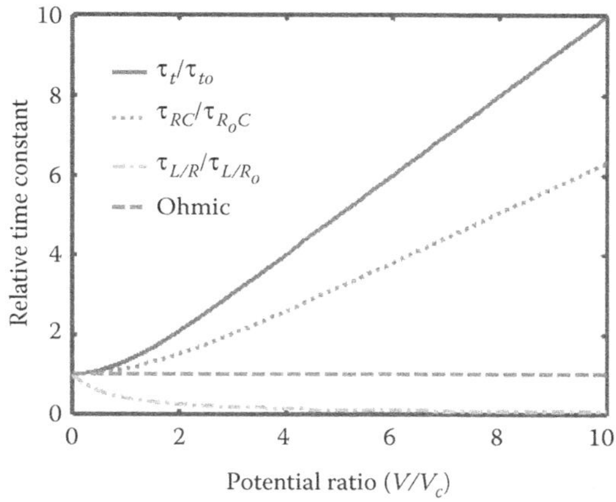

FIGURE 5.17
Relative time constants as a function of normalized voltage. Also shown is the ohmic limit ($\tau_{L/R}/\tau_{L/Ro} = 1$ and $\tau_{RC}/\tau_{R_oC} = 1$) for comparison. *RC* time constant dominates transit time and *L/R* delay. τ_t is in fact constant, however τ_{to} decreases giving impression of ratio τ_t/τ_{to} rising.

It is shown that the transit time and inductive delay will decrease and *RC* time constants will enhance due to the resistance surge effect as resistive channels are being scaled down to the nanometer scale. The rise in τ_t/τ_{to} is not attributed to the rise in τ_t as Figure 5.17 indicates. In fact, for a given length L, $\tau_t = L/v_{sat}$ saturates. As Equation 5.54 clearly indicates, the rise in τ_t/τ_{to} is due to τ_{to} that decreases with the applied voltage. This decrease in τ_{to} may result in the wrong prediction of the high-frequency cutoff $f_o = 1/2\pi\tau_{to}$; the cutoff frequency increases as voltage increases. The limiting value of frequency due to velocity saturation is $f_{sat} = v_{sat}/2\pi L$, independent of the voltage when $V \gg V_c$. Similarly, when *RC* time constant is used to extract parasitic capacitance utilizing the cutoff frequency expression $f = 1/2\pi RC$, $C = 1/2\pi f R_o$ will be higher when $R = R_o$ is utilized. A replacement of R_o with R will give smaller capacitance than estimated from ohmic transport. Resistance surge effect is thus of paramount importance in the extraction of parameters and performance evaluation of high-frequency response. The switch to low-resistivity materials (e.g., copper) and low-*k* dielectric will certainly decrease the value of $\tau_{R_oC} = R_oC$ and hence $\tau_{RC} = (R/R_o)\tau_{R_oC}$. However, the intrinsic enhancement of τ_{RC} due to R/R_o enhancement cannot be ignored as it increases τ_{RC} several fold.

As explained by Hu [18], the cutoff frequency for the unity gain and maximum oscillation frequency for unity power gain are affected by RC time constants. The input resistance consists of gate-electrode resistance and the intrinsic input resistance. As both are of nanoscale, surge effect is bound to take place for both. The power-frequency product that is the figure of merit in VLSI circuits will be likely affected. As shown above, the power consumed in a VLSI circuit is a linear function of the applied voltage across a resistor [5].

5.10 Voltage and Current Division

As current–voltage characteristics become nonlinear and resistance is no longer a constant, it is natural that the familiar voltage and current division rules may not apply. The length of a resistor plays a predominant role in transforming I–V and resistive behavior. Two resistors of length 5 μm ($s_{L1} = 5$) and 10 μm ($s_{L2} = 10$) of the same ohmic value ($R_o = 67.2\ \Omega$) are utilized in the voltage and current divider circuits to follow.

In a potential divider circuit of Figure 5.18, an applied voltage V will not be equally divided as predicted by Ohm's law when the two resistors (R_1 and R_2) are connected in series. The dc voltage (with signal source reduced to zero) across the smaller-length resistor V_1 is obtained from

$$V_{c1} \tanh\left(\frac{V_1}{V_{c1}}\right) = V_{c2} \tanh\left(\frac{V - V_1}{V_{c2}}\right) \quad (5.63)$$

with $V_{c1} = 1.9$ V for the 5-μm resistor and $V_{c2} = 3.8$ V for the 10-μm resistor. The scaling factors for the width are proportionately scaled so that $W/L = 10$. Therefore, $s_{W1} = 50$ and $s_{W2} = 100$.

The voltage V_1 and V_2 appearing across each resistor is shown in Figure 5.19. The 5-μm resistor gets a larger share of the input voltage in the nonohmic regime with $V > V_c$, while leaving the predictions of Ohm's law intact for $V < V_c$. As voltage goes beyond its critical value, most of the applied potential

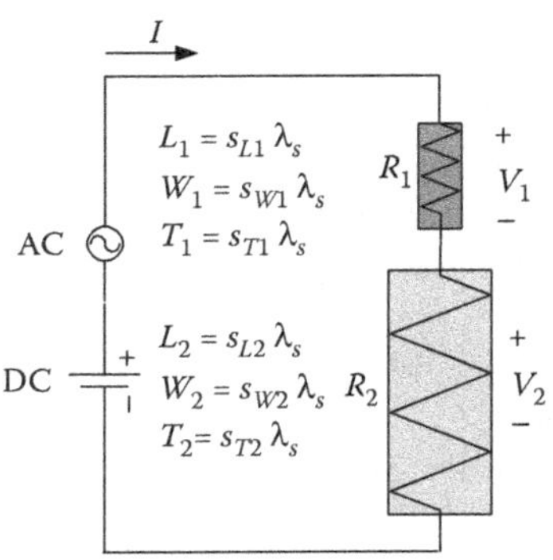

FIGURE 5.18
Voltage divider circuit with two microresistors ($R_{o1} = R_{o2} = 67.2\ \Omega$) of equal $W/L = 10$.

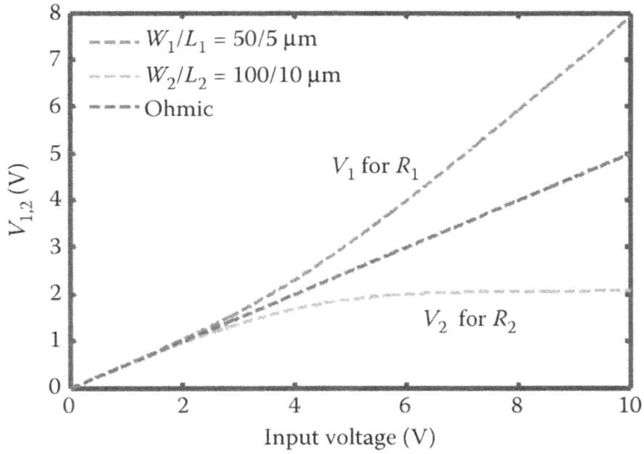

FIGURE 5.19
Voltage division as a function of the applied circuit voltage across two microresistors with the same ohmic value ($R_{o1} = R_{o2} = 67.2\ \Omega$) and $W/L = 10$ connected in series.

goes to the smaller resistor with minuscule incremental change in the voltage across the larger resistor. The circuit designers may advantageously utilize this effect for particular applications in designing VLSI circuits with micro-/nanoscale-conducting channels. Analog and digital signal resistances may transform similarly depending on the bias point of the device. As signal resistance blows up much faster than the dc resistance, virtually all the signals may appear across the smaller-length resistor as shown in Figure 5.20. Therefore, a carefully designed channel may make the effect of parasitic regions negligible for the propagating signal.

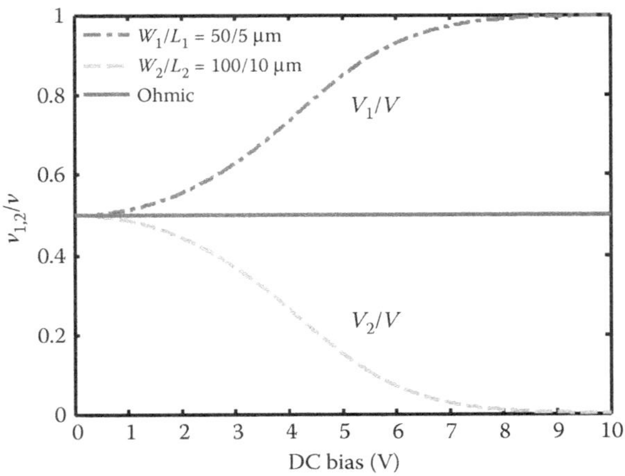

FIGURE 5.20
Signal fraction appearing across each resistor in a series circuit.

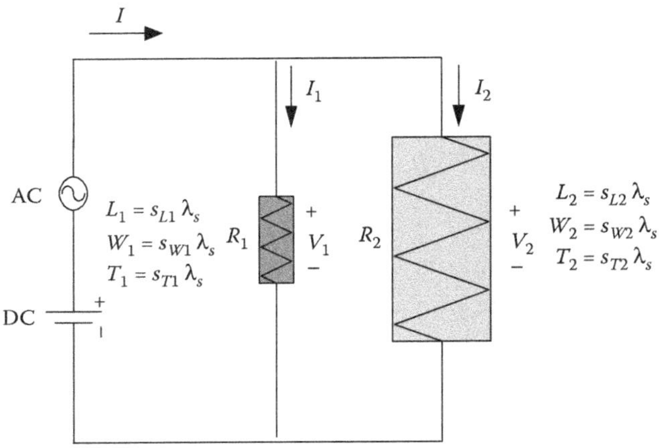

FIGURE 5.21
Current divider circuit with two microresistors ($R_{o1} = R_{o2} = 67.2\ \Omega$).

Figure 5.21 shows a current divider circuit where two resistors of same resistance ($R_{o1} = R_{o2} = 67.2\ \Omega$) are connected in parallel. Figure 5.22 shows that the resulting current in each resistor is substantially below its ohmic value. As V increases beyond V_c, the maximum current that can be drawn from the voltage source is 85 mA when compared to 300 mA predicted from Ohm's law at $V = 10$ V; the larger resistor is less resistive and draws more current. With the current per unit width of 565 μA/μm, the saturation current for resistor 1 is $I_{sat1} = 28$ mA and that for resistor 2 is $I_{sat2} = 56$ mA. The two resistors share equal currents in the ohmic domain ($V < V_c$). However, the current

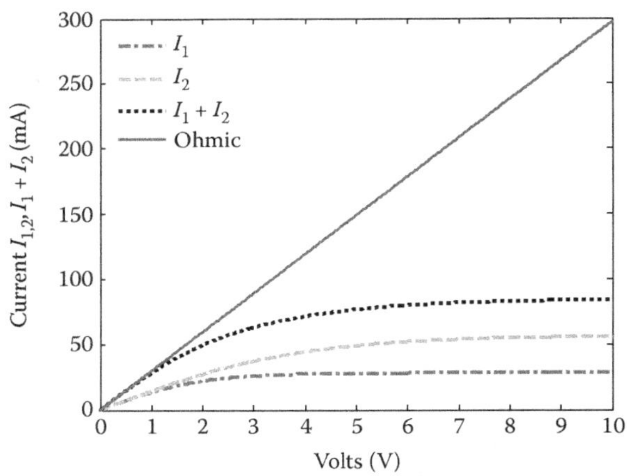

FIGURE 5.22
Ohmic and nonohmic currents in a current divider circuit with two microresistors ($R_{o1} = R_{o2} = 67.2\ \Omega$) and $W/L = 10$ connected in parallel.

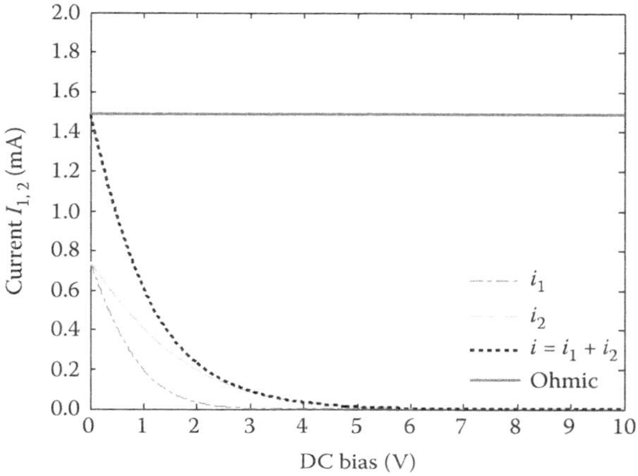

FIGURE 5.23
Signal current in each resistor and the total signal current drawn from the source for two resistors connected in the parallel configuration.

$I_{sat} = V_c/R_o = n_2 q v_{sat} W$ in the saturation region ($V \gg V_c$) is proportional to the width of each resistor. Here, $n_2 = n_3 T$ (m^{-2}) is the surface density of the sheet resistor and v_{sat} is the saturation velocity that relies on temperature and carrier concentration depending on the degeneracy of the resistor, as stated above. The signal current will similarly be affected by this transformation. The signal may pass only through the larger resistor with a negligible portion going through the smaller resistor as shown in Figure 5.23.

EXAMPLES

E5.1 Calculate the resistivity for a uniformly doped bulk silicon sample with 10^{17} donors per cubic centimeter. The mobilities are $\mu_n = 6.3 \times 10^2$ for the majority carriers (electrons) at and $\mu_p = 4.6 \times 10^2$ for minority holes.

$$\sigma_3 = q\mu_n n_3 + q\mu_p p_3$$

The material is n-type, and not degenerately doped; so, $n_3 = 10^{17}$ and $p_3 = n_i^2/n_3 = (1.08 \times 10^{10}\,\text{cm}^{-3})^2/10^{17}\,\text{cm}^{-3} = 1.2 \times 10^3\,\text{cm}^{-3}$

$$\begin{aligned}\sigma_3 &= q\mu_n n_3 + q\mu_p p_3 \\ &= 1.6 \times 10^{-19}(630)(10^{17}) + 1.6 \times 10^{-19}(460)(1.2 \times 10^3) \\ &= 10\ (\Omega\text{-cm})^{-1}\end{aligned}$$

The resistivity ρ is $\rho = 1/\sigma = 0.1\ \Omega$-cm.

Note that the contribution of holes to the conduction is negligible.

E5.2 A voltage of 2.5 V is applied to a sample of silicon whose cross-sectional area is 0.1×1 μm. The length of the path is 0.1 μm. If the material is doped n-type with $N_D = 10^{18}$ cm^{-3}, what is the current in the sample? What is the current density? These dimensions could represent the channel of a field-effect transistor. The majority carrier mobility is $\mu_n \approx 230$ cm^2/V · s.

The cross-sectional area is $A = 0.1 \times 1.0$ μm $= 10^{-9}$ cm^2, and the length is $L = 0.1$ μm $= 10^{-5}$ cm.

The current is given by $I = JA$, where J is the current density per unit area in a 3D semiconductor and $J = \sigma E$. To find σ, we need to find n_3 and p_3. $n_3 = N_D = 10^{18}$ cm^{-3}, and p_3 is negligible so that it would not contribute to conductivity.

$$\sigma = q\mu_n n_3 = (1.6 \times 10^{-19}\,\mathrm{C})(230)(10^{18}) = 37\ (\Omega\text{-cm})^{-1}$$

Thus

$$J = \sigma \mathcal{E} = \sigma \frac{V}{L} = 37\ (\Omega\text{-cm})^{-1} \cdot \frac{2.5\ \mathrm{V}}{0.1 \times 10^{-4}\,\mathrm{cm}} = 9 \times 10^6\ \mathrm{A/cm^2}$$

and

$$I = JA = 9 \times 10^6\ \mathrm{A/cm^2} \cdot 10^{-9}\,\mathrm{cm^2} = 9\ \mathrm{mA}$$

E5.3 Calculate the electron drift velocity for $N_D - N_A = 10^{16}$ cm^{-3} in a bar of Si of cross-sectional area 1.0 mm^2 for a current of 50 mA.

$$I = JA, \quad \text{or} \quad J = \frac{I}{A} = \frac{50 \times 10^{-3}\,\mathrm{A}}{1\,\mathrm{mm^2}(1\,\mathrm{cm^2}/100\,\mathrm{mm^2})} = 5\ \mathrm{A/cm^2}$$

The carrier density is $n = N_D = 10^{16}$ cm^{-3}.

$$v = \frac{J_{n(drift)}}{qn} = \frac{5\ \mathrm{A/cm^2}}{1.6 \times 10^{-19}\,\mathrm{C}\ (10^{16}\,\mathrm{cm^{-3}})} = 3.1 \times 10^3\ \mathrm{cm/s}$$

E5.4 In a layered structure of an InGaAs nanotransitor, the critical value of the electric field is identified to be $\mathcal{E}_c = 0.38$ V/μm and the saturation current per unit gate width is identified to be $I_{sat}/W = 565$ μA/μm. From this information, find the ohmic resistance of a microresistor with $W = 50$ μm and $L = 5$ μm.

$$I_{sat} = V_c/R_o = n_s q v_{sat} W \quad \text{gives} \quad R_o = V_c/I_{sat} = \mathcal{E}_c L/(I_{sat}/W)W = 67.2\ \Omega$$

PROBLEMS

P5.1 For a silicon conductor of length 5 μm, doped n-type at 10^{15} cm^{-3}, calculate the current density for an applied voltage of 2.5 V across the length. How about for a voltage of 2.5 kV? $\mu_n = 1500$ cm^2/V · s, $\mu_p = 500$ cm^2/V · s, and $v_{sat(p)} = 10^5$ m/s.

P5.2 A p-type sample with hole mobility $\mu_{op} = 250$(cm^2/V · s) is doped with $N_A = 10^{17}$ cm^{-3} acceptors. What is the resistivity of the sample?

P5.3 Show the electron drift velocity in pure Si ($\mu_n = 1400$(cm^2/V · s) = 0.14(m^2/V · s)) for (100 V/cm) is less than v_{th3} at room temperature. Comment on the electron drift velocity for 10^8 V/m.

P5.4 Calculate the speed of an electron in Si with kinetic energy 0.013 eV. Draw an equilibrium energy band diagram for silicon and indicate where this electron will be. Compare your calculated thermal speed to the typical drift velocities $v_n = \mu_{on}\mathcal{E}$ obtained, say, for 5 V applied across a 1-cm resistor. How does it compare to typical saturation velocities?

P5.5 Calculate the resistivity for a uniformly doped silicon sample with 10^{17} donors per cubic centimeter. The mobility for the doped silicon at 10^{17} per cubic centimeter density is 630 cm^2/V · s.

P5.6 A voltage of 2.5 V is applied to a sample of silicon whose cross-sectional area is 0.1 × 0.1 μm. The length of the path is 0.1 μm. If the material is doped n-type with $N_D = 10^{18}$ cm^{-3}, what is the current in the sample? What is the current density? These dimensions could represent the channel of a field-effect transistor.

P5.7 Calculate the electron drift velocity for $N_D - N_A = 10^{16}$ cm^{-3} in a bar of Si of cross-sectional area 1.0 mm^2 for a current of 50 mA.

P5.8 Compare the mean free time between collisions for electrons and for holes in intrinsic GaAs. How do these values compare to those for silicon? Using the thermal velocity v_{th3}, estimate the mfp of electrons and holes in GaAs and Si.

P5.9 A test resistor of the following figure, in the form of a sheet, is of length $L = 50$ μm, width $W = 5$ μm, and thickness $T = 2$ μm. It is doped with donor impurities $N_D = 2 \times 10^{17}$ cm^{-3}. All donors are ionized.

a. When 1 V is applied across the length of the resistor, a current of 0.5 mA flows through it. What is the resistance R of the resistor?

b. What is its resistivity ρ and conductivity σ?

c. Estimate the electron mobility in the sample.

d. Estimate the drift velocity of the electrons when current as in (a) is flowing through it.

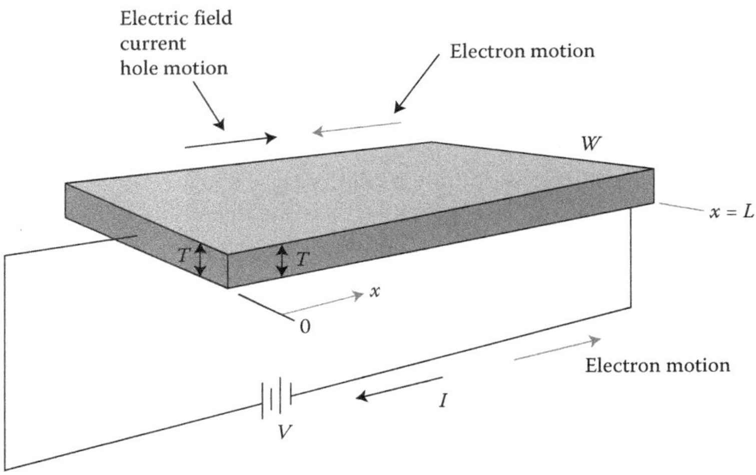

Test register of Problem 5.9.

P5.10 a. It is well known that Ohm's law breaks down when the potential V across a resistor is larger than the critical electric field $V_c = (V_t/\ell_o)L$. Calculate the critical voltage (in V) for electrons in a silicon resistor of length $L = 1$ μm with the mfp $\ell_o = 0.1$ μm at room temperature.

b. Calculate the current in a resistor of ohmic value $R_o = 10\ \Omega$ when a voltage of 10 V is applied across a 1-μm resistor assuming Ohm's law is valid.

c. Recalculate the current assuming Ohm's law is not valid.

CAD/CAE PROJECTS

C5.1 *I–V Characteristics.* In this project, you will read the paper on Velocity Saturation in the Extrinsic Device: A Fundamental Limit in HFETs, by David R. Greenberg and Jesus A. del Alamo published in the *IEEE Transactions on Electron Devices,* Vol. 41, No. 8, August 1994. You will also plot *I–V* characteristics using the empirical Equation 5.8 along with the theoretical Equation 5.24. Also, discuss the approximations in the linear (ohmic) limit and in the saturation limit. Plot these two limiting expressions on the same graph. You can use material parameters for Si with reasonable estimates or InGaAs as determined by Greenberg and del Alamo.

C5.2 *Resistance surge for a microcircuit.* The critical field for an $In_{0.15}Ga_{0.85}As$ conducting channel has been identified to be $\mathcal{E}_c = 3.8$

kV/cm and saturation current is $I_{sat}/W = 565\ \mu A/\mu m$, as outlined in a paper titled, Velocity Saturation in the Extrinsic Device: A Fundamental Limit in HFETs, by David R. Greenberg and Jesus A. del Alamo published in the *IEEE Transactions on Electron Devices*, Vol. 41, No. 8, August 1994. There is an empirical relation stated where $\gamma = 2.8$. The effective mass can be extracted by linear interpolation between InAs and GaAs.

a. Plot I/I_{sat} as a function of V/V_c with theoretical and empirical results along with curves for linear and saturation limits.
b. Plot normalized plots of R/R_o versus V/V_c (and another plot vs. I/I_{sat}) along with the ohmic value, both for direct and incremental resistance. Also, plot the graph of ohmic range when $V/V_c < 1$ and for the regime when $V/V_c \gg 1$. Discuss the functional behavior by a comparative study of theoretical and empirical relationships.
c. Take some reasonable value of a microresistor, say $W = 1000$ nm and $L = 100$ nm. Plot the graphs of resistance R of a nanoresistor versus voltage V and R versus the channel current.
d. Plot power P consumed by this resistor along with its counterpart of what is obtained from Ohm's law. Discuss your findings.

C5.3 *Kirchhoff's current law (KCL) and Kirchhoff's voltage law (KVL) applications.* Taking the parameters from C5.2, continue with 2–3 resistor circuits. You have to find the voltage drop and current flowing through each resistor as a function of source voltage. Discuss how resistance surge can affect dc and ac currents flowing through the resistors. Take a reasonable value, say for one $W_1 = 1000$ nm, $L_1 = 100$ nm and the other $W_2 = 1000$ nm, $L_2 = 200$ nm, and the third one may be $W_3 = 2000$ nm, $L_3 = 200$ nm. What differences do you see if you connect two resistors in series or parallel to the voltage source of $V = 0$–10 V. How do dc and signal values change? What is the amount of current drawn for the source? What is the power consumed by each resistor as a function of V?

C5.4 *RC, L/R, and transit time delay.* In this chapter, you discussed the transit time delay and RC delay when a nanoresistor of C5.2 is connected in series with a capacitor of say 1 ps. Plot these time constants as a function of voltage along with their ohmic and asymptotic values for high voltages. What conclusions do you draw regarding the effect of resistance surge on the time constants? Does Ohm's law hold? In what range? And so on.

Appendix 5A: Derivation of the *L*/*R* Time Constant

The transient response is expressed as

$$V = v_R(t) + v_L(t) = V_c \tanh^{-1}\left(\frac{i(t)}{I_{sat}}\right) + \mathcal{L}\,\frac{di}{dt} \tag{5A.1}$$

In terms of the time constant $\tau_{L/R_o} = \mathcal{L}/R_o$, the differential equation can be written as

$$\tau_{L/R_o}\frac{di}{dt} + I_{sat}\tanh^{-1}\left(\frac{i}{I_{sat}}\right) = \frac{V}{R_o} \tag{5A.2}$$

with the initial condition

$$i(0) = 0, \quad i(\infty) = I_{sat}\tanh\left(\frac{V}{V_c}\right) \tag{5A.3}$$

Using the approximation of $i(t) = I_{sat}\tanh\left(\frac{v_R(t)}{V_c}\right) \approx I_{sat}\frac{(v_R(t)/V_c)}{1+(v_R(t)/V_c)}$, we get

$$v_R(t) = V_c\frac{(i(t)/I_{sat})}{1-(i(t)/I_{sat})} \tag{5A.4}$$

Using Equation 5A.4 in Equation 5A.1 gives

$$\mathcal{L}\,\frac{di}{dt} + V_c\frac{(i/I_{sat})}{1-(i/I_{sat})} = V \tag{5A.5}$$

Substituting $x = (i/I_{sat})$ transforms Equation 5A.5 into

$$I_{sat}\,\mathcal{L}\,\frac{dx}{dt} + V_c\frac{x}{1-x} = V \tag{5A.6}$$

Rewriting in the form of a differential equation yields

$$\frac{dx}{dt} + \frac{V_c}{I_{sat}\mathcal{L}}\frac{x}{1-x} = \frac{V}{I_{sat}\mathcal{L}} \tag{5A.7}$$

Substituting $a = (V_c/I_{sat}\mathcal{L})$ and $b = (V/I_{sat}\mathcal{L})$ yields the equation in a compact form

$$\frac{dx}{dt} + a\frac{x}{1-x} = b \tag{5A.8}$$

Separation of variables and integration yields

$$\frac{dx}{b - a(x/1-x)} = dt \quad \text{or} \quad \int_0^{x(t)} \frac{dx}{b - a(x/1-x)} = \int_0^t dt \tag{5A.9}$$

The reorganization of the integral leads to

$$\int_0^{x(t)} \frac{(1-x)dx}{b-(a+b)x} = t \tag{5A.10}$$

It can be broken into two terms yielding

$$\int_0^{x(t)} \frac{1}{b-(a+b)x}dx - \int_0^{x(t)} \frac{x}{b-(a+b)x}dx = t \tag{5A.11}$$

Integration yields

$$-\frac{1}{a+b}\ln(b-(a+b)x) + \frac{1}{a+b}\left[x - b\frac{\ln(b-(a+b)x)}{-(a+b)}\right] = t \tag{5A.12}$$

Further simplification yields

$$-\frac{a}{(a+b)^2}\ln\left[1 - \frac{a+b}{b}x\right] + \frac{x}{a+b} = t \tag{5A.13}$$

Rewriting a and b gives

$$a = \frac{V_c}{I_{sat}\mathcal{L}} = \frac{R_o}{\mathcal{L}_{ind}} = \frac{1}{\tau_{L/R_o}}, \quad b = \frac{V}{I_{sat}} = \frac{R_o}{\mathcal{L}}\frac{V}{V_c} = \frac{1}{\tau_{L/R_o}}\frac{V}{V_c}, \quad \text{and} \quad \frac{b}{a} = \frac{V}{V_c} \tag{5A.14}$$

With these values of a and b, Equation 5A.13 transforms into

$$-\tau_{RLo}\frac{1}{(1+(V/V_c))^2}\ln\left[1-\left(1+\frac{V_c}{V}\right)\frac{i(t)}{I_{sat}}\right]+\tau_{RLo}\frac{(i(t)/I_{sat})}{1+(V/V_c)}=t \tag{5A.15}$$

Now, substituting at $t=\tau_{L/R}$

$$i(\tau_{L/R})=(1-e^{-1})I_{sat}\frac{(V/V_c)}{1+(V/V_c)} \tag{5A.16}$$

gives linking of $\tau_{L/R}$ to τ_{L/R_o} by

$$\begin{aligned}\tau_{L/R}&=-\tau_{L/R_o}\frac{1}{(1+(V/V_c))^2}\ln\left[1-(1-e^{-1})\left(1+\frac{V_c}{V}\right)\frac{(V/V_c)}{1+(V/V_c)}\right]\\&\quad+\tau_{L/R_o}(1-e^{-1})\frac{(V/V_c)}{(1+(V/V_c))^2}\end{aligned} \tag{5A.17}$$

A further simplification yields

$$\frac{\tau_{L/R}}{\tau_{L/R_o}}=\frac{1+(1-e^{-1})(V/V_c)}{(1+(V/V_c))^2} \tag{5A.18}$$

References

1. M. L. P. Tan, T. Saxena, and V. K. Arora, Resistance blow-up effect in micro-circuit engineering, *Solid State Electronics*, 54, 1617–1624, Dec. 2010.
2. D. R. Greenberg and J. A. d. Alamo, Velocity saturation in the extrinsic device: A fundamental limit in HFETs, *IEEE Transactions of Electron Devices*, 41, 1334–1339, 1994.
3. I. Saad, M. L. P. Tan, I. Hui Hii, R. Ismail, and V. K. Arora, Ballistic mobility and saturation velocity in low-dimensional nanostructures, *Microelectronics Journal*, 40, 540–542, 2009.
4. M. T. Ahmadi, M. L. P. Tan, R. Ismail, and V. K. Arora, The high-field drift velocity in degenerately-doped silicon nanowires, *International Journal of Nanotechnology*, 6, 601–617, 2009.
5. T. Saxena, D. C. Y. Chek, M. L. P. Tan, and V. K. Arora, Microcircuit modeling and simulation beyond Ohm's law, *IEEE Transactions on Education*, 54, 34, Feb. 2011.
6. V. K. Arora, Theory of scattering-limited and ballistic mobility and saturation velocity in low-dimensional nanostructures, *Current Nanoscience*, 5, 227–231, May 2009.

7. V. K. Arora, High-electric-field initiated information processing in nanoelectronic devices, in *Nanotechnology for Telecommunications Handbook*, S. Anwar, Ed. Oxford, UK: CRC/Taylor & Francis Group, pp. 309–334, 2010.
8. V. K. Arora, M. L. P. Tan, and C. Gupta, High-field transport in a graphene nanolayer, *Journal of Applied Physics*, 112, 114330, 2012.
9. V. K. Arora and A. Bhattacharyya, Cohesive band structure of carbon nanotubes for applications in quantum transport, *Nanoscale*, 5, 10927, 2013.
10. Z. Yao, C. L. Kane, and C. Dekker, High-field electrical transport in single-wall carbon nanotubes, *Physical Review Letters*, 84, 2941–2944, 2000.
11. V. K. Arora, Ballistic transport in nanoscale devices, presented at the *MIXDES 2012:19th International Conference MIXED Design of Integrated Circuits and Systems*, Warsaw, Poland, 2012.
12. V. K. Arora, M. S. Z. Abidin, S. Tembhurne, and M. A. Riyadi, Concentration dependence of drift and magnetoresistance ballistic mobility in a scaled-down metal–oxide–semiconductor field-effect transistor, *Applied Physics Letters*, 99, 063106, 2011.
13. P. H. S. Wong and D. Akinwande, *Carbon Nanotube and Graphene Device Physics*. Cambridge: Cambridge University Press, 2011.
14. L. S. Tan, S. J. Chua, and V. K. Arora, Velocity-field characteristics of selectively doped $GaAs/Al_xGa_{1-x}As$ quantum-well heterostructures, *Physical Review B*, 47, 13868–13871, 1993.
15. V. K. Arora, D. C. Y. Chek, M. L. P. Tan, and A. M. Hashim, Transition of equilibrium stochastic to unidirectional velocity vectors in a nanowire subjected to a towering electric field, *Journal of Applied Physics*, 108, 114314–114318, 2010.
16. M. L. P. Tan, V. K. Arora, I. Saad, M. T. Ahmadi, and R. Ismail, The drain velocity overshoot in an 80 nm metal–oxide–semiconductor field-effect transistor, *Journal of Applied Physics*, 105, 074503, Apr. 1 2009.
17. V. K. Arora, M. L. P. Tan, I. Saad, and R. Ismail, Ballistic quantum transport in a nanoscale metal–oxide–semiconductor field effect transistor, *Applied Physics Letters*, 91, 103510, 2007.
18. C. C. Hu, *Modern Semiconductor Devices for Integrated Circuits*. Upper Saddle River, NJ: Pearson/Prentice-Hall, 2010.

6

Nano-MOSFET and Nano-CMOS

The advances in ULSI technology are heavily based on downscaling of the minimum feature size of a metal-oxide-semiconductor field-effect-transistor (MOSFET) [1,2]. NMOS (n-type MOSFET) and PMOS (p-type MOSFET) pair comprises a CMOS inverter circuit. The recent evolution of nanotechnology may provide challenges and opportunities for novel devices, such as single-electron devices, carbon nanotubes, Si nanowires, and new materials. The utilization of quantum effects and ballistic transport characteristics may also provide novel functions for silicon-based devices. Among various candidate materials for nanometer-scale devices, silicon nanodevices are particularly promising because of the existing silicon process infrastructure in semiconductor industries, the compatibility to CMOS circuits, and a nearly perfect interface between the natural oxide and silicon. Nanoscale MOSFETs are particularly of great interest to discover fundamental physics as well as applications to the technology for product development. MOSFETs are grown on (100)-substrate for a better silicon dioxide (SiO_2) interface.

6.1 Primer

The mobility and saturation velocity are the two important parameters that control the charge transport in a conducting MOSFET channel [3]. The mobility is degraded by both the gate electric field and the channel electric field; the former is due to quantum confinement and the latter is due to streaming of velocity vectors in the intense driving electric field. The saturation velocity in the channel is ballistic, which is limited to the thermal velocity for nondegenerate carriers and to the Fermi velocity for degenerate carriers in the inversion regime. The drain-end carrier velocity is always smaller than the ultimate saturation velocity due to the presence of the finite electric field at the drain. The popular channel pinchoff assumption is reexamined for either a long or SC. Channel conduction beyond quasi-pinchoff arises from an increase in the drain velocity as a result of the enhanced electric field as the drain voltage is increased, giving an alternative description of the channel-length modulation. The current–voltage characteristics of a nanoscale MOSFET in the inversion regime is the subject of this chapter.

The MOSFET of Figure 6.1 is a vehicle for the design of an integrated circuit both for digital and analog applications. It has a long history of

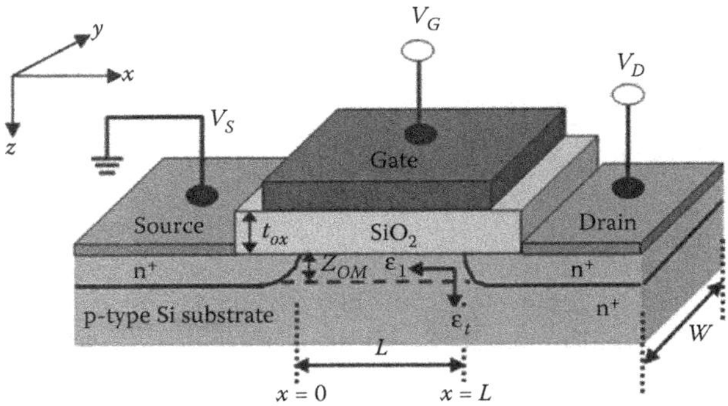

FIGURE 6.1
Basic structure of an n-channel MOSFET with electrons removed from the Si/SiO_2 interface due to the quantum-confinement effect.

channel length being scaled down that is now in the deca–nanometer regime. The fundamental processes that control the performance of the MOSFET channel continue to elude physicists and engineers alike. Figure 6.1 demonstrates that in a nanoscale channel, where oxide thickness is a few nm, the separation z_{QM} of electrons from the interface due to the quantum-confinement effect cannot be ignored. The gate electric field E_t does not heat electrons as it is not an accelerating field; rather, it is a confining electric field that makes an electron a quantum entity described by the wave character with discrete (digitized) energy levels. The wavefunction vanishes at the Si/SiO_2 interface and peaks at a distance approximately z_{QM} away from the interface. This alters the gate capacitance and hence the carrier density in the channel.

MOSFET channel, like any other conducting channel, is created by the application of gate voltage V_{GS} that is above the threshold voltage for the channel to grow as the overdrive voltage $V_{GT} = V_{GS} - V_T$ is increased. Care should be taken not to confuse V_T with thermal voltage V_t that also appears in the subthreshold swing. This is the principle of MOS-capacitor (MOS-C) discussed in the next section. The charge in the channel flows due to the drain voltage applied between the gate and the source. Normally, the source is put at the ground potential ($V_S = 0$). V_T is adjustable by applied voltage to the body of the MOSFET with respect to the source. For a carefully designed MOSFET for threshold voltage, the source and the body are tied together internally, making MOSFET a three-terminal device (source, drain, and gate).

6.2 MOS Capacitor

The metal (or polysilicon)–oxide–semiconductor sequence without source and drain contacts forms an MOS-C capacitor (C is for capacitor not to be

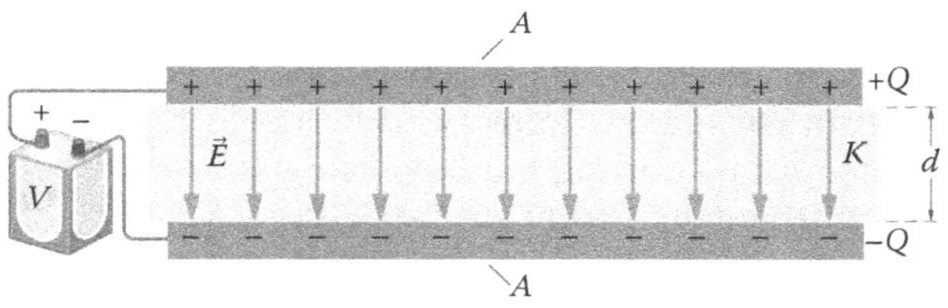

FIGURE 6.2
A MOSFET capacitor depicted as the top plate (polysilicon/metal) as a gate with induced electron gas on the lower end. The spacing $d \approx t_{ox} + (\varepsilon_{ox}/\varepsilon_{si})z_{QM}$ (t_{oxeff} below) consists of two components: oxide thickness t_{ox} and quantum mechanical separation corrected for the difference in the dielectric constant of Si and SiO_2.

confused with CMOS to come later) where charge is induced in the substrate channel through the application of the gate voltage V_{GS} that is larger than the threshold voltage V_T. A simplified model of the capacitor is shown in Figure 6.2 [4].

Considering Figure 6.2, the charge Q on each side ($Q_- = Q_+ = Q$) of the capacitor is

$$Q = CV, \quad C = \frac{\varepsilon_s A}{d} = \frac{\varepsilon_o \kappa A}{d} \tag{6.1}$$

where C is the capacitance of the capacitor, A is the area of the charge-carrying sheets, $\kappa = \varepsilon_r = \varepsilon/\varepsilon_o$ is the dielectric constant (relative permittivity), and d is the spacing between the charge sheets. $\varepsilon_r = \varepsilon/\varepsilon_o$, the relative permittivity, is the modern way of describing the dielectric constant in SI units. ε_o is the permittivity of the free space. The charge per unit area $Q_A = Q/A$ is preferred for use in the MOS-C because it is related to charge density per unit area ($Q_A = n_2 q$). Equation 6.1 for charge per unit area is then

$$Q_A = C_A V, \quad C_A = \frac{C}{A} = \frac{\varepsilon_s}{d} = \frac{\varepsilon_o \kappa}{d} \tag{6.2}$$

In an NMOS channel, the substrate is p-type that gets depleted of holes leaving behind a negatively charged depletion layer on which the itinerant (mobile) electrons float in the form of a sheet. This is called the inversion layer as polarity is inverted. The inversion layer floats at a distance z_Q from the interface as will be discussed below. In that case, the gate capacitance $C_{Aox} = \varepsilon_{ox}/t_{ox}$ is in series with quantum capacitance $C_{AQ} = \varepsilon_{si}/z_{QM}$ giving areal gate capacitance

$$C_{AG} = \frac{C_{Aox} C_{AQ}}{C_{Aox} + C_{AQ}} = \frac{\varepsilon_{ox}}{d}, \quad d = t_{oxeff} = t_{ox} + \frac{\varepsilon_{ox}}{\varepsilon_{Si}} z_{QM} \approx t_{ox} + \frac{1}{3} z_{QM} \tag{6.3}$$

For a polysilicon gate, there is a penetration of the wave function on the gate side as well. Considering that the polysilicon gate is heavily doped, perhaps, that quantum capacitance effect can be neglected. Otherwise, it can be easily accounted for by adding the increasing effective z_{QM} by wavefunction penetration in the gate. Quantum capacitance should not be confused with the depletion layer capacitance $C_{AD} = \varepsilon_{Si}/W_D$ due to the formation of the depletion layer in the substrate as it does not affect the charge in the channel, but could be important for the signal through the substrate. In a nano-MOS, since gate oxide t_{ox} is very small, the quantum thickness z_{QM} cannot be neglected.

The channel in a nano-scale MOSFET is indeed a quantum one that is constrained by the gate electric field forming an approximately linear quantum well (see Figure 6.3). Contrary to the belief of many naïve researchers, no heating of the electrons by the gate is possible. Rather, the gate confines the electron to length z_{QM} comparable to the de Broglie wavelength. The energy spectrum is digital (quantum) in the z-direction perpendicular to the gate, while the other two Cartesian directions with lengths L_x and L_y are analog (or classical). This confinement makes the channel quasi-2-dimensional (Q2D) [5–7]:

$$\varepsilon_{ki} = \frac{\hbar^2 k_x^2}{2m_1^*} + \frac{\hbar^2 k_y^2}{2m_2^*} + \varepsilon_i \tag{6.4}$$

with

$$\varepsilon_i = \xi_i E_o \approx \left[\frac{\hbar^2}{2m_3}\right]^{1/3} \left[\frac{3\pi q}{2} \mathscr{E}_t \left(i + \frac{3}{4}\right)\right]^{2/3}, \quad i = 0,1,2,3 \tag{6.5}$$

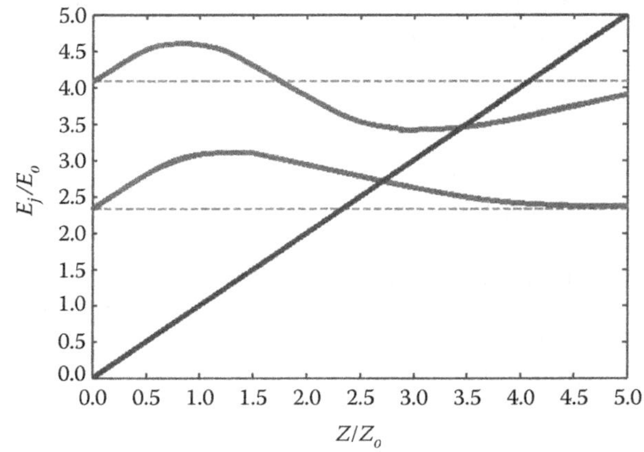

FIGURE 6.3
Electron distribution in the first two quantized levels.

$$E_o = \left(\frac{\hbar^2 q^2 \mathcal{E}_t^2}{2m_3^*}\right)^{1/3} \tag{6.6}$$

where $k_{x,y}$ is the momentum vector in the analog 2D x–y-plane and ε_i is the quantized energy in the digitized z-direction.

As shown in Figure 6.4, $m_{1,2}^*$ is the effective mass in the x–y-plane of the Q2D channel and m_3^* is the effective mass in the z-direction for a given conduction valley, assuming (100) Si/SiO_2 interface. ξ_i are the zeros of the Airy function ($Ai(-\xi_i) = 0$) with $\xi_0 = 2.33811$, $\xi_1 = 4.08795$, and $\xi_2 = 5.52056$. For (100) Si MOSFET, the conduction band energy surfaces are six ellipsoids with a longitudinal direction along $\pm x,y,z$ in k-space. The two valleys have $m_3^* = m_\ell = 0.916m_o$ and four valleys with $m_3^* = m_t = 0.198m_o$. $\varepsilon_o = \xi_0 E_o$ is the ground-state energy corresponding to $i = 0$ with $m_3 = 0.916\ m_o$ for the two valleys. The other four valleys are not occupied in the quantum limit when all electrons are in the lowest energy state. $m_{1,2}^* = 0.198m_o$ is the conductivity effective mass in the x–y-plane of the QTD channel for the lower two valleys (two vertical valleys with circular projection on x–y plane in Figure 6.4). E_t is the electric field generated by the gate, which in the strong inversion regime is given by [8]

$$\mathcal{E}_t \approx \frac{V_{GT} + V_T}{6t_{ox}} \tag{6.7}$$

Here, $V_{GT} = V_{GS} - V_T$ is the overdrive gate voltage above the threshold voltage V_T and t_{ox} is the thickness of the gate oxide. Eigenfunctions corresponding to the eigenvalues in Equation 6.1 are given by

$$\psi(x, y, z) = \frac{1}{\sqrt{L_x L_y}} e^{j(k_x x + k_y y)} Z_i(z) \tag{6.8a}$$

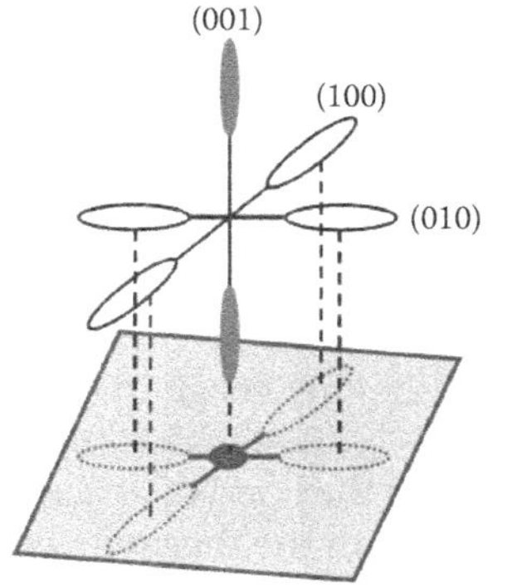

FIGURE 6.4
The populated valley (filled) with quantum confinement in the z-direction with projection on the x–y-plane with an isotropic effective mass $m_t = 0.19m_o$.

$$Z_i(z) = \frac{1}{Ai'(-\xi_i)z_o^{1/2}} Ai\left(\frac{z}{z_o} - \xi_i\right) \tag{6.8b}$$

$$z_o = \frac{E_o}{q\mathcal{E}_t} \tag{6.8c}$$

The confined eigenfunction $Z_i(z)$ is for the standing waves in the approximate triangular quantum well ($V(z) = qE_t z$) at the gate. Its corresponding eigenstates are shown in Figure 6.3 for the first two digitized energy levels. Appendix 6A gives some essential properties of the Airy function.

The average distance of the electron from the interface ($z = 0$) for electrons [6,7] in the ground state is

$$z_{QM} = \frac{2}{3}\frac{E_o}{q\mathcal{E}_t} \tag{6.9}$$

Threshold voltage V_T is an important parameter for VLSI design. MOSFET characterization involves extracting V_T from experiments. From a design perspective, it is useful to know the material characteristics that can be varied to obtain a desired V_T. As the inversion layer is formed, the bands near the Si/SiO_2 interface band downward so that the Fermi level is closer to the conduction band at the threshold of inversion, as shown in Figure 6.5.

As is clear from Figure 6.5, the surface potential energy at the interface is given by

$$q\phi_s = E_g - E_B - (E_F - E_{vbulk}) \tag{6.10}$$

The threshold of inversion is defined as the gate voltage when the surface electron concentration n_{3s} is equal to n_{3bulk} far from the interface. Naturally, this means $(E_c - E_F)_{surface} = (E_F - E_v)_{bulk}$ or equivalently $(E_F - E_i)_{surface} = (E_i - E_F)_{bulk} = \phi_B$ for background concentration N_B of the substrate. Equivalently, this means

$$q\phi_s = 2q\phi_B = 2qV_t \ln\left(\frac{N_B}{n_i}\right), \quad V_t = \frac{k_B T}{q} \tag{6.11}$$

Another effect comes from the difference in the work function of the metal and semiconductor $qV_{FB} = q\phi_m - q\phi_s$ that makes the band across the oxide tilted. The flat-band voltage V_{FB} makes the bands flat across the oxide. The threshold voltage V_T (not to be confused with the thermal voltage V_t) is given by

$$V_T = V_{FB} + 2\phi_B + \left(\frac{qN_B W_{dep}}{C_{ox}}\right), \quad W_{dep} = \sqrt{\frac{2\varepsilon_s}{q} 2\phi_B \frac{1}{N_B}} \tag{6.12}$$

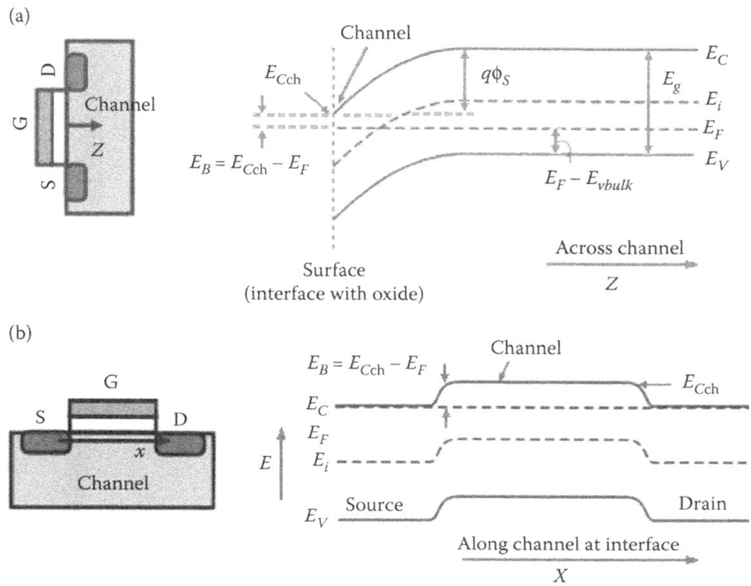

FIGURE 6.5
The energy band diagram of the NMOS (a) across the channel and (b) along the channel when no source-to-drain voltage exists.

where W_{dep} is the depletion layer width in the substrate that depends on the background concentration N_B of the p-type acceptors in the substrate.

The induced areal charge qn_2 in the channel in equilibrium is thus given by

$$qn_2 = C_{AG}(V_{GS} - V_T) = C_{AG}V_{GT} \tag{6.13}$$

where V_{GT} is a shorthand notation for overdrive gate voltage beyond threshold.

6.3 *I–V* Characteristics of Nano-MOSFET

With a channel established with the gate voltage $V_{GS} > V_T$, the current will flow in the channel if a potential difference exists between the source and the drain of the channel. Normally, the source is connected to the substrate and grounded; voltage V_{DS} is applied to the drain as shown in Figure 6.6.

The induced carrier concentration along the length of the channel (x-direction) varies as

$$qn_2(x) = C_{AG}[V_{GS} - V_T - V(x)] = C_{AG}[V_{GT} - V(x)] \tag{6.14}$$

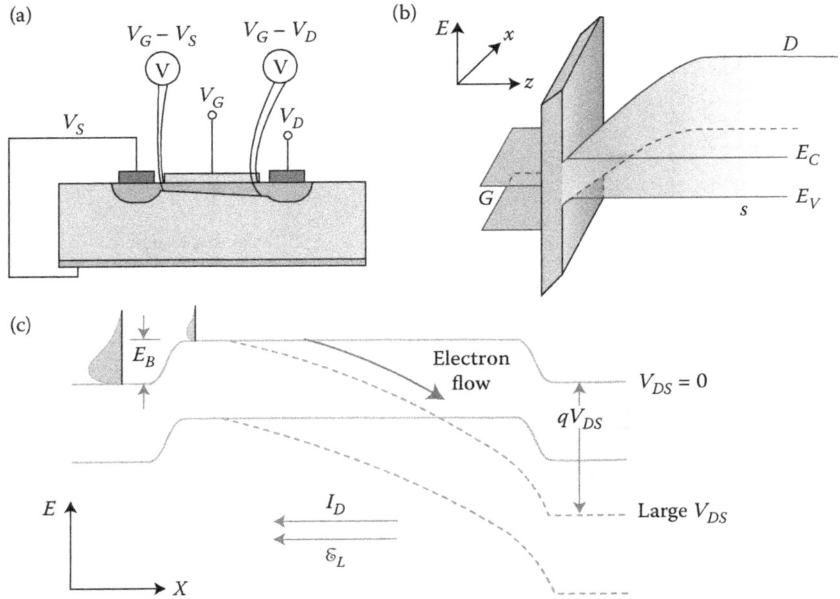

FIGURE 6.6
(a) The drain voltage $V_{DS} > 0$, not only the carrier concentration but also the depletion layer varies with position. (b) The energy band diagrams normal to the gate at the source and at the drain. (c) The energy band diagram along the channel both with and without the drain voltage.

where $V(x)$ is the potential across the length of the channel with $V(0) = 0$ and $V(L) = V_{DS}$. The channel does not exist when $V_{GT} < 0$. The drain current as discussed in Chapter 5 is given by

$$I_D = qn_2(x)v(x)W \tag{6.15}$$

The drain current I_D is continuous as there are no sources or sinks in the channel. The fact that neither $n_2(x)$ nor $v(x)$ function is known *a priori*, analytical work can be performed only with some simplifying assumptions. The velocity-field relationship established before gives

$$v(x) = v_{sat}(x)\tanh\left[\frac{\mathcal{E}(x)}{\mathcal{E}_c}\right] \tag{6.16}$$

A suitable approximation for $\tanh(x)$ is $\tanh(x) \approx x/(1+x)$ giving

$$v(x) \approx \frac{\mu_{ef}\mathcal{E}}{1+(\mathcal{E}/\mathcal{E}_c)}, \quad \mathcal{E}_c = \frac{v_{sat}}{\mu_{ef}} \tag{6.17}$$

v_{sat} the saturation velocity and $\mu_{\ell f}$ the low-field mobility are assumed constant with their experimental values utilized:

$$\mu_{\ell f} \approx 500\,\text{cm}^2/\text{V}\cdot\text{s}, \quad v_{sat} \approx 0.6 \times 10^5\,\text{m/s} \tag{6.18}$$

Another assumption that simplifies the solution is that all carriers at the drain end are exiting with the saturation velocity at the onset of current saturation. This assumption automatically implies that the electric field at the drain end is infinity [9,10]. The saturation current at the drain end beyond the saturation point for drain voltage is then given by

$$I_{Dsat} = C_{AG}(V_{GT} - V_{Dsat})v_{sat}W, \quad V_{DS} \geq V_{DSsat} \tag{6.19}$$

The use of Equation 6.17 in Equation 6.15 gives

$$I_D = C_{AG}(V_{GT} - V(x))\frac{\mu_{\ell f}\mathcal{E}}{1 + (\mathcal{E}/\mathcal{E}_c)}W \tag{6.20}$$

With the electric field existing from the drain to the source in the $-x$-direction, the electric field as a negative gradient of the potential is given by

$$\mathcal{E}(x) = -(-)\frac{dV(x)}{dx} = \frac{dV(x)}{dx} \tag{6.21}$$

Multiplying Equation 6.20 by the denominator on each side, applying Equation 6.21 and integrating over the length of the channel from $x = 0$ ($V(0) = 0$) to $x = L$ ($V(0) = V_{DS}$) yields

$$I_D\int_0^L\left(1 + \frac{1}{\mathcal{E}_c}\frac{dV}{dx}\right)dx = \int_0^L C_G(V_{GT} - V(x))\mu_{\ell f}\frac{dV}{dx}W\,dx \tag{6.22}$$

Equation 6.22 further reduces to

$$I_D\int_0^L dx + \int_0^{V_{DS}}\frac{1}{\mathcal{E}_c}dV = \mu_{\ell f}C_{AG}W\int_0^{V_{DS}}(V_{GT} - V)dV \tag{6.23}$$

Integration is simple and results in

$$I_D\cdot\left(L + \frac{1}{\mathcal{E}_c}V_{DS}\right) = \mu_{\ell f}C_G W\left[V_{GT}V_{DS} - \frac{1}{2}V_{DS}^2\right] \tag{6.24}$$

With the use of critical voltage $V_c = \mathcal{E}_c L = (v_{sat}/\mu_{\ell f})L$, Equation 6.24 transforms into

$$I_D = \frac{\mu_{\ell f} C_{AG} W}{L} \frac{\left[V_{GT}V_{DS} - (1/2)V_{DS}^2\right]}{1 + (V_{DS}/V_c)}, \quad 0 \le V_{DS} \le V_{DSsat} \tag{6.25}$$

Reconciling Equations 6.19 and 6.25 at the saturation point $V_{DS} = V_{DSsat}$ yields

$$\frac{\mu_{\ell f} C_{AG} W}{L} \frac{\left[V_{GT}V_{DSsat} - (1/2)V_{DSsat}^2\right]}{1 + (V_{DSsat}/V_c)} = C_{AG}(V_{GT} - V_{Dsat})v_{sat}W \tag{6.26}$$

Equation 6.26 is further reduced to

$$\left[V_{GT}V_{DSsat} - \frac{1}{2}V_{DSsat}^2\right] = (V_{GT} - V_{DSsat})\left(\frac{v_{sat}}{\mu_{\ell f}}L\right)\left(1 + \frac{V_{DSsat}}{V_c}\right) \tag{6.27}$$

resulting in

$$V_{GT}V_{DSsat} - \frac{1}{2}V_{DSsat}^2 = V_{GT}V_c + V_{GT}V_{DSsat} - V_{DSsat}V_c - V_{DSsat}^2 \tag{6.28}$$

which can be rearranged as a quadratic equation in V_{DSsat}

$$V_{DSsat}^2 + 2V_cV_{DSsat} - 2V_{GT}V_c = 0 \tag{6.29}$$

The solution of Equation 6.29 yields

$$V_{DSsat} = \frac{-2V_c \pm \sqrt{(2V_c)^2 + 4 \times 2V_{GT}V_c}}{2} \tag{6.30}$$

which after simplification gives

$$V_{DSsat} = V_c\left[-1 + \sqrt{1 + \frac{2V_{GT}}{V_c}}\right] = V_c\left[\sqrt{1 + \frac{2V_{GT}}{V_c}} - 1\right] \tag{6.31}$$

This expression was originally given by Arora and Das [9,10] and now appears in a number of textbooks [2]. I_{Dsat} at the onset of current saturation is easier to calculate either from Equation 6.19 or from Equation 6.25 by inserting $V_{DS} = V_{DSsat}$. Perhaps, it is easier to use Equation 6.19 after noticing from Equation 6.29 that

$$2V_c(V_{GT} - V_{DSsat}) = V_{DSsat}^2 \tag{6.32}$$

This substitution in Equation 6.19 gives

$$I_{Dsat} = \frac{1}{2} C_{AG} v_{sat} W \frac{V_{Dsat}^2}{V_c} = \frac{1}{2} C_{AG} v_{sat} W \frac{V_{Dsat}^2}{(v_{sat}/\mu_{eff})L} = \frac{1}{2} \frac{C_{AG} W \mu_{eff}}{L} V_{Dsat}^2 \qquad (6.33)$$

Figure 6.7 shows the current–voltage characteristics of the FET for both NMOS and PMOS. When $V_{GT} \gg V_c$, the current steps in the saturation region tend to be equidistant, whereas the steps tend to show a quadratic rise in long channels (LCs) where $V_{GT} \ll V_c$. Clearly, if device-packing density continues to rise, effects such as velocity saturation (of electrons in a given material) and wave–particle duality must be taken into account. The carrier mobility in a MOSFET is distinctly different from that in bulk silicon. In the earlier version for electron mobility, the mobility is given by [11]

$$\mu_{efn} = \mu_{on} - k_{1n} V_{GT}, \quad \mu_{on} = 625\ \text{cm}^2/\text{V}\cdot\text{s}, \quad k_{1n} = 25\ \text{cm}^2/\text{V}^2\cdot\text{s} \qquad (6.34)$$

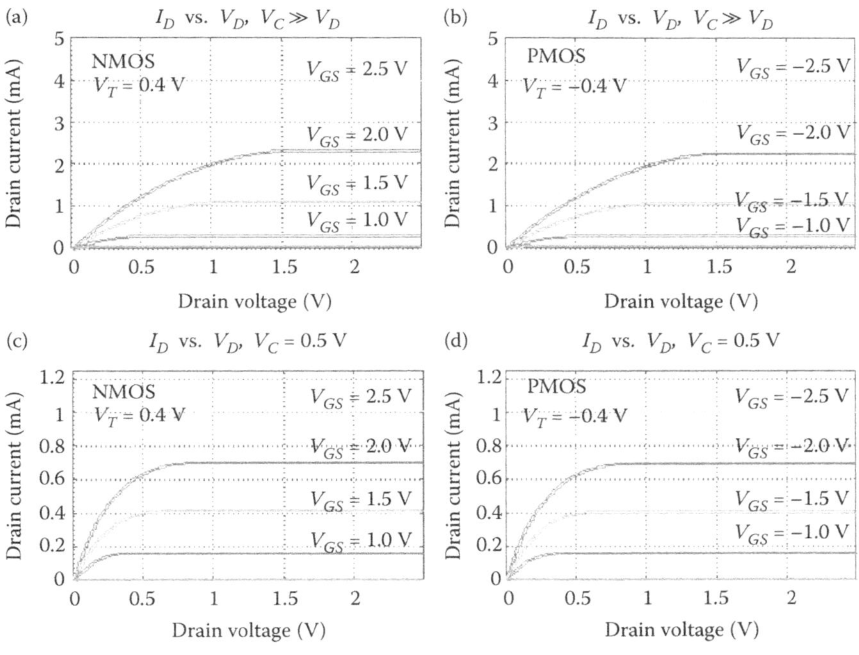

FIGURE 6.7
I–V characteristics of MOSFET for various values of gate voltage. (a) NMOS with velocity saturation documented. (b) PMOS with velocity saturation documented. (c) Ohmic NMOS. (d) Ohmic PMOS.

The hole mobility is given by [11]

$$\frac{1}{\mu_{\ell fp}} = \frac{1}{\mu_{op}} - \frac{1}{k_{1p}}(V_{GS} + 2V_T), \quad \mu_{op} = 279\,\text{cm}^2/\text{V}\cdot\text{s}, \quad k_{1p} = 1508\,\text{cm}^2/\text{V}^2\cdot\text{s} \tag{6.35}$$

The mobility dependence on gate voltage is being reexamined and changing from time to time as ballistic transport processes are understood.

An observation of Equation 6.25 shows that it is the net injection of current in the channel from source-to-drain I_{SD} and drain-to-source I_{DS} [1]

$$I_D = I_{SD} - I_{DS} = \frac{1}{2}\frac{\mu_{\ell f} C_{AG} W}{L}\left[\frac{V_{GT}^2}{1 + (V_{DS}/V_c)} - \frac{(V_{GT} - V_{DS})^2}{1 + (V_{DS}/V_c)}\right] \tag{6.36}$$

In the original paradigm, the saturation condition arose from the fact that $I_{DS} = 0$ requires $V_{Dsat} = V_{GT}$. This is the condition that is commonly referred to as channel pinchoff when the charge at the onset of current saturation vanishes

$$qn_{2D} = C_{AG}[V_{GT} - V_{DSsat}] = 0 \quad \text{(pinchoff condition)} \tag{6.37}$$

As it is shown in the next section, channel pinchoff was a mirage [9,10] as velocity saturation exists even in the LC as the electric field is extremely high on the drain end.

The complete electric field and voltage profile is evaluated by repeating derivation (6.22) except the limit of the integral is changed from $x = 0$ to x and for the potential from $V = 0$ to $V(x)$. The result is [10]

$$V(x) = \left[V_{GT} - \frac{I_D}{C_{AG} v_{sat} W}\right] - \left[\left\{V_{GT} - \frac{I_D}{C_{AG} v_{sat} W}\right\}^2 - \frac{2I_D}{C_{AG}\mu_{\ell f} W}x\right]^{1/2} \tag{6.38}$$

$$\mathcal{E}(x) = \frac{dV}{dx} = \frac{I_D}{C_{AG}\mu_{\ell f} W}\left[\left\{V_{GT} - \frac{I_D}{C_{AG} v_{sat} W}\right\}^2 - \frac{2I_D}{C_{AG}\mu_{\ell f} W}x\right]^{-1/2} \tag{6.39}$$

Equations 6.38 and 6.39, at the onset of the saturation point ($I_D = I_{Dsat}$), simplify to give

$$V(x) = V_{DSsat} - V_{DSsat}\left(1 - \frac{x}{L}\right)^{1/2} \tag{6.40}$$

$$\mathcal{E}(x) = \frac{V_{DSsat}}{2L}\frac{1}{\left(1-(x/L)\right)^{1/2}} \tag{6.41}$$

6.4 Long- (LC) and Short-Channel (SC) MOSFET

The results above are applicable for channels of arbitrary length. In the twentieth-century MOSFET, Ohm's law with linear current–voltage relationship was extensively used for *I–V* characteristics. Ohm's law is certainly valid; so far, the applied voltage remains below the critical value ($V_D < V_c$). For a microscale MOSFET of the twentieth century, certainly this assumption is valid. In that extreme, the MOSFET channel is considered long. In the LC ($V_D \ll V_c$), Equation 6.25 transforms into

$$I_D = \frac{\mu_{ef} C_{AG} W}{L}\left[V_{GT}V_{DS} - \frac{1}{2}V_{DS}^2\right], \quad 0 \le V_{DS} \le V_{DSsat} \tag{6.42}$$

$V_{DSsatLC}$ in the LC approximation is attained by use of the binomial theorem

$$(1+x)^n \approx 1 + nx + \frac{n(n-1)}{2!}x^2 + \cdots, \quad x \ll 1 \tag{6.43}$$

With $x = 2V_{GT}/V_c$, this approximation when applied to Equation 6.31 infers

$$V_{DSsatLC} \approx V_c\left[1 + \frac{1}{2}\frac{2V_{GT}}{V_c} - \frac{1}{8}\left(\frac{2V_{GT}}{V_c}\right)^2 + \cdots - 1\right] \tag{6.44}$$

Equation 6.44 gives $V_{DSsat} = V_{GT}$, with only the first-order term in V_{GT} considered, as assumed in the twentieth-century model. However, when the second-order term in Equation 6.44 is also included, the results are

$$qn_{2DLC} = \frac{1}{2}C_{AG}\frac{V_{GT}^2}{V_c} \quad \text{(no channel pinchoff!)} \tag{6.45}$$

The channel pinchoff condition of Equation 6.37 is a direct result of not retaining the second-order term in Equation 6.44. The drain current as obtained from Equation 6.33, instead of vanishing at the pinchoff point, is finite

$$I_{DsatLC} \approx \frac{1}{2}\frac{C_{AG}W\mu_{ef}}{L}V_{GT}^2 = qn_{2S}\mu_{ef}\mathcal{E}_S W \tag{6.46}$$

which is the same as expected from the application of Ohm's law at the source end with the source-end electric field $\mathcal{E}_S = V_{Dsat}/2L$. Thus, the source electric field also saturates as indicated in Reference [2]. Equations 6.45 and 6.46 are consistent with the current continuity as the current on the drain end at saturation, using Equation 6.45 and $V_c = (v_{sat}/\mu_{\ell f})L$, is the same as in Equation 6.46

$$I_{DsatLC} = qn_{2D}v_{sat}W = \frac{1}{2}\frac{C_{AG}W\mu_{\ell f}}{L}V_{GT}^2 \tag{6.47}$$

What happens beyond pinchoff in a nanochannel will be discussed when we invoke the condition that the carriers travel at the drain with velocity fraction α of the v_{sat} as discussed in Reference [12]. The comparison of Equations 6.46 and 6.47 clearly indicates that the current is ohmic at the source end and saturation limited on the drain end. Equation 6.45 for the carrier concentration at the drain end is

$$n_{2DLC} = n_{2SLC}\frac{V_{GT}}{2V_c} = \frac{I_{Dsat}}{qv_{sat}W} \tag{6.48}$$

As the channel length enters into the nanoregime, the source-end velocity will become saturation limited with the appropriate value of α changing with the length of the channel and approaches unity $\alpha \approx 1$ in the LC.

In the SC limit ($L \to 0$), $V_D \gg V_c$ as V_c is lower, as low as a fraction of a volt in nanoscale channels. In this limit, $2V_{GT}/V_c \gg 1$. Equation 6.31 then yields

$$V_{DsatSC} = V_c\left[\sqrt{1+\frac{2V_{GT}}{V_c}}-1\right] \approx V_c\left[\frac{2V_{GT}}{V_c}\right]^{1/2} = \left[2V_{GT}V_c\right]^{1/2} \tag{6.49}$$

With the application of this approximation to current I_{DsatSC} as obtained from Equation 6.33, it is given by

$$I_{DsatSC} = \frac{1}{2}\frac{C_G W\mu_{\ell f}}{L}2V_{GT}V_c = C_G V_{GT}v_{sat}W = n_{2S}qv_{sat}W \tag{6.50}$$

The comparison of Equations 6.47 and 6.50 vividly indicates that both V_{Dsat} and I_{Dsat} reduce considerably in the SC limit:

$$\frac{V_{DsatSC}}{V_{DsatLC}} = \frac{\left[2V_{GT}V_c\right]^{1/2}}{V_{GT}} = \left[\frac{2V_c}{V_{GT}}\right]^{1/2} \ll 1 \tag{6.51}$$

$$\frac{I_{DsatSC}}{I_{DsatLC}} = \frac{2V_c}{V_{GT}} \ll 1 \tag{6.52}$$

This reduction is clearly visible in Figure 6.7. In Figure 6.7a and b, the saturation current is larger as well as the step spacing is quadratic with gate voltage LC behavior. Figure 6.7c and d depicts SC behavior. The saturation current is smaller in Figure 6.7c and d. Also, the steps are equally spaced.

6.5 Model Refinements for Nano-CMOS Application

The low-field mobility $\mu_{\ell f}$ along the channel is really not constant and strongly depends on the gate electric field that varies along the length of the channel. Fairus and Arora [5] calculated the mobility for a MOSFET channel and found that the ratio $\mu_{\ell f}/\mu_b$ of the channel mobility $\mu_{\ell f}$ to μ_b in a bulk material is proportional to z_{QM}/λ_D, where z_{QM} is the length of confinement and λ_D is the de Broglie wavelength. In this model, $\mu_{\ell f} \propto \mathcal{E}_t^{-1/3}$. This relationship is valid for the nondegenerate statistics. Rothwarf [13] with quantum mechanical arguments gave a formula in agreement with an empirical formula derived from experiments of Cooper and Nelson [14]. In this framework, $\mu_{\ell f}$, as a function of the transverse gate electric field $\mathcal{E}_t$, is given by

$$\mu_{\ell f} = \frac{0.1105\ \mathrm{m}^2/\mathrm{V}\cdot\mathrm{s}}{1+(\mathcal{E}_t/(30.5\ \mathrm{V}/\mu\mathrm{m}))^{0.657}} \tag{6.53}$$

A number of empirical formulas exist in the literature describing $\mu_{\ell f}$ as a function of the gate electric field. Equations 6.34 and 6.35 are one possible way to express the decline of low-field mobility with the gate voltage. Anderson [2] gives another option that is the function of voltage across the gate $V_{GT} - V(x)$ and is given by

$$\mu_{\ell f} = \frac{\mu_S}{1+\vartheta(V_{GT} - V(x))} \tag{6.54}$$

where μ_S is the channel mobility at the source ($V(0) = 0$) at the threshold of inversion ($V_{GT} = 0$). $\vartheta = 0.03\text{–}0.2\ \mathrm{V}^{-1}$ is an empirical parameter. The data extracted from experiments by Tan and Ismail [15] on an 80-nm MOSFET follow an exponential trend

$$\mu_{\ell f} = 0.0700 \frac{\mathrm{m}^2}{\mathrm{V}\cdot\mathrm{s}} e^{-(V_{GS}/1.33)} \tag{6.55}$$

Equation 6.55 clearly establishes the nature of ballistic processes in the nanoscale channel that need further investigation. It is possible that the mobility in a quantum channel is barrier-limited and is controlled by processes distinctly different from a microchannel. One such possibility is the reflection of quantum waves from contacts that may yield an exponential behavior.

The correct identification of v_{satD} at the drain is another factor that needs careful investigation, especially when an inverted carrier gas in the channel is degenerate. v_{satD} is then limited by the intrinsic velocity v_{i2} discussed earlier. In fact, when electrons stream in an electric field, quantum states in the forward direction (opposite to the electric field) are occupied, making the unidirectional velocity v_{u2} appropriate for twice the number filling quantum states or equivalently, at the saturation, only half the quantum states are available in the forward direction. In that case, $v_{u2} = \sqrt{2}\, v_{i2}$. The degenerate drain velocity v_{satD} is then given by

$$v_{satD} = v_{u2} = \frac{2}{3}\frac{\hbar}{m_t^*}\sqrt{4\pi n_2} = \frac{2}{3}\frac{\hbar}{m_t^*}\sqrt{\frac{4\pi C_G(V_{GT} - V_{Dsat})}{q}} \tag{6.56}$$

As expected, V_{Dsat} itself depends on the saturation velocity, necessitating an iterative solution. An iterative process perhaps can be initiated with the nondegenerate saturation velocity $v_{sat} = v_{th2} = 1.96 \times 10^5$ m/s as the seed value that is always independent of the carrier concentration or gate voltage.

The drain velocity is always smaller than that expected from an infinite electric field at the drain ($V_{Dsat} = V_{Dsat1}$ for $\alpha = 1$). The calculated v_{sat} as a function of gate voltage from the iterative solution is shown in Figure 6.8. The solid

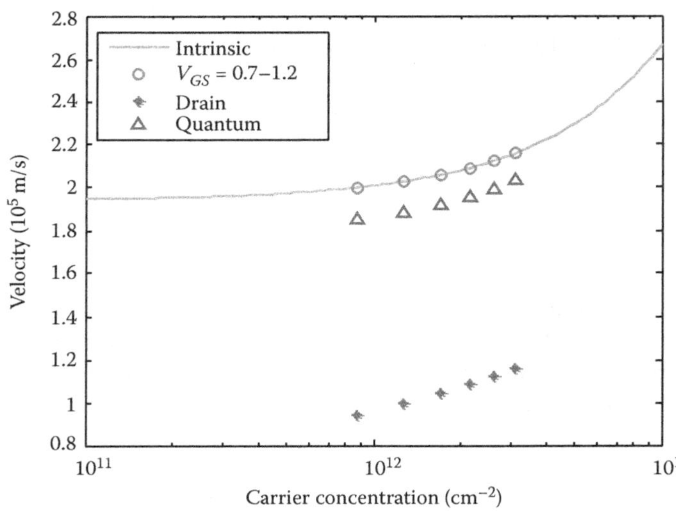

FIGURE 6.8
The ultimate saturation velocity as a function of gate voltage. The saturation velocity increases as carrier concentration increases with the gate voltage.

line is the intrinsic velocity (the ultimate saturation velocity for an infinite electric field) obtained from Equation 6.56 in the absence of quantum emission for $T = 300$ K. The intrinsic velocity is independent of carrier concentration for low values and rises as a square root for high concentration. When calculated for other temperatures (not shown), the intrinsic velocity in the degenerate domain is independent of temperature as given by Equation 6.56. This is a significant shift from the mindset in the published literature where saturation velocity is written in simulation programs as $v_{sat} = 0.6 \times 10^5\,\text{m/s}$. The quantum emission lowers the saturation velocity as shown by triangles (▲). However, the drain velocity (open circles o) with which carriers leave the drain is lower due to the presence of a finite electric field.

Chen et al. [16] brought an important aspect of the FET that indicates that it is impossible for velocity to reach the ultimate saturation value at the drain. The drain velocity is $v_{Drain} = \alpha v_{sat}$ because of the finite electric field at the drain. Yes, in the LC, $\alpha \approx 1$ in the limit of infinite length that itself is a fiction. As channels are entering the nanoregime, such corrections are warranted as implemented in Reference [12]. Equation 6.19 with v_{sat} replaced by αv_{sat} modifies to

$$I_{Dsat} = \alpha C_{AG}(V_{GT} - V_{Dsat})v_{sat}WV_{DS} \geq V_{DSsat\alpha} \tag{6.57}$$

Reconciliation of Equations 6.57 to 6.25 at the saturation point $V_{DS} = V_{DSsat\alpha}$ yields

$$V_{Dsat\alpha} = \frac{1}{(2\alpha - 1)}\left[(s - \alpha)V_c - (1 - \alpha)V_{GT}\right] \tag{6.58}$$

$$I_{Dsat\alpha} = \frac{\alpha}{(2\alpha - 1)}\frac{C_G\mu_{ef}W}{L}V_c\left[\alpha V_{GT} - (s - \alpha)V_c\right] \tag{6.59}$$

with

$$s = \sqrt{\left[\alpha + (1 - \alpha)\frac{V_{GT}}{V_c}\right]^2 + 2\alpha(2\alpha - 1)\frac{V_{GT}}{V_c}} \tag{6.60}$$

As expected, $V_{DSsat\alpha} < V_{DSsat1} = V_{Dsat}$ with V_{Dsat} given in Equation 6.31. α is related to the finite electric field $\mathcal{E}_{Dr}$ at the drain through

$$\alpha = \frac{\mathcal{E}_{Dr}/\mathcal{E}_c}{1 + \mathcal{E}_{Dr}/\mathcal{E}_c} \tag{6.61}$$

As stated earlier, an iterative process determines v_{sat}, $\mathcal{E}_{Dr}$, and α. The electric field as a function of distance along the channel for an 80-nm MOSFET

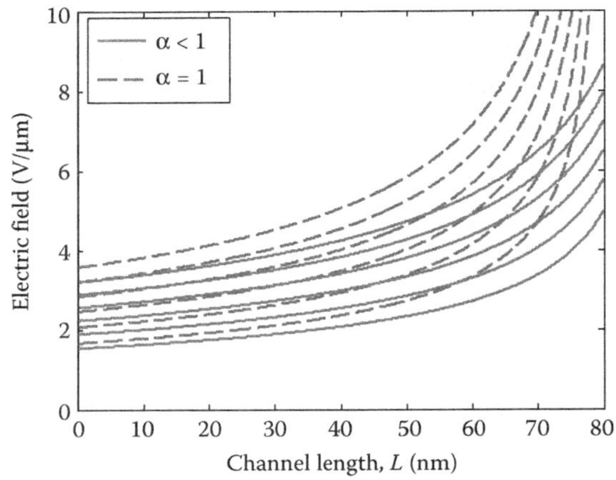

FIGURE 6.9
Electric field profile (solid) with partial saturation ($\alpha \leq 1$) and that (dotted) with full saturation ($\alpha = 1$).

is shown in Figure 6.9. As expected, the electric field rises to infinity in the $\alpha = 1$ model. However, it is finite when $\alpha < 1$.

The drain-to-saturation-velocity ratio $\alpha = v_D/v_{sat}$ is plotted in Figure 6.10 as a function of drain voltage V_D starting from the saturation point $V_{Dsat\alpha}$ to $V_D = 1.0$ V. α and hence the associated drain velocity rises from its value at quasi-saturation current approaching 1, as the electric field increases due

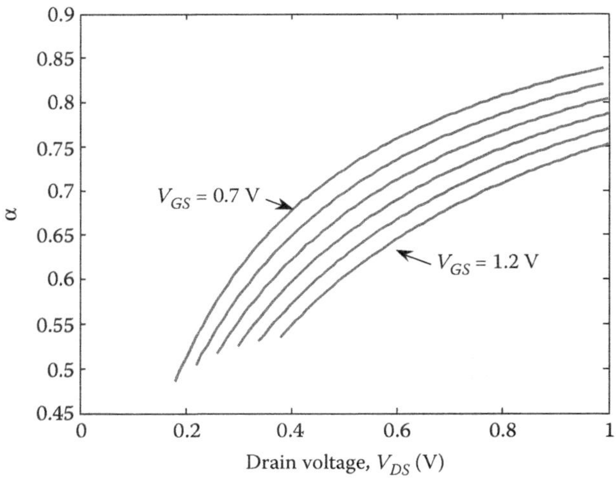

FIGURE 6.10
The ratio $\alpha = v_D/v_{sat}$ of the drain velocity as a function of drain voltage beyond the onset of saturation $V_D > V_{Dsat}$ for $V_{GS} = 0.7$ (the topmost curve), 0.8, 0.9, 1.0, 1.1, and 1.2 (bottom curve).

to increase in the drain voltage beyond saturation. α is higher for lower V_{GS} values. This is due to the fact that the SC effect arises when V_{GS} is larger than the critical voltage $V_c = E_c L$. V_{GS} being smaller is equivalent to the channel length being larger as the ratio V_{GS}/V_c appears in the expression for I_{Dsat} and V_{Dsat} as shown below. Therefore, for LC transistors, the value of α is almost closer to 1. This feature is clearly visible in LC *I–V* characteristics.

Another argument that is often given is that the source controls the current as the electric field there reaches a constant value as given by Equation 6.41 with $x = 0$. This observation is made on the assumption of ohmic transport where current is proportional to the voltage applied, consistent with the drift velocity proportional to the electric field. The current continuity demands that the current is controlled by both the source and the drain as indicated in Equations 6.46 and 6.47. The reason that the velocity is larger at the drain end is due to a larger electric field compensated by the lower carrier concentration; the product $n_2 v$ remaining the same throughout the channel. Beyond the saturation point, the gradual channel approximation breaks down. Its continual use in the saturation domain also leads to varied interpretations of the simulated results.

As Figure 6.11 shows, the concentration-dependent saturation velocity replicates the observed behavior well in a nanochannel without using any fitting parameters. The pseudo-channel-length modulation in the absence

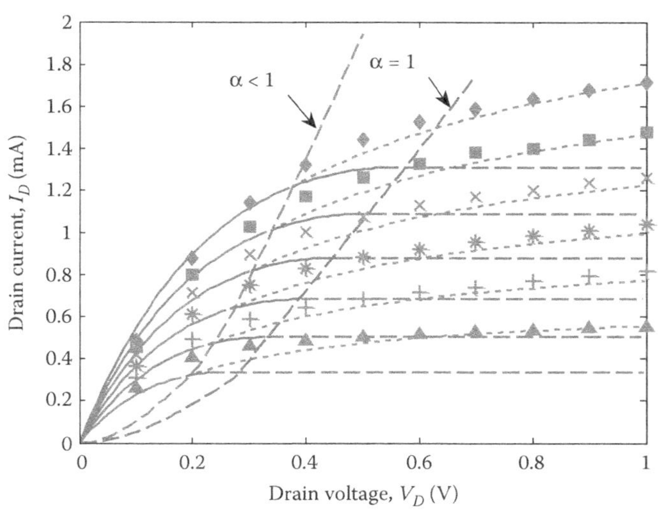

FIGURE 6.11
I–V characteristics of an 80-nm MOSFET for gate voltage $V_{GS} = 0.7, 0.8, 0.9, 1.0, 1.1$, and 1.2. Solid lines are from the triode equation in the range $0 \le V_D \le V_{Dsat}$. The dotted lines are extensions for $V_D \ge V_{Dsat}$ as the rise of an electric field brings the drift velocity closer to the saturation velocity with the rising value of α. The dashed lines are for the limit $\alpha = 1$.

of a quasi-pinchoff and finite output conductance in *I–V* characteristics at the onset of drain saturation are well explained by the velocity-field model based on the archival work of Arora [17]. In many ways, NEADF is similar to what is presented by Buttiker [18]. A series of ballistic channels, each of length ℓ (mfp) comprise a macrochannel of length L. The behavior is well understood as we consider the ballistic low-field mobility where ℓ is replaced by L [19–21] and implemented by Mugnaini and Iannaccone [22,23]. However, it does not affect the velocity saturation in a high electric field that is always ballistic. Thus, the ends of each mfp can be considered Buttiker's thermalizing virtual probes, which can be used to describe transport in any regime. In a high electric field, the electrons are in a coordinated relay race, each electron passing its velocity to the next electron at each virtual probe. The saturation velocity is thus always ballistic whether or not the length of the device is smaller or larger than the mfp. The ballistic saturation velocity is independent of the scattering-limited low-field mobility that may be degraded by the gate electric field. The relation between mobility and the mfp has deep consequences on the understanding of the transport in any nanoscale device.

Natori [24,25] raised an inherent possibility of ballistic saturation velocity in his work. Our results are consistent with Natori's mindset. The fundamental difference lies in how saturation velocity is calculated. Similar to Buttiker probes mentioned above, Natori considers the net exchange of electrons that are injected into the channel from the source and the drain. The major difference arises from the unidirectional nature of saturation velocity in a high electric field as considered here and only half the injected electrons going into the channel from either the source or the drain are taken to be a reservoir of electrons. Good agreement with the experiment without the use of any artificial parameters is indicative of the fact that the vision of Natori has been extended to streaming electrons in a high electric field initiated by the electric field.

Another feature that is visible in Figure 6.11 is the almost linear rise of $V_{Dsat\alpha}$ with the gate electric field and steps are almost equally spaced. Obviously, channel conductance $g_d = \partial I_{Dsat}/\partial V_{DS}$ is not zero as in the LC model. This is not due to channel modulation as in a simulation program, but due to the rise of the saturation velocity as the drain field is increased beyond the saturation point. The small-signal channel conductance g_d and transconductance g_m in the triode region $(0 \le V_D \le V_{Dsat})$ are obtained as

$$g_d = \left.\frac{\partial I_D}{\partial V_{DS}}\right|_{V_{GT}=const} = \frac{1}{2}\frac{C_{AG}\mu_{ef}W}{L}\frac{2(V_{GT}-V_{DS})-(V_{DS}^2/V_c)}{(1+(V_{DS}/V_c))^2} \tag{6.62}$$

$$g_m = \left.\frac{\partial I_D}{\partial V_{GT}}\right|_{V_{DS}=const} = g_{mo}\frac{V_D/V_c}{1+(V_D/V_c)}, \quad g_{mo} = C_{AG}Wv_{sat} \tag{6.63}$$

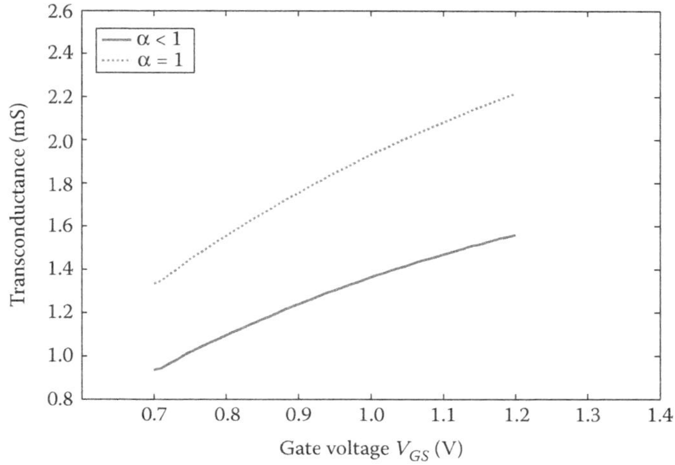

FIGURE 6.12
The channel transconductance as a function of gate voltage for channel current at full saturation with $\alpha = 1$ (dashed line) and partial saturation $\alpha = 1 < 1$ (solid curve).

The transconductance as a function of gate voltage is given in Figure 6.12. Owing to the lower drain velocity in $\alpha < 1$ model, the transconductance tends to be lower than predicted from the full velocity saturation ($\alpha = 1$) model. g_m is below its limiting value g_{mo}. The transconductance in the saturation regime g_{msat} is

$$g_{msat} = g_{msato}\left[1 - \frac{1}{\{1 + (2V_{GT}/V_c)\}^{1/2}}\right] \tag{6.64}$$

The channel conductance in the linear (triode) region is shown in Figure 6.13. With $\alpha < 1$, no carrier has reached the ultimate intrinsic velocity at the saturation point giving $g_d \neq 0$. In a model where $\alpha = 1$, the carriers are traveling at saturation velocity with an intrinsic assumption that the drain electric field is infinite. The channel conductance is $g_{d1} = 0$ where the dashed curve terminates in V_{Dsat1} in Figure 6.13. However, when $\alpha < 1$, the solid curves of Figure 6.13 terminate with a finite value of g_d at $V_D = V_{Dsat.}$

6.6 CMOS Design

A complementary MOS circuit consists of a pair of NMOS and PMOS channels connected in series as shown in Figure 6.14 that invert the logical value

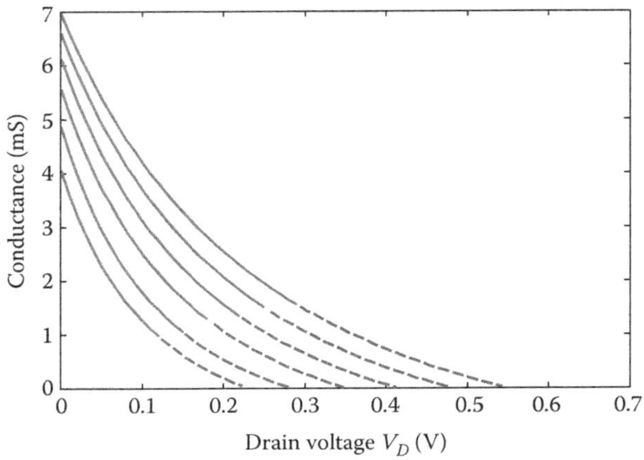

FIGURE 6.13
The channel conductance as a function of drain voltage $0 \le V_D \le V_{Dsat}$ for $V_{GS} = 0.7, 0.8, 0.9, 1.0, 1.1$, and 1.2. The conductance is finite at the onset of saturation (solid line). The dashed lines cover the region $V_{Dsat} \ge V_D \ge V_{Dsat1}$.

of the voltage applied to its input terminal. In a CMOS inverter, the gates of n- and p-channel MOSFETs are tied together. A positive gate bias, which turns the NMOS on, turns off the PMOS, and vice versa, and hence, the output is connected to V_{SS} (i.e., when the input is high, the output is low). This reverses when the input is low and then the output is connected to V_{DD} (high). Both stable states correspond to a very low current consumed from the power supply.

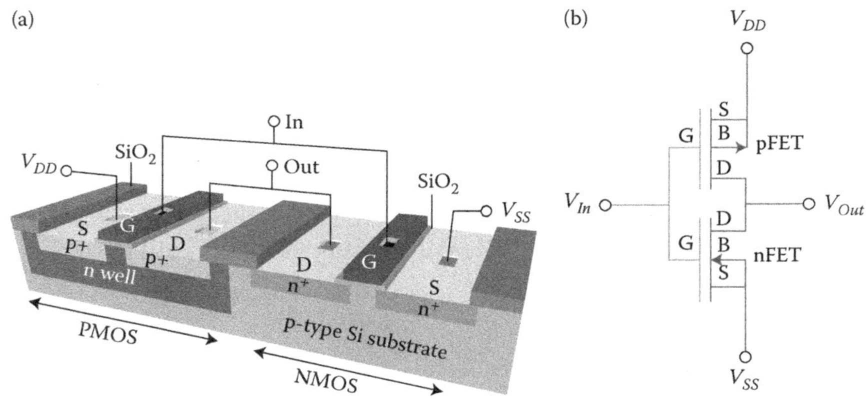

FIGURE 6.14
(a) A 3D view of CMOS with a complementary pair of NMOS and PMOS in an *n*-well process. (b) The circuit schematics of a CMOS inverter.

In a CMOS circuit design, switching speed is enhanced when the current $I_{Dn(p)}$ in NMOS and PMOS transistors is the same. In the fabrication process, the oxide thickness t_{ox} can be made the same for both transistors and threshold voltage can be adjusted so that $V_{Tn} = V_{Tp}$. In a traditional design based on Ohm's law ($V_{DS} \ll V_c$), Equation 6.42 is valid. To have the same current in the n- and p-MOS channels requires from Equation 6.42

$$\frac{\mu_{\ell fn} C_{AGn} W_n}{L_n} = \frac{\mu_{\ell fp} C_{AGp} W_p}{L_p} \tag{6.65}$$

The oxide thickness with a small quantum capacitance effect makes gate capacitance the same for both channels $C_{AGn} = C_{AGp}$. As length is the smallest dimension in the photolithographic process, the designers tend not to sacrifice the smallest dimensions attained and hence, keep the channel length the same ($L_n = L_p$) for both channels. The width is scaled so that $(\mu_{\ell f} W)_{n(p)}$ product is the same for current to be identical in both channels. To account for the lower mobility of PMOS, the width ratio of p-channel to n-channel is scaled as

$$\frac{W_p}{W_n} = \frac{\mu_{\ell fn}}{\mu_{\ell fp}} \tag{6.66}$$

The above design is represented in Figure 6.15. As expected, LC currents match (topmost curve). However, with this design, the SC currents do not match (bottom two curves). Since $W_p \approx 2W_n$, the current in PMOS is twice

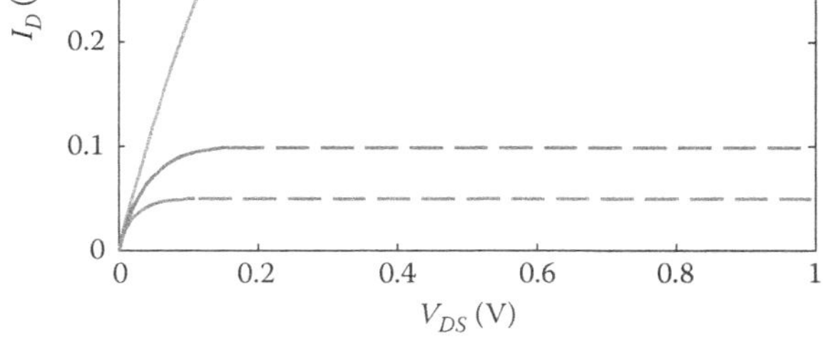

FIGURE 6.15
I–V characteristics of an 80-nm NMOS for SC and LC. In an LC model (top curve), the current is the same for NMOS and PMOS. In an SC model, the current not only drops, but also is proportional to the width of the channel ($W_p \approx 2W_n$).

as large as that in NMOS. The reduction in V_{Dsat} and I_{Dsat} in the SC model is clearly visible. Also, the saturation current is proportional to the width of the channel with this design.

The above design is based on the validity of Ohm's law that obviously does not hold good for nanoscale channels. When velocity saturation is considered, the saturation current of Equation 6.19 requires (Wv_{sat}) to be matched for both n- and p-channel. In the triode region $0 \le V_D \le V_{Dsat}$, the current expression of Equation 6.25, including V_c, must also match to complement I–V characteristics of both channels. This leads to

$$\frac{W_p}{W_n} = \frac{v_{satn}}{v_{satp}}, \quad \frac{L_p}{L_n} = \frac{\mu_{\ell fp}}{\mu_{\ell fn}} \tag{6.67}$$

The length of the channel is scaled proportional to the channel mobility and the width is scaled inversely proportional to the saturation velocity of the channel. The appropriate effective mass for a PMOS channel due to heavy holes is $m_{hh}^* = 0.48m_o$, the light-hole subband is lifted higher in energy, making its population highly unlikely. Hence, it is appropriate to calculate the saturation velocity with this effective mass. The new design matches well for both the LC and SC characteristics. Figure 6.16 implements this design with a perfect match for the current both for LCs and SCs. Hence, this is most appropriate and generalized, eliminating the need to distinguish between SCs and LCs.

The presence of ballistic mobility as the channel length goes below the scattering-limited mfp presents another complication in the design process.

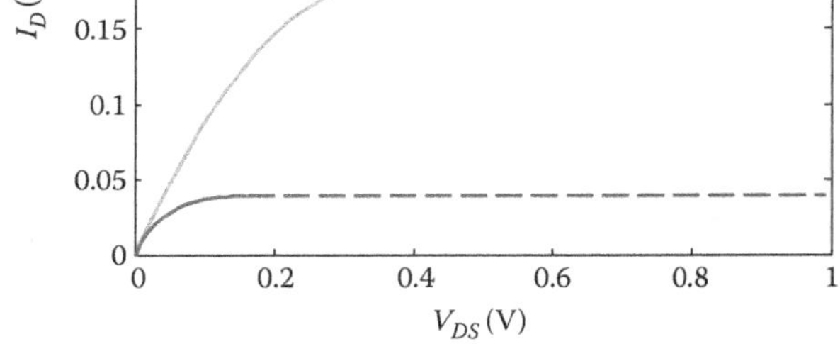

FIGURE 6.16
I–*V* characteristics of an 80-nm MOSFET for SC and LC on a new design based on the velocity saturation model. (Adapted from B. L. Anderson and R. L. Anderson, *Fundamentals of Semiconductor Devices*. New York, NY: McGraw-Hill, 2005.)

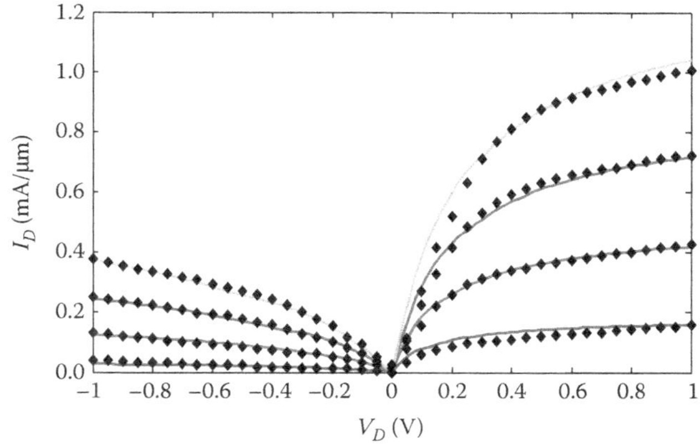

FIGURE 6.17
Analytical 45-nm PMOS and NMOS (solid lines) drain current characteristics with comparison to 45-nm IBM [26] experimental data (symbol). Initial V_G for MOSFET is 0.4 V with 0.2 V increment.

Figure 6.17 shows the output from a 45-nm simulation for NMOS and PMOS. As expected with the mobility from the literature, there is a very good agreement between theory and experiment for both NMOS and PMOS. However, the current in PMOS is much smaller than that in NMOS. The ballistic mobility required for the fit is much smaller than that in the LC MOSFET.

The nondegenerate intrinsic velocity is $v_{th2n} = 1.94 \times 10^5$ m/s for electrons and $v_{th2p} = 1.22 \times 10^5$ m/s that is used as a seed value in the program. The comparison with the 45-nm experimental data indicates that the compact model presented above is able to accurately predict the drain current not only from Taiwan Semiconductor Manufacturing Company (TSMC) but also from The International Business Machines Corporation (IBM) data. The design parameters are given in Table 6.1. This match is indicated in Figure 6.18.

The ballistic nature of the mobility depending on the length of the channel as discussed in Chapter 4 requires an iterative solution to find the appropriate length to match the low drain voltage component and appropriate width

TABLE 6.1
Device Model Specification for NMOS and PMOS

Parameter	NMOS	PMOS
Effective oxide thickness (nm)	1.2	
Insulator thickness, t_{ox} (nm)	2.1	1.2
Channel length, L (nm)	45	
Low-field mobility, $\mu_{\ell f}$ ($m^2/V \cdot s$)	0.036	0.005

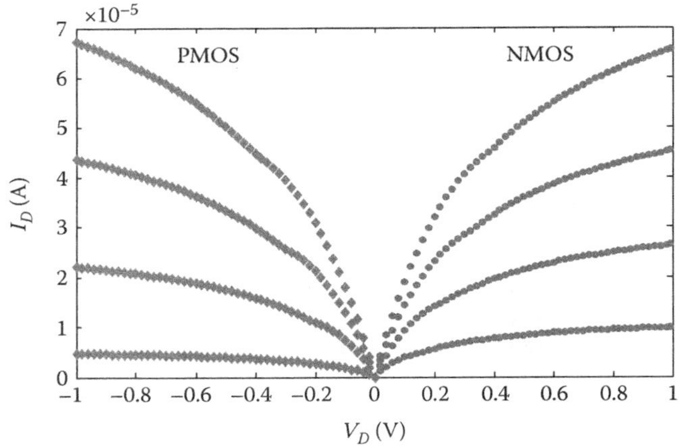

FIGURE 6.18
Analytical CMOS drain current characteristics. The initial V_G for MOSFET is 0.4 V with 0.2 V increment. Both devices ratio are able to provide similar drain current following the mobility and velocity scaling.

for the saturation-limited component. Table 6.2 lists the design parameters that give the same current in both complementary channels. Matched *I–V* characteristics of NMOS and PMOS are given in Figure 6.18.

As discussed in Chapter 5, the resistance is a function of current in the resistor or voltage across the resistor [27], and the normalized resistance surges as current in the resistor approaches the saturation value. To avoid this resistive enhancement, the current in the parasitic region should be kept much lower than its saturation value. This should be taken into account while designing parasitic regions, for example, the lightly doped drain and other related design concepts for submicron devices. Alternative designs are needed to ameliorate hot-electron effects and resistance surge effect.

TABLE 6.2
Device Model Specification for CMOS

Parameter	NMOS	PMOS
Channel length, L (nm)	90	22
Channel width, W (nm)	90	108
Length ratio, L_p/L_n	0.24	
Width ratio, W_p/W_n	1.20	
Ballistic mobility ratio, μ_p/μ_n	0.11	
Saturation velocity ratio, v_{satn}/v_{satp}	1.53	

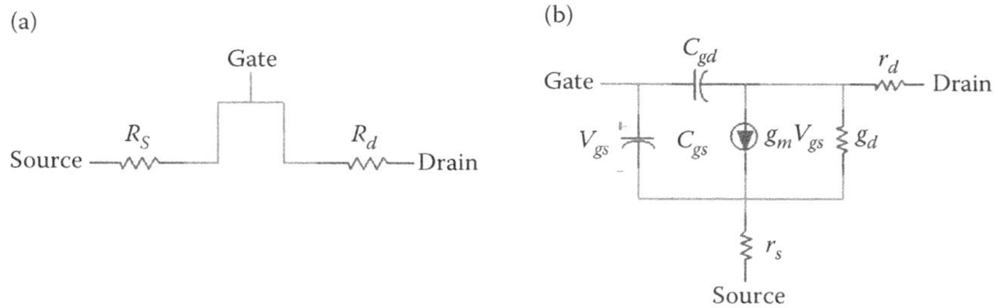

FIGURE 6.19
(a) Equivalent circuit of MOSFET with parasitic source and drain series resistances. (b) Simplified MOSFET small-signal equivalent circuit model used in device analysis.

The fundamental limits on device performance set by the impact of the extrinsic source and drain can be understood in terms of the simplified small-signal equivalent circuit shown in Figure 6.19 [27,28].

The extrinsic small-signal source and drain resistances, r_s and r_d, play an important role in the performance of a CMOS. Normally, the extrinsic source and drain are modeled as simple linear resistances with values obtained from low-current ohmic measurements. In this model, r_s and r_d are equal to their low-current values, R_{s0} and R_{d0}, respectively, regardless of the current. It is important to utilize potential divider principles laid out in Chapter 5 to see what voltage appears across the parasitic regions. The resistance surge because of the smaller value of critical voltage V_c in parasitic regions when compared to that in the channel region can considerably degrade the transconductance and other small-signal parameters. When blown-up r_s and r_d are phased in with the intrinsic transconductance g_{mo}, the external transconductance appears as

$$g_m = \frac{g_{mo}}{1 + g_{mo} r_s + g_{do}(r_s + r_d)} \tag{6.68}$$

where g_{mo} and g_{do} are the intrinsic transconductance and drain conductance respectively. Similarly, the unity-gain cutoff frequency f_T is given by

$$f_T = \frac{g_m}{2\pi(C_{GS} + C_{GD})} = \frac{g_m}{2\pi C_G} = \frac{g_m}{2\pi C_{AG} WL} \tag{6.69}$$

For a transistor operating in the saturation region, g_{msat} of Equation 6.64 is appropriate.

The propagation delay of a switching circuit depends on the change in the input to be reflected at the output [29]. This delay can be estimated from charge $C_L V_{DD}$ added to the load capacitor or drawn back from the load capacitor. The load capacitor charges through the PMOS capacitor when the input is low. The average charge $Q = C_L V_{DD}/2$ flowing during the charging process gives $I_{Dsatp} = C_L V_{DD}/2t_{dr}$, where t_{dr} is the propagation delay. t_{dr} as estimated from $I_{Dsatp} = C_L V_{DD}/2t_{dr}$ when the load capacitor is charged is given by

$$t_{dr} = \frac{C_L V_{DD}}{2I_{Dsatp}} \tag{6.70}$$

Similarly, the estimated propagation time when the load capacitor discharges through the NMOS transistor shorted through the ground is given by

$$t_{df} = \frac{C_L V_{DD}}{2I_{Dsatn}} \tag{6.71}$$

The propagation delay is defined as the average of the two time delays

$$t_d = \frac{t_{dr} + t_{df}}{2} = \frac{1}{2}\left[\frac{C_L V_{DD}}{2I_{Dsatn}} + \frac{C_L V_{DD}}{2I_{Dsatp}}\right] \tag{6.72}$$

For matched transistors, $I_{Dsatp} = I_{Dsatn}$ and hence, the delay time is either of the times of Equation 6.70 or 6.71. This has not taken into account the RC timing delay due to the resistance surge in the channel.

Another figure of merit is the product of the energy frequency known as dynamic power. The energy stored in the capacitor is $E_{capacitor} = (1/2)C_L V_{DD}^2$. This energy is dissipated both in the rise and fall of the signal giving the total energy $E_{cycle} = C_L V_{DD}^2$. The dynamic power during switching is the rate of change of energy that for one cycle is

$$P_{dynamic} = \frac{E_{cycle}}{T} = C_L V_{DD}^2 f \tag{6.73}$$

EXAMPLES

E6.1 An NMOS has a substrate doping $N_A = 10^{16}$ cm^3.

a. Evaluate the band-bending E_B when the surface concentration of inverted electrons at the surface is at least equal to the number of holes in substrate, that is, $n_{3s} = p_{3B} = N_A = 10^{16}$ cm^3.
b. Reevaluate for a realistic maximum value $n_{3s} = 2 \times 10^{18}$ cm^3.
c. Find the flat-band surface potential at the threshold

Answer

a. $E_B = qV_t \ln\frac{N_C}{n_{3s}} = 0.0259\,\text{eV} \ln\frac{2.9\times 10^{19}\,\text{cm}^{-3}}{10^{16}\,\text{cm}^{-3}} = 0.207\,\text{eV}$

b. $E_B = qV_t \ln \frac{N_C}{n_{3s}} = 0.0259\,\text{eV} \ln \frac{2.9 \times 10^{19}\,\text{cm}^{-3}}{2 \times 10^{18}\,\text{cm}^{-3}} = 0.069\,\text{eV}$

The change ΔE_B in is $E_B \Delta E_B = 0.207 - 0.069 = 0.138\,\text{eV}$

c. $\phi_s = 2qV_t \ln \frac{N_A}{n_i} = 2 \times 0.0259\,\text{eV} \ln \frac{10^{16}\,\text{cm}^{-3}}{1.08 \times 10^{10}\,\text{cm}^{-3}} = 0.357\,\text{eV}$

which is much larger than the change ΔE_B affected by the increase in carrier concentration.

E6.2 An NMOS has the following design parameter: $W/L = 5$, $\vartheta = 0.13$ V^{-1}, $t_{ox} = 5$ nm, $V_T = 1$ V, and $\mu_S = 480$ (cm²/V · s). Calculate (a) C_{Aox} and (b) μ_{ef} for gate voltage $V_{GS} = 0,1,2,3$, and 4.

$$C_{Aox} = \frac{\varepsilon_{ox}}{t_{ox}} = 6.9 \times 10^{-3} \frac{\text{F}}{\text{m}^2}$$

$$\mu_{ef} = \frac{\mu_S}{1 + \vartheta V_{GT}} = 480, 425, 380, 345, 316 \frac{\text{cm}^2}{\text{V} \cdot \text{s}}$$

PROBLEMS

P6.1 The threshold voltage of an NMOS capacitor on a p-type substrate is $V_T = 1.0$ V. The gate oxide thickness is 15 nm. Its channel length is $L = 1$ μm and the channel width is $W = 100$ μm. The applied gate voltage is 5 V.

a. What is the oxide capacitance per unit area (in F/cm²)?

b. What is the inverted charge in C/cm²?

c. What is the inverted charge (C) in the inverted sheet?

d. How many electrons are in the sheet of charge in (c)?

P6.2 An n-MOSFET is made with $t_{ox} = 4$ nm, $L = 0.5$ μm, $W = 2$ μm, $V_T = 1$ V, $\mu_{lf} = 500$ cm²/V · s, and $v_{sat} = 5 \times 10^6$ cm/s.

a. What is the total charge (in C) in the channel when a gate voltage of 5 V is applied and the drain voltage is 0 V?

b. Find the critical voltage V_C for this MOSFET.

c. Find the value of V_{Dsat} and I_{Dsat} for the applied gate voltage of 5 V. You are free to use any approximation, provided you state so.

P6.3 An n-channel MOSFET has the following characteristics:

$$\mu_{ef} = 400 \frac{\text{cm}^2}{\text{V} \cdot \text{s}}, \quad v_{sat} = 0.6 \times 10^7 \frac{\text{cm}}{\text{s}}, \quad C_{AG} = 2.23 \times 10^{-8} \frac{\text{F}}{\text{cm}^2}$$

$$L = 0.4\,\mu\text{m}, \quad W = 10\,\mu\text{m}, \quad V_T = 1\,\text{V, and } V_{GS} = 3\,\text{V}$$

a. Calculate the critical voltage for the onset of nonohmic behavior.

b. Calculate the channel transconductance.

c. Calculate the transistor cutoff frequency.

P6.4 A complementary pair of an ideal n-and p-channel MOSFET is to be designed to produce the same *I–V* characteristics when they are equivalently biased. The low-field mobility of the n-channel is $\mu_{\ell fn} = 600(\text{cm}^2/\text{V}\cdot\text{s})$ and its saturation velocity is $v_{satn} = 0.6\times10^7$ (cm/s). For the p-channel, the parameters are $\mu_{\ell fp} = 220(\text{cm}^2/\text{V}\cdot\text{s})$ and $v_{satp} = 0.4\times10^7(\text{cm/s})$. The maximum gate width for either the n- or p-channel is 20 μm. Photolithographic limits do not allow fabrication below 100 nm. Design the transistor by specifying the lengths and widths of each channel (n and p). Explain all steps.

P6.5 An enhancement NMOS FET has gate capacitance $C_G = 8.6\times10^{-7}$ F/cm², low-field mobility $\mu_{\ell f} = 500$ cm²/V·s, threshold voltage $V_T = 1$ V, channel length of $L = 1$ μm, and a width of $W = 5$ μm. Considering velocity saturation, with $v_{sat} = 5\times10^6$ cm/s, find the current I_D for

a. $V_{GS} = 0$ V, $V_{DS} = 1$ V

b. $V_{GS} = 2$ V, $V_{DS} = 1$ V

c. $V_{GS} = 3$ V, $V_{DS} = 1$ V

P6.6 A CMOS inverter drives a load that consists of the gate of another FET. The gate area is 0.2 × 5 μm.

a. Find the dynamic power dissipation if the clock frequency is 350 MHz, the supply voltage is $V_{DD} = 2.5$ V, and the oxide thickness is 5 nm.

b. If an MSI circuit (medium-scale integrated circuit) has 1000 transistors, what is the power consumption just due to switching? (Neglect the feedthrough current.)

c. Determine the propagation delay time for the CMOS inverter driving the load of another FET given $I_{Dsarn} = I_{Dsarp} = 1$ mA.

P6.7 In a velocity saturation model, the current is reduced by a factor $1 + (V_{DS}/V_c)$ from its ohmic counterpart. Find the length of NMOS and PMOS that will reduce the current by a factor of 2 for an applied drain voltage of 2 V. The following material properties are given

$$\mu_{\ell fn} = 500\,\text{cm}^2/\text{V}\cdot\text{s} = 0.05\,\text{m}^2/\text{V}\cdot\text{s},\quad \mu_{\ell fp} = 200\,\text{cm}^2/\text{V}\cdot\text{s} = 0.02\,\text{m}^2/\text{V}\cdot\text{s}$$

$$v_{satn} = v_{satp} = 0.4\times10^5\,\text{m/s}$$

P6.8 Consider two silicon MOSFETs, one n-channel and the other p-channel, with substrate dopings of 10^{16} cm^{-3}. The NMOS has an n$^+$ gate and the PMOS has a p$^+$ gate, both doped to 10^{19} cm^{-3}. Find the barrier energy E_B for each as the carrier moves from heavily doped contacts to the channel.

P6.9 Design a capacitor (find an area and geometry) of 1 pF from an NMOS process with the oxide thickness of 4 nm.

P6.10 An NMOS is made with $t_{ox} = 4$ nm, $L = 1$ μm, $W = 2$ μm, $V_T = 1$ V, and $\mu_{lf} = 500 \text{cm}^2/\text{V}\cdot\text{s}$. If the LC model is used, what should the width of the PMOS be to get the same characteristics? Let the low-field mobility for holes be 200 cm^2/V · s.

P6.11 An NMOS and a PMOS are made on the same chip, using the same process. The NMOS has $C'_{ox} = 8.6 \times 10^{-7}$ F/cm^2, $t_{ox} = 4$ nm, $L = 0.2$ μm, $W = 15$ μm, $V_T = 1.5$ V, and $\mu_{lf} = 500$ cm^2/V · s. If the p-channel field-effect transistor (PFET) is identical except for its mobility (200 cm^2/V · s) and width, W.

a. What should W be for the PMOS to make the characteristics the same as for the NMOS, as predicted by the LC model?

b. Find V_{DSsat} and I_{Dsat} for $V_{GS} - V_T = 1$ V.

c. If velocity saturation is considered, how different are V_{DSsat} and I_{Dsat} for the NMOS and the PMOS compared to the simple model results? Express your result as a ratio (e.g., $I_{DsatNMOS}/I_{DsatPMOS}$ and $V_{DSsatNMOS}/V_{DSsatPMOS}$).

P6.12 a. Find W_p/W_n needed to match I_{Dsat} for CMOS transistors if $\mu_{lfn} = 500$ cm^2/V · s, $\mu_{lfp} = 200$ cm^2/V · s, $L = 0.5$ μm, and $V_{GS} - V_T = 2.6$ V according to the LC model.

b. Find V_{DSsat} for the NMOS and the PMOS in the velocity saturation model assuming that $v_{sat} = 4 \times 10^6$ cm/s.

c. Adjust the length of the NMOS to equalize the V_{DSsat}'s in the velocity saturation model. What should the new W_p/W_n be to keep the I_{Dsat}'s equal?

P6.13 A CMOS inverter has $W_p/W_n = 1.5$. The channel lengths are $L = 1$ μm, and $W_n = 10$ μm. Find the propagation delay time if the load capacitance is 1 pF. Let $t_{ox} = 4$ nm, $V_{DD} = 2.5$ V, and $V_{GS} - V_T = 2$ V.

P6.14 According to Gauss's Law, the electric field lines terminating on a sheet of charge is given by $\mathcal{E} = \mathcal{E}\sigma/\varepsilon_s = qn_2/2\varepsilon_s$ giving $n_2 = 2\varepsilon_s\mathcal{E}/q$.

a. Calculate the Fermi energy of Si MOSFET assuming the electron gas is strongly degenerate for a gate electric field of $\mathcal{E} = 5 \times 10^7 V/m$.

b. Repeat for GaAs HEMT.

CAD/CAE PROJECTS

C6.1 *Nano-CMOS*

a. You will plot *I–V* characteristics of a nano-MOSFET with reasonable values of length and width, for example, $W = 1000$ nm, $L = 100$ nm. You will compare the ohmic model (LC) and velocity saturation models. Choose reasonable values of drain and gate voltages.

b. Discuss the transconductance behavior over the whole range of the applied voltage with plots of g_m versus gate and drain voltages.

c. Present a reasonable design format for nano-CMOS as found in the literature using Ohm's law (twentieth-century paradigm). How would you respond to the changes in the paradigm as you encounter the saturation velocity? Discuss the timing delays, power consumption, and high-frequency behaviors.

C6.2 *I–V characteristics using tanh function.* The degenerate characteristics offer complex mathematics in terms of the Fermi–Dirac function. Use the degeneracy temperatures defined in the earlier chapters to see if the tan$h(x)$ function offers any advantage for degenerately induced channels. Also, compare the *I–V* characteristics obtained in the chapter with the approximate form tan$h(x) \approx x/(1+x)$ that does meet the asymptotic limits, but may be different in intermediate values. Plot tan$h(x)$ and its approximate form $x/(1+x)$ to see any advantage in describing *I–V* characteristics.

C6.3 *NMOS operation for signal processing.* Considering the Reference 27 paper by D. R. Greenberg and J. A. d. Alamo, titled Velocity saturation in the extrinsic device: A fundamental limit in HFETs, published in the *IEEE Transactions of Electron Devices*, Vol. 41, pp. 1334–1339, 1994, evaluate the role of source and drain contacts in degrading the signal parameters (unity frequency cutoff, transconductance, etc.) to evaluate the functioning characteristics of a nanoscale MOSFET. Search the literature and make the comparison.

Appendix 6A: Properties of Airy Function

The Schrodinger equation for a triangular potential well is given by

$$\frac{d^2X}{dx^2} + \frac{2m_1^*}{\hbar^2}(E - q\mathscr{E}_T x) = 0 \quad (6A.1)$$

Its solution is the Airy function given by

$$X_n(x) = \frac{1}{Ai'(-\xi_n)\, x_o^{1/2}} Ai\left(\frac{x}{x_o} - \xi_n\right) \tag{6A.2}$$

where ξ_n's are zeros of the Airy function with

$$x_o = \frac{\varepsilon_0}{q\mathcal{E}_T}, \quad \varepsilon_o = \left(\frac{\hbar^2 q^2 \mathcal{E}_T^2}{2m_1^*}\right) \tag{6A.3}$$

Other properties of the Airy function are given by

$$\int_x^\infty Ai^2(x')dx' = -x\,Ai^2(x) + Ai'^2(x) \tag{6A.4}$$

$$\int_x^\infty x'\,Ai^2(x')dx' = \frac{1}{3}\left[-x^2 Ai^2(x) + x\,Ai'^2(x) - Ai(x)Ai'(x)\right] \tag{6A.5}$$

$$\int_x^\infty x'^2\,Ai^2(x')dx' = \frac{1}{5}\left[-x^3 Ai^2(x) + x^2\,Ai'^2(x) - 2x\,Ai(x)Ai'(x) + Ai^2(x)\right] \tag{6A.6}$$

The average distance from the interface and square of the distance is given by

$$x_n = \langle x \rangle_n = \frac{\int x\,X^2(x)dx}{\int X^2(x)dx} = \frac{2}{3}\frac{E_n}{q\mathcal{E}_T} \tag{6A.7}$$

$$\langle x^2 \rangle_n = \frac{6}{5} x_n^2 \tag{6A.8}$$

References

1. S. Oda and D. Ferry, eds., *Silicon Nanoelectronics*. Boca Raton, FL: Taylor and Francis, 2006.
2. B. L. Anderson and R. L. Anderson, *Fundamentals of Semiconductor Devices*. New York, NY: McGraw-Hill, 2005.

3. M. L. P. Tan, V. K. Arora, I. Saad, M. Taghi Ahmadi, and R. Ismail, The drain velocity overshoot in an 80 nm metal–oxide–semiconductor field-effect transistor, *Journal of Applied Physics*, 105, 074503, 2009.
4. J. Walker, *Physics*, 4th ed. Upper Saddle River, NJ: Pearson/Prentice-Hall, 2010.
5. A. T. M. Fairus and V. K. Arora, Quantum engineering of nanoelectronic devices: The role of quantum confinement on mobility degradation, *Microelectronics Journal*, 32, 679–686, 2001.
6. F. Stern, Self-consistent results for n-type Si inversion layers, *Physical Review B*, 5, 4891, 1972.
7. F. Stern and W. E. Howard, Properties of semiconductor surface inversion layers in the electric quantum limit, *Physical Review*, 163, 816, 1967.
8. Y. Taur and T. H. Ning, *Fundamentals of Modern VLSI Devices*. Cambridge, UK: Cambridge University Press, 1998.
9. V. K. Arora and M. B. Das, Role of velocity saturation in lifting pinchoff condition in long-channel MOSFET, *Electronics Letters*, 25, 820–821, 1989.
10. V. K. Arora and M. B. Das, Effect of electric-field-induced mobility degradation on the velocity distribution in a sub-μm length channel of InGaAs/AlGaAs heterojunction MOSFET, *Semiconductor Science and Technology*, 5, 967–973, 1990.
11. K. Lee, M. Shur, T. A. Fjeldly, and T. Ytterdal, *Semiconductor Device Modeling for VLSI*. Englewood Cliffs, NJ: Prentice-Hall, 1993.
12. I. Saad, M. L. P. Tan, R. Ismail, and V. K. Arora, Ballistic saturation velocity of quasi-2D low-dimensional nanoscale field effect transistor, *Nanoscience and Nanotechnology*, AIP Conference.
13. A. Rothwarf, A new quantum mechanical channel mobility model for Si MOSFETs, *IEEE Electron Device Letters*, EDL-8, 499–502, 1987.
14. J. F. Cooper and D. F. Nelson, High-field drift velocity of electrons at the Si–SiO_2 interface as determined by a time-of-flight technique, *Journal of Applied Physics*, 54, 1445–1456, 1983.
15. M. L. P. Tan and R. Ismail, Modeling of nanoscale MOSFET performance in the velocity saturation region, *Jurnal Elektrika*, 9, 37–41, 2007.
16. J.-W. Chen, M. Thurairaj, and M. B. Das, Optimization of gate-to-drain separation in submicron gate-length modulation doped FETs for maximum power gain performance, *IEEE Transactions on Electron Devices*, 41, 465–475, 1994.
17. V. K. Arora, High-field distribution and mobility in semiconductors, *Japanese Journal of Applied Physics, Part 1: Regular Papers and Short Notes*, 24, 537–545, 1985.
18. M. Büttiker, Role of quantum coherence in series resistors, *Physical Review B*, 33, 3020, 1986.
19. M. S. Shur, Low ballistic mobility in submicron HEMTs, *IEEE Electron Device Letters*, 23, 511–513, Sep. 2002.
20. I. Saad, M. L. P. Tan, I. Hui Hii, R. Ismail, and V. K. Arora, Ballistic mobility and saturation velocity in low-dimensional nanostructures, *Microelectronics Journal*, 40, 540–542, 2009.
21. J. Wang and M. Lundstrom, Ballistic transport in high electron mobility transistors, *IEEE Transactions on Electron Devices*, 50, 1604–1609, Jul. 2003.
22. G. Mugnaini and G. Iannaccone, Physics-based compact model of nanoscale MOSFETs—Part II: Effects of degeneracy on transport, *IEEE Transactions on Electron Devices*, 52, 1802–1806, Aug. 2005.

23. G. Mugnaini and G. Iannaccone, Physics-based compact model of nanoscale MOSFETs—Part I, *IEEE Transactions on Electron Devices*, 52, 1795–1801, 2005.
24. K. Natori, New solution to high-field transport in semiconductors: II. Velocity saturation and ballistic transmission, *Japanese Journal of Applied Physics*, 48.3R, 034504, 2009.
25. K. Natori, Current–voltage characteristics of silicon-on-insulator metal–oxide–semiconductor field-effect transistors in ballistic mode, *Japanese Journal of Applied Physics Part 1—Regular Papers Short Notes and Review Papers*, 33, 554–557, Jan. 1994.
26. V. Chan, R. Rengarajan, N. Rovedo, J. Wei, T. Hook, P. Nguyen, C. J. C. E. Nowak et al., High speed 45 nm gate length CMOSFETs integrated into a 90 nm bulk technology incorporating strain engineering, in *IEEE International Electron Devices Meeting 2003*, Washington, DC, 2003, pp. 3.8.1–3.8.4.
27. D. R. Greenberg and J. A. d. Alamo, Velocity saturation in the extrinsic device: A fundamental limit in HFETs, *IEEE Transactions of Electron Devices*, 41, 1334–1339, 1994.
28. M. Shur, *Introduction to Electronic Devices*. New York: John Wiley, 1996.
29. G. Samudra, A. K. F. Yong, T. K. Lee, and V. K. Arora, The role of velocity saturation on switching delay of RC-loaded inverter, *Semiconductors Science and Technology*, 9, 1108–1116, 1994.

7

Nanowire Transport

In this chapter, we cover the application of high-field distribution previously discussed in Chapters 4 and 5 to charge transport in a nanowire. The equilibrium stochastic velocity vectors randomly oriented in and opposite to the quasi-free direction of a nanowire (NW) are shown to streamline in the presence of an extremely high electric field. The complete velocity-field characteristics are acquired. The ultimate directed drift velocity in a towering field is shown to be limited to the appropriately averaged Fermi velocity in the strongly degenerate limit where only half of the quantum states are accessible to electrons. This unidirectional velocity does not depend on the low-field ohmic mobility. The emission of a quantum in the form of a phonon or photon lowers the saturation velocity from its ultimate unidirectional limit.

7.1 Primer

Recent developments in nanofabrication technology have allowed researchers in laboratories around the world to fabricate structures as small as a fraction of a monolayer. One might wonder how much fundamental difference there is by downsizing from bulk to an NW or even to a quantum dot form [1]. In a quantum transport theory for electric conductivity by Arora [1], the ratio of NW mobility to that of a bulk semiconductor was shown to be proportional to A/λ_D^2, where λ_D is the de Broglie wavelength and A is the NW area of the cross-section. In the ultrathin quantum limit ($A \ll \lambda_D^2$), the mobility degrades due to the quantum effect that is sometimes mistaken to be caused by the enhanced scattering from the interfaces. The diameter (and hence area)-dependent performance metrics of InAs NW FETs are investigated by Abul Khayer and Lake [2] using an analytical two-band semiclassical ballistic transport model. A study [3] on size-dependent subband structures of silicon NWs aligned along the [100]-direction, ranging from 0.77 to 2.69 nm in width, is performed by the first-principles calculations. The results presented indicate the importance of size-dependent injection velocity that strongly depends on the Fermi level measured from the conduction band edge when carriers are degenerate. The size-dependent subband structures of NWs are shown to have a serious effect on performance. When compared to a bulk semiconductor, Kim et al. [4] find the potential for up to 50% improvement in performance. Yang et al. [5] provide a perspective with a critical look at the research progress within the NW community for the past decade. These observations are a few of the many good surprises, resulting from strong photon, phonon, and electron confinement within a semiconductor NW.

A historical perspective on growth and transport studies in InAs NWs is provided by Dayeh [6]. Experimental measurements indicating the enhanced probability of a ballistic transport on InAs NWs demonstrate a large electron mfp and correlate with the high electron mobility measured on similar NWs, consistent with the observations of Arora [7]. NW transistor performance and superiority of InAs NWs for high-on currents and high-speed applications is demonstrated where saturation velocity plays a predominant role. Gnani et al. [8] observed mobility degradation at large gate voltages predominantly due to carrier degeneracy, which is particularly severe in 1D nanostructures, much more than an enhanced scattering rate. In fact, a methodology for the extraction of the average momentum-relaxation length from experimental measurements of the drain conductance, independent of the carrier density has been identified. These and other works [9] justify the need to study NWs under strongly degenerate conditions where quantum states are filled up to the Fermi level. As electrons transfer to higher states in streamlined velocity vectors, only half of the phase space is accessible for electrons. The Fermi level then moves higher and gives enhanced Fermi velocity that gives higher saturation velocity. The physics of carrier backscattering in 1D and 2D transistors has been examined analytically and by numerical simulations in a number of works. Kim and Lundstrom [9] derive an analytical formula for the backscattering coefficient for elastic scattering in a 1D channel. The backscattering appears to be identical to the rearrangement of velocity vectors in a strong electric field. An analytical formula for the backscattering coefficient is derived for elastic scattering in a 1D channel. For an inelastic process, the critical length is determined from the phonon energy for both 1D and 2D channels. A limited study of inelastic scattering length limited by the emission of an optical phonon is, therefore, desirable and is included in the results to follow [9].

The issues on the discovery of fundamentally new phenomena versus performance benchmarking for many of the NW applications were originally discussed by Sakaki [10,11]. Sakaki proposed fabrication of an NW transistor [12] from a layered GaAs/AlGaAs heterostructure by V-groove etching. These and other applications are often related to the nanostructure's optical, electrical, and magnetic properties, where one can find related reports and discussion for their bulk counterparts in a separate research community. For example, field broadening $q\mathcal{E}\lambda_D$ caused by the finite extent of the de Broglie wavelength λ_D, where q is the electronic charge, overwhelms the collision broadening in a high electric field $\mathcal{E}$, highlighting the importance of ballistic transport [13,14]. Infrequent comparisons between nanostructured and bulk materials partly reflects the undisputed reality that there is indeed some disconnect between those who are working on semiconductor nanostructures and those who are working on more traditional bulk semiconductors. The connection between NW with 3D bulk is important if we want to have the biggest impact from NW technology. The motivation behind this kind of work is to connect the NWs to bulk semiconductors [13,15] using the high-field transport framework that has been so successful for bulk (3D) semiconductors. As most nanostructures are degenerate in their carrier concentration, an application of the Fermi–Dirac (FD) statistics is the most appropriate as opposed to Maxwell–Boltzmann (MB) statistics applicable only in the ND limit. The degeneracy enhances in a high electric field [16] as higher quantum states are populated.

A NW is a 1D nanoconductor where quasi-ballistic transport with propagating waves is present only in the quasi-free Cartesian direction, while the other two directions are quantum confined with standing waves [17]. The applicability of familiar transport theories is questionable as conducting channels embrace nanoscale dimensions. The drift-diffusion theories based on Monte Carlo simulations and Boltzmann framework are in vogue, but the understanding of the fundamental processes is lost in the computational maze. Reduced channel length L is the key to ballistic transport for three reasons: (1) reduced L below the scattering-limited long-channel (LC) mfp $\ell_{o\infty}$ reduces the probability of collision [18–20]; (2) the presence of a high electric field $\mathcal{E} = V/L$ for any reasonable applied voltage V makes Ohm's law invalid [21,22]; and (3) the proximity of the contacts across L from which the high-energy electrons with Fermi velocity of the contacts are launched [19]. The industry is already making transistors with channel lengths as small as 45 nm [23]. However, the focus in research labs is on devices with acceptable transistor characteristics in the sub-10-nm regime, where classical laws of drift diffusion are invalid [24]. Carbon-based devices are in high gear [25–28] and are on a landscape where energy momentum is linear.

The avenues for studying nanoelectronics transport are many. NEADF [15] is the most direct with natural growth starting from equilibrium carrier statistics as given by the FD distribution. It is also consistent with Boltzmann transport framework as well as equilibrium statistics for Dirac fermions and bosons. NEGF is another approach [29] similar to the density matrix formalism or super operator theory of Arora and Al-Massári [30]. It requires a density matrix in Hilbert space that is much more complicated. The ballistic transport in an NW covers a variety of situations such as: (1) adopting a new device structure; (2) using a different material other than silicon with better transport properties; and (3) exploring a different carrier transport mechanism for current conduction.

7.2 Ballistic Quantum Conductance

Figure 7.1 shows a ballistic conductor with channels through which electrons traverse ballistically. At extremely undersized dimensions L ($L < \ell_{o\infty}$), smaller than the scattering-limited LC mfp $\ell_{o\infty}$, it is natural to surmise that in traversing a carrier path from the source to the drain electrode, the carriers would not undergo any scattering, eliminating the resistance. Contrary to the expectation of collision-free transport, Arora et al. [18,19,31,32] show the probabilistic nature of the collisions that give a finite probability for scattering even in short channels leading to nonunity ballisticity [19] that takes into account ballistic injection from the contacts. The length-limited ballistic mobility μ_{oL} as discussed in Chapter 4 is given by [19]

$$\mu_{oL} = \mu_{o\infty}(1 - e^{-(L/\ell_B)}) \tag{7.1}$$

The ballistic mfp $\ell_B = \ell_{o\infty}(v_{inj}/v_{m1})$ differs from the traditional low-field (o) LC (∞) mfp $\ell_{o\infty}$ by a factor v_{inj}/v_{m1}, where v_{inj} is the velocity of the injected carriers from the contacts and v_{m1} is the channel mobility velocity. v_{m1} is essentially the same as intrinsic velocity in 1D NW except for a numerical factor $\pi/2$:

$$v_{m1} = \left(\frac{\pi}{2}\right)v_{i1} = \left(\frac{\pi}{2}\right)v_{th1}\frac{\Im_o(\eta_o)}{\Im_{1/2}(\eta_o)} \tag{7.2}$$

with

$$v_{th1} = \sqrt{\frac{\pi k_B T}{2m^*}} \tag{7.3}$$

$\Im_j(\eta_o)$ is the FD integral of order j [33], $\eta_o = (E_F - E_c)/k_BT$ is the zero-field-reduced Fermi energy $E_F - E_c$ in terms of thermal energy k_BT, and m^* is the electron effective mass. Ballistic mfp ℓ_B is much larger than its counterpart

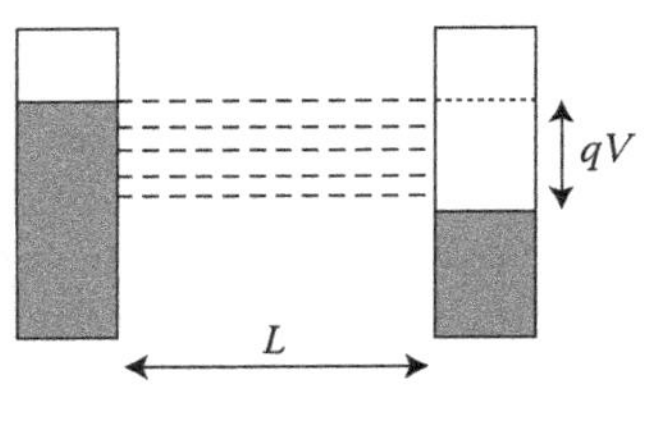

FIGURE 7.1
A ballistic conductor of length L connected between two metallic reservoirs.

$\ell_{o\infty}$ in a LC as $v_{inj} \gg v_{i1}$. The LC mfp $\ell_{o\infty}$ is obtainable from experimentally measured $\mu_{o\infty}$ as given by

$$\ell_{o\infty} = \frac{m^* v_{m1}}{q} \mu_{o\infty} \tag{7.4}$$

At low temperatures, electrons are distinctly degenerate with $v_{inj} = v_{F1}$, the Fermi velocity [33,34] is given by

$$v_{F1} = \frac{\pi\hbar}{2m^*} n_1 = \frac{h}{4m^*} n_1 \text{ (degenerate)} \tag{7.5}$$

where n_1 is the line concentration per unit length for a given subband, normally called a channel in the published literature. The number of channels, as shown in Figure 7.1, depends on the gate voltage that induces a finite number of electrons in the channel. The driving voltage makes it possible for the injected electron from the left reservoir to traverse the channel length L ballistically only if the corresponding state at the injected energy is empty in the right reservoir in conformity with the Pauli Exclusion Principle. The total number of injected electrons is $(Nn_1/2)$ with n_1, the channel concentration per unit length for each channel, and N is the number of channels; the factor of 2 is accounting for $k > 0$ electrons only. In the strongly degenerate regime appropriate for $T = 0$ K, only states with energy $E_{FL} > E > E_{FR}$ contribute to the total current, where E_{FL} or E_{FR} is the Fermi level on the left (L) or right (R) contact region.

Figure 7.2 shows the normalized length-limited mobility $\mu_{oL}/\mu_{o\infty}$ as a function of normalized length L/ℓ_B. The transition from linear ballistic to LC constant mobility takes place at $L = \ell_B$. In the linear $L \ll \ell_B$ regime, the ballistic mobility μ_B is given by

$$\mu_B = \mu_{o\infty} \frac{L}{\ell_B} = \frac{qL}{m^* v_{inj}} = \frac{4qL}{hn_1} \tag{7.6}$$

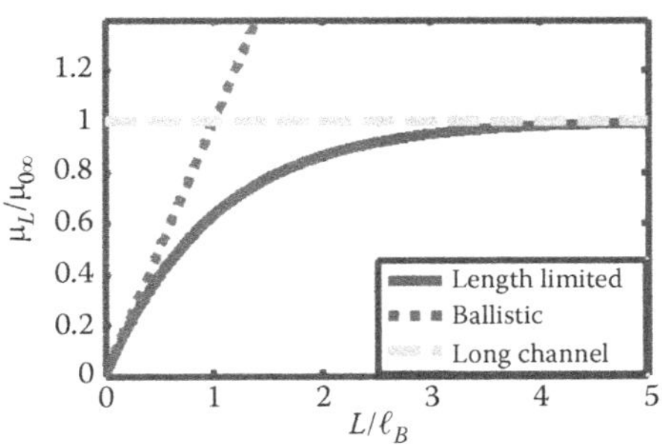

FIGURE 7.2
Normalized mobility versus normalized length. The solid curve is the general curve with the extreme ballistic case and LC constant mobility.

The ballistic resistance R_B of an NW channel with resonant $v_{inj} = v_{F1}$ is given by [35]

$$R_B = \frac{1}{(Nn_1/2)q\mu_B} L = \frac{h}{2q^2N} = \frac{R_Q}{N} \tag{7.7}$$

where $R_Q = (h/2q^2)$ is the resistance quantum. It is worth noting that the number of electrons is very few in an NW. For example, in a 10-nm NW, the line density is $n_1 = 10^8$ m^{-1} with only one electron as the charge cannot be less than $q = 1.6 \times 10^{-19}$ C. This is much higher than the 1D effective density of states at low temperatures. That is why a quantum grain of charge plays a distinct role in an NW as compared to 2D and 3D nanochannels, similar to what is noted in a quantum Hall effect [36–38]. The Pauli Exclusion Principle plays a predominant role with fewer electrons present in an NW. As channels are induced, the first plateau appears at $R_Q = h/2q^2$ and the next one will appear when two levels are induced and so on. In a silicon NW, the degeneracy of the valleys ($g_v = 2$) makes an observable quantum to $h/4q^2$. The quantum of resistance should also be observable in a scaled-down bulk channel with a longitudinal magnetic field present parallel to the electric field. The ballistic formalism thus applies equally to all 1D configurations. This is also the case for a carbon nanotube.

The resistance R/R_B of an NW channel as compared to that of a ballistic channel is given by

$$\frac{R}{R_B} = \frac{(L/\ell_B)}{(1 - e^{-(L/\ell_B)})} \tag{7.8}$$

Figure 7.3 gives R/R_B as a function of L/ℓ_B, converging to unity as $L \to 0$ and is a linear function of L as $L \gg \ell_B$. This type of scaling and electron mfp has been demonstrated by Purewal et al. [39] in a number of experiments on metallic and semiconducting nanotubes.

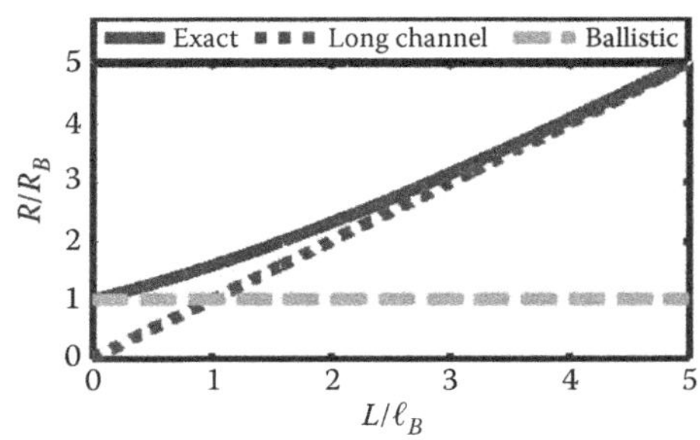

FIGURE 7.3
The resistance in an NW as a function of channel length (solid line). The dotted line is applicable for LC behavior. The flat line is the ballistic quantum resistance.

7.3 Quantum Emission

In an NW, the electrons exhibit a wave-like character with the standing-wave pattern in two of the three Cartesian directions, while the third one is quasi-free with propagating quantum waves. The classical (analog) character exists only in the quasi-free direction parallel to the NW's length, making it a distinct quasi-1D entity. An example of this structure is Sakaki's transistor [40], fabricated as shown in Figure 7.4. Sakaki grew a layered structure of AlGaAs/GaAs/AlGaAs using the molecular beam epitaxy (MBE). An isotropic etch was then used to create a V-groove in the layered structure. An insulting layer was grown and a metal gate was deposited. When a positive gate voltage is applied, an electron gas confined to nanoscale dimensions is induced. When confined to the GaAs nanolayer, the application of the gate voltage induces an NW that has a quasi-free direction only in one direction (parallel to the length of the gate). The electrons are scattered or realign only in the forward or opposite direction depending on the direction of the electric field parallel to the NW. In fact, Sakaki showed that an extremely high mobility is possible at low temperatures where only impurity scattering is predominant. However, no serious work has been done to demonstrate the interrelationship of the ultimate velocity giving saturation to the low-field ohmic mobility. A high mobility may or may not lead to a higher saturation velocity.

NWs of the rectangular, triangular, or circular cross-section can either be easily fabricated or induced in an FET prototype of Figure 7.5. An NW arises when two of the three Cartesian directions are smaller than the de Broglie wavelength of the confined carriers. The energy of an electron confined to an NW has an analog or classical character only in one dimension (taken to be x-axis) and is given by

$$E_k = E_{c1} + \frac{\hbar^2 k_x^2}{2m^*} \tag{7.9}$$

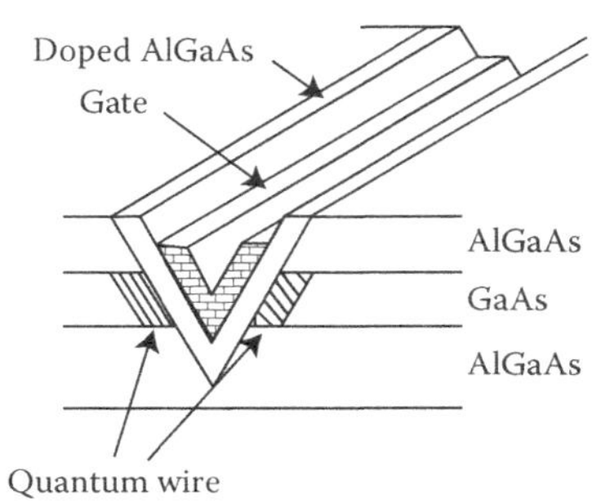

FIGURE 7.4
Field-induced quantum well wires in a V-groove of a layered structure of AlGaAs/GaAs/AlGaAs.

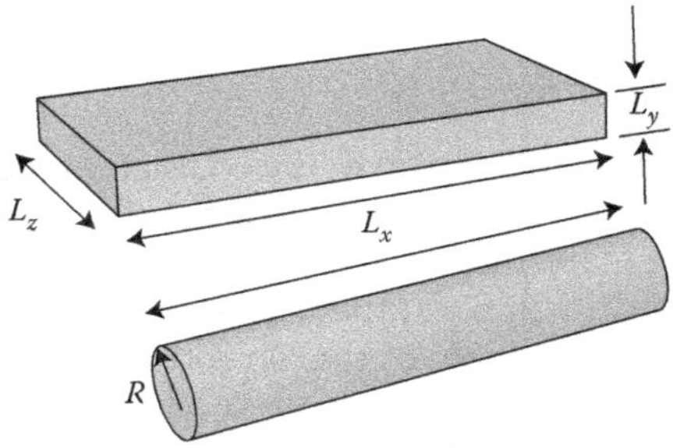

FIGURE 7.5
A prototype NW with $L_{y,z} \ll \lambda_D$ and $L_x \gg \lambda_D$ for a rectangular cross-section and $R \ll \lambda_D$ and $L \gg \lambda_D$ for a circular cross-section.

where $E_{c1} = E_{co} + \varepsilon_o$ is the raised conduction band edge by zero-point energy ε_o, due to quantum confinement in an NW from its level E_{co} in a bulk semiconductor [41]. An NW approaches a quantum limit when the confinement span in the transverse direction is smaller than the de Broglie wavelength, as shown in Figure 7.5. Only the lowest level with energy $E_{co} + \varepsilon_o$ is appreciably populated, making it a distinct 1D entity [17,42]. The analog-type momentum vector $-\infty > k_x > \infty$ is directed along the length of the wire in the x-direction and m^* is the effective mass. The stochastic nature of the electronic motion in equilibrium directs an equal number of electrons along the +ve or −ve direction. An electron in an NW thus behaves very similar to an electron confined to a strong magnetic field. In a magnetic field, the quantized energy is $\varepsilon_o = \hbar\omega_c/2$, where $\omega_c = qB/m^*$ is the cyclotron frequency of the electron orbit perpendicular to the magnetic field B applied in the x-direction.

The theory of quantum transport in an NW in the ohmic regime is well established [17,42]. The original thrust behind the development of NWs was the fact that the ionized impurity scattering is considerably reduced and hence high mobility could be obtained especially at low temperatures [12,43–46]. The mobility in an NW is shown to exceed 10^6 cm^2/V · s at liquid helium temperatures and can be optimized at liquid nitrogen temperatures. The question that remains unanswered in the published literature is: Does a high mobility lead to a higher saturation velocity or higher operational speed? The answer lies in the transition from the scattering-limited ohmic regime to a ballistic regime [7].

To obtain a high-speed and high-frequency performance of such a transistor, the length of the wire should necessarily be small. At such small lengths, high electric fields along the length of the wire are anticipated. The understanding of these high-field effects is crucial to the design of an NW transistor. All scattering interactions change the carrier momentum. However, only some of them, inelastic scattering events change the energy of the carriers. Usually, several scattering events are followed by one inelastic scattering usually by generating a phonon (acoustic or optical). The transition between quantized levels by emitting a photon is also possible. The quantum scattering length ℓ_Q is the length during which a drifting electron in response to an applied electric field $\mathcal{E}$ emits a quantum of average energy $E_Q = \hbar\omega_o(N_o + 1)$ where $\hbar\omega_o$ is the energy of a quantum emitted and $(N_o + 1)$ is the probability of emission. It is given by

$$\ell_Q = \frac{E_Q}{q\mathcal{E}}, \quad E_Q = \hbar\omega_o(N_o + 1) \tag{7.10}$$

with

$$N_o = \frac{1}{[\exp(\hbar\omega_o/k_BT) - 1]} \tag{7.11}$$

where E_Q is the average energy of a quantum causing inelastic scattering. It could be a photon emitted from subband transitions or phonon following Bose–Einstein statistics. The absorption of a quantum is also possible with absorption probability equal to N_o. However, it does not play a significant role in high-field transport. To suppress the emission of an optical phonon, Sakaki [10] suggested an NW array and a box structure. With the development of such arrays, the width and spacing of minibands can be tailored to suppress the emission of an optical phonon. This will enhance the saturation velocity and hence provide a prototype structure for high-speed applications. Unfortunately, such arrays are difficult to fabricate and implement in high-speed circuits. As Equation 7.10 indicates, the quantum emission is inherently impossible in a low electric field as $\ell_Q \gg \ell_{o\infty}$. The quantum emission comes into play only in a high electric field when $\ell_Q \ll \ell_o$. In the limit of low quantum energy, $\hbar\omega_o \ll k_BT$, $E_Q \approx k_BT$.

7.4 Stochastic to Streamlined Unidirectional Velocity

As stated in Chapter 3, the equilibrium carrier statistics in a semiconductor of arbitrary degeneracy is well known to be the FD distribution function given by

$$f(E_k) = \frac{1}{\exp\{(E_k - E_F)/k_BT\} + 1} \tag{7.12}$$

where E_F is the Fermi energy that is above the conduction band edge E_c for a degenerate carrier concentration and in the forbidden bandgap for an ND

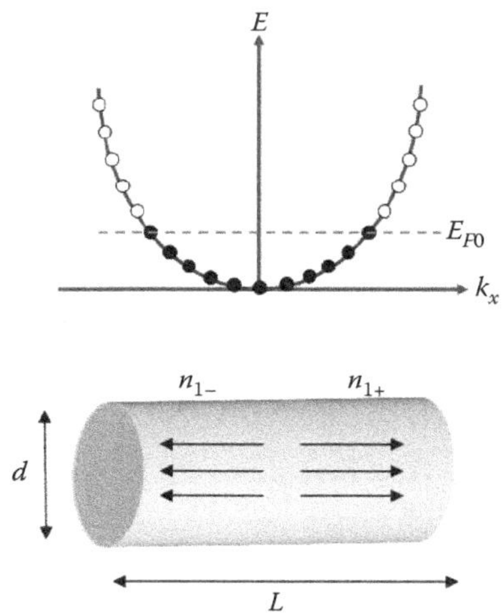

FIGURE 7.6
Stochastic velocity vectors in equilibrium with equal number of electrons in each of the $\pm k_x$ directions filling the states up to the Fermi level E_{Fo} in equilibrium.

concentration. The energy E_k for a quantum NW, as given in Equation 7.12, is the same for carrier momentum $\pm k_x$. The number of carriers with a velocity (or momentum) vector pointing either in the $+x$- or $-x$-direction is equal, canceling the velocity vectors in equilibrium, as expected, with no net drift in either direction. This equilibrium symmetrical distribution of the carriers with net zero drift is shown in Figure 7.6 for degenerate statistics where all energy states are filled up to the Fermi level E_{Fo}. However, the magnitude of the velocity vectors is not zero. The average of the magnitude $|v|$ of the velocity vector, the intrinsic velocity v_{i1} [7,47] is easily calculated by multiplying FD distribution and the density of quantum states [47]. v_{i1} for an NW is given by [7,47] (see Chapter 3)

$$v_{i1} = v_{th1}\left[\frac{\Im_0(\eta_o)}{\Im_{-1/2}(\eta_o)}\right] \tag{7.13}$$

with

$$v_{th1} = \frac{v_{th}}{\sqrt{\pi}}, \quad v_{th} = \sqrt{\frac{2k_BT}{m^*}} \tag{7.14}$$

$$\eta_o = \frac{E_{Fo} - E_{c1}}{k_BT} \tag{7.15}$$

$$\Im_j(\eta) = \frac{1}{\Gamma(j+1)} \int_0^\infty \frac{x^j}{e^{x-\eta}+1}\,dx \tag{7.16}$$

η_o is the reduced Fermi energy with respect to the band edge that depends on the carrier concentration and temperature and can be obtained from

$$n_1 = N_{c1}\Im_{-1/2}(\eta_o), \quad N_{c1} = 2\left(\frac{m^* k_B T}{2\pi\hbar^2}\right)^{1/2} \tag{7.17}$$

$\Im_j(x)$ is the FD integral of order j, n_1 is the carrier density per unit length of the wire, and N_{c1} is the effective density per unit length of quantum states. In a strongly degenerate state appropriate for Figure 7.6, all quantum states up to the Fermi energy E_{Fo} are occupied and all states above E_{Fo} are empty.

With the electric field applied (say in the $-x$-direction) as shown in Figure 7.7, the electron population is depleted in the $-x$-direction as electrons populate the higher quantum states in the $+x$-direction in the k_x-phase space. In the strong degenerate limit, the Fermi level is lifted higher than its equilibrium value due to the Pauli Exclusion Principle. These directed velocity vectors are consistent with the directed moments reported by Lundstrom and Guo [48].

In a towering electric field, as shown in Figure 7.8, virtually all quantum states are depleted in the $-k_x$-phase space and those in the $+k_x$-phase space

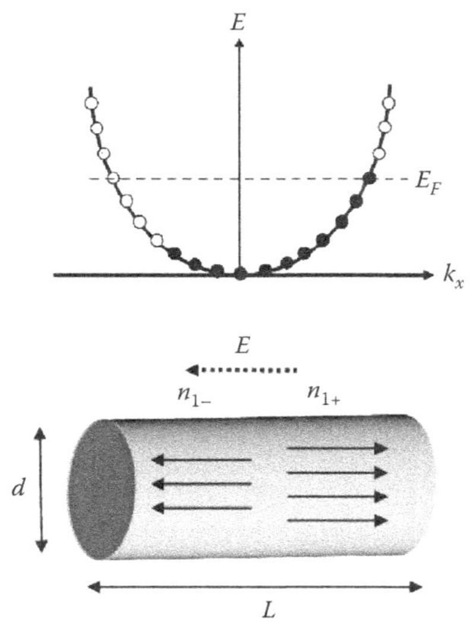

FIGURE 7.7
Nonequilibrium velocity vectors with electrons oriented in the +ve x-direction outweighing those in the $-x$-direction lifting the Fermi level E_F higher ($E_F > E_{Fo}$).

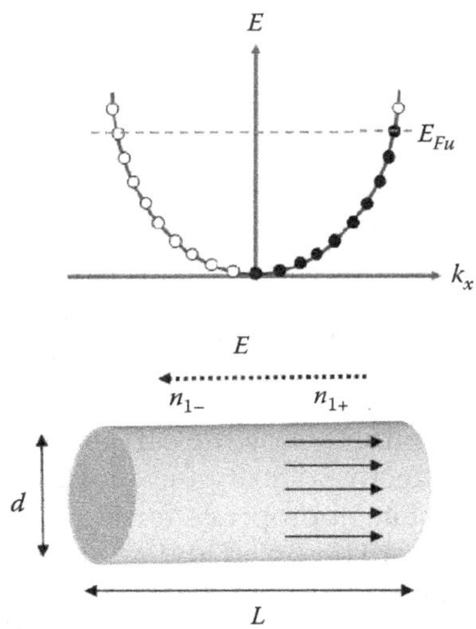

FIGURE 7.8
The extreme nonequilibrium when all velocity vectors are streamlined in the $+x$-direction, with the Fermi level E_{Fu} in the ultimate position.

get saturated with all electrons. In this limit, the unidirectional reduced Fermi energy $\eta_u = (E_{Fu} - E_{c1})/k_BT$ is appropriate to the available $N_{c1}/2$ quantum states only in the forward $+x$-direction. Equivalently, electron population in the available quantum states is twice than that in the equilibrium state.

Figure 7.9 shows the reduced Fermi energy η_o and η_u as a function of the degeneracy level $u_1 = n_1/N_{c1}$. The ratio η_u/η_o saturates to 4 in extreme degeneracy as η is proportional to n_1^2. The unidirectional intrinsic velocity

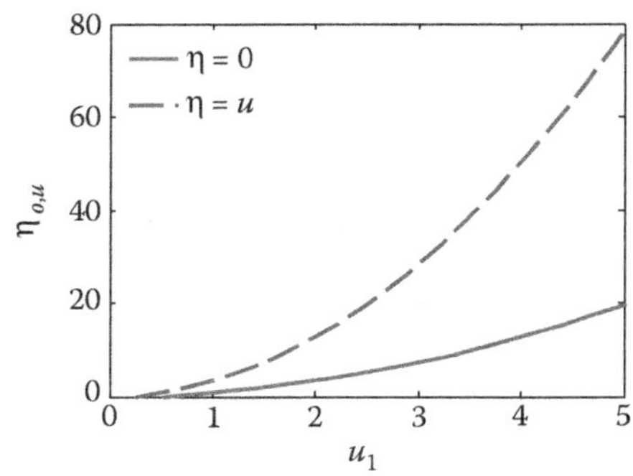

FIGURE 7.9
Comparison of reduced Fermi energy in equilibrium and extreme nonequilibrium states as a function of ID degeneracy $u_1 = n_1/N_{c1}$.

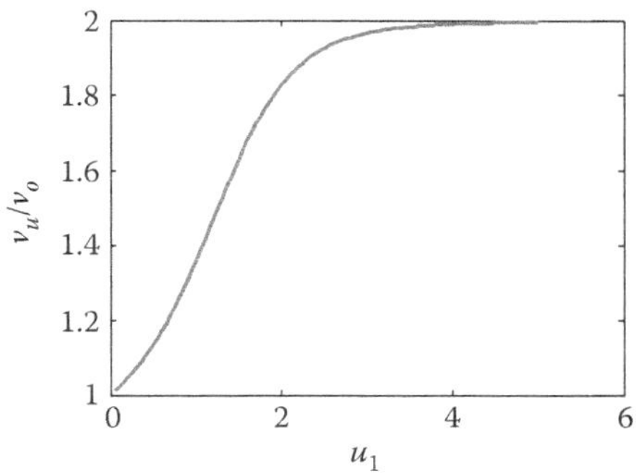

FIGURE 7.10
The ratio of unidirectional intrinsic velocity v_u to that of equilibrium intrinsic velocity v_o as a function of 1D degeneracy $u_1 = n_1/N_{c1}$.

v_{u1} in extreme nonequilibrium of the directed velocity vectors is twice that of the equilibrium intrinsic velocity v_{i1} as shown in Figure 7.10. v_{u1} is appropriate to the Fermi energy η_u. This reorientation of the velocity vectors arises due to transition of the equilibrium FD carrier statistics to the nonequilibrium one as described by NEADF [15] that is the subject of the discussion to follow.

7.5 NEADF Application to NW

NEADF arises as a direct extension of equilibrium FD statistics by inclusion of anisotropy introduced by the energy $q\vec{\mathscr{E}} \cdot \vec{\ell}$ gained/absorbed in and opposite to the electric field in an mfp [34]. For 1D nanowire, NEADF is given by

$$f(E_k) = \frac{1}{\exp(x - \eta \pm \delta_o) + 1} \tag{7.18}$$

with

$$x = \frac{E_k - E_c}{k_B T}, \quad \eta = \frac{E_F - E_c}{k_B T}, \quad \delta_o = \frac{\mathscr{E}}{\mathscr{E}_{co}} \tag{7.19}$$

$E_k - E_c$ is the energy of the electron along the NW with respect to the band edge, $E_F - E_c$ is the Fermi energy, $\mathscr{E}$ is the electric field, and $\mathscr{E}_{co} = V_t/\ell_{o\infty}$ is the critical voltage below which ohmic transport is valid. $\mathscr{E}_{co}$ thus marks the transition from ohmic linear to nonohmic saturated behavior.

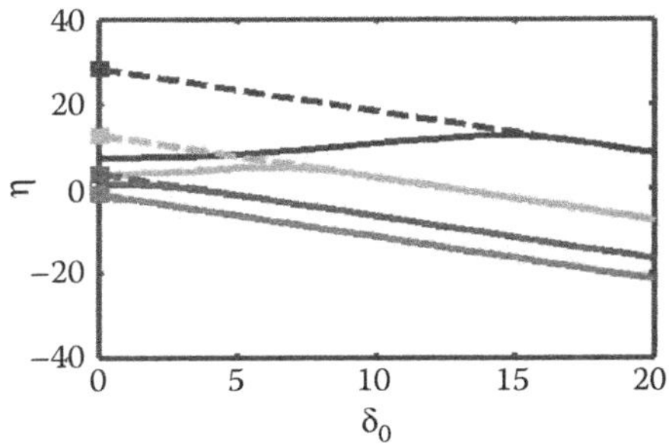

FIGURE 7.11
The reduced Fermi energy as a function of reduced electric field. The solid lines are for η as a function of the electric field with $u_1 = 3$ (top), 2, 1, and 0.1 (bottom). The dashed lines extrapolate to the unidirectional η_u with marked squares.

The Fermi energy is a function of the electric field as electrons with velocity vectors (or mfps) antiparallel to the electric field are favored over those with mfps in the parallel direction. The normalized Fermi energy η in the presence of an electric field is shown in Figure 7.11. As the electric field surpasses its critical value $\delta_o = 1$, the Fermi energy is a linear function of the electric field as given by

$$\eta = \eta_u - \delta_o \tag{7.20}$$

Here, η_u is the Fermi energy appropriate for $2n$ electrons. Figure 7.11 shows this linear behavior in a high-field domain. In extreme nonequilibrium, the electric field is extremely high. Realignment of the velocity vectors depletes phase space parallel to the electric field. Hence, there are $2n$ electrons in one direction and none in the other, causing a unidirectional transport resulting in saturation velocity that is the unidirectional intrinsic velocity v_{u1} for 2n electrons. This lifts the Fermi energy to η_u as quantum states already occupied cannot accommodate realigned electrons and must move to higher quantum states, thereby lifting the Fermi level.

The drift velocity is the average of velocity v with the product of the density of states and the distribution function. Following the same procedures as in Chapter 3 and 4, drift velocity is given by [34]

$$v_D = \frac{v_{th1}}{2u_1} \ln\left[\frac{1 + e^{\eta+\delta_o}}{1 + e^{\eta-\delta_o}}\right] \tag{7.21}$$

where $u_1 = n_1/N_{c1}$ is the normalized concentration with an effective density of states. The normalized drift velocity v_D/v_{u1} as a function of $\delta_o = \mathcal{E}/\mathcal{E}_{co}$ is shown in Figure 7.12 for $u_1 = 0.1$, 1, 2, and 3. The graphs overlap with that obtained from the realignment of electrons with a net ratio $\Delta n_1/n_1$ moving

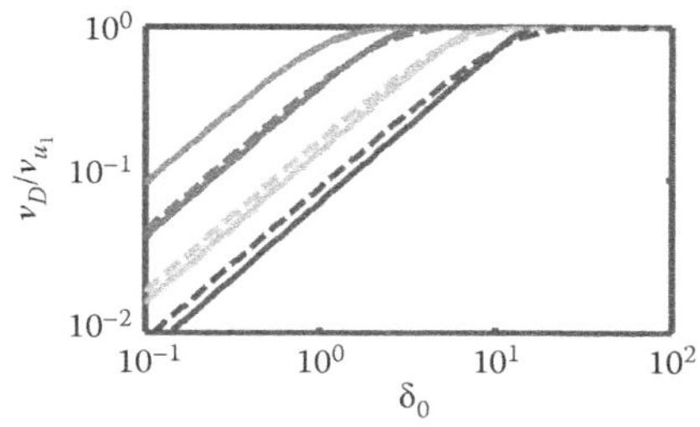

FIGURE 7.12
The normalized drift velocity v_D/v_{u1} as a function of normalized electric field reaching saturation ($v_D/v_{u1} = 1$) in a very high electric field. The dashed curves are $\Delta n_1/n_1$ that are nearly identical to v_D/v_{u1} indicating the carrier realignment in an electric field.

in the antiparallel (+ve direction) to the electric field applied along the –ve direction. $\Delta n_1/n_1$ is given by

$$\frac{\Delta n_1}{n_1} = \frac{(n_{1+} - n_{1-})}{(n_{1+} + n_{1-})} = \frac{v_D}{v_{u1}} \tag{7.22}$$

with

$$n_{1\pm} = N_{c1}\Im_{-1/2}(\eta \pm \delta_o) \tag{7.23}$$

Here, $\Im_{-1/2}(\eta \pm \delta_o)$ is an FD integral of order –1/2.

The quantum emission with a photon, phonon, or transition from one quantum state to the other is accounted for by the replacement of δ_o in Equation 7.23 with δ given by [41]

$$\delta = \delta_o(1 - e^{-(\delta_Q/\delta_o)}), \quad \delta_Q = \frac{E_Q}{k_B T} \tag{7.24}$$

with

$$E_Q = \hbar\omega_q(N_{BE} + 1), \quad N_{BE} = \frac{1}{[\exp(\hbar\omega_q/k_B T) - 1]} \tag{7.25}$$

where $\hbar\omega_q$ is the energy of a quantum and $(N_{BE} + 1)$ is the probability of emission with Bose–Einstein distribution. N_{BE} is applicable to phonons/photons. $N_{BE} \approx 0$ when $\hbar\omega_q \gg k_B T$ depressing the importance of quantum emission. In the other extreme, when $\hbar\omega_q \ll k_B T$, for example, for an acoustic phonon, $N_{BE} \approx k_B T/\hbar\omega_q \gg 1$, the probability of emission is very large making $\delta_Q = 1$. The effect of quantum emission is shown in Figure 7.13 for $\delta_Q = 1$, 5, and 10. The emission of a quantum does not alter in anyway the ohmic region; it affects only the saturation velocity decreasing with the reduced value of δ_Q.

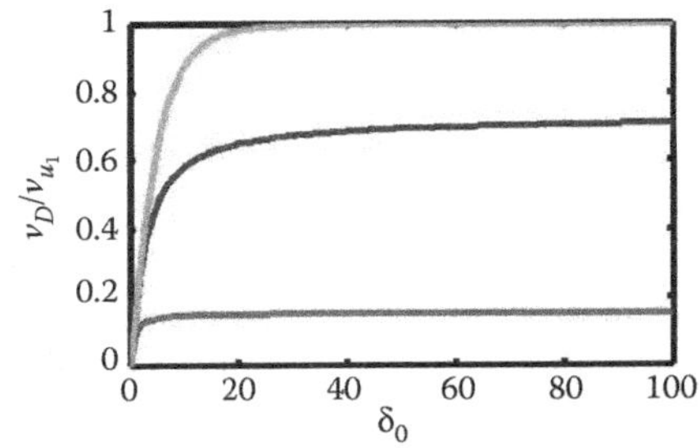

FIGURE 7.13
The normalized drift velocity as a function of electric field lowering the saturation velocity due to the emitted quantum for $\delta_Q = 10$ (top), $\delta_Q = 5$ (middle), and $\delta_Q = 1$ (bottom).

The hot-electron term is normally associated with enhanced energy with the application of an applied electric field. Temperature is, in fact, the measure of disorder or chaos. The temperature in thermodynamics is described by an increase in energy with respect to the Fermi level η that decreases with the electrical field as given in Equation 7.20 and Figure 7.11. In this scenario, an apparent rise in temperature $\Delta T/T = \Delta\eta/\eta_o = \delta_o$ is a linear function of the electric field. The fact and fiction of hot electrons is discussed by Arora [49]. Since the unidirectional electron system is less chaotic in the presence of an electric field, it is inappropriate to think in terms of hot electrons. The apparent increase in the temperature can be attributed to the increase in average energy E_m that can be equated to $(1/2)k_BT_m$. The mean degeneracy temperature T_{m1} arises when the average of energy $E - E_c$ with respect to the conduction band edge is calculated with the FD distribution, leading to the relative temperature

$$\frac{T_{m1}}{T} = \frac{\Im_{1/2}(\eta_o)}{\Im_{-1/2}(\eta_o)} \tag{7.26}$$

This is compared in Figure 7.14 to the relative intrinsic temperature $T_{i1}/T = (v_{i1}/v_{th1})^2$ with a relative increase in kinetic energy. The unidirectional velocity further enhances the degeneracy. Hence, the complete unidirectional degeneracy temperature can be defined as

$$\frac{T_{u1}}{T} = \left(\frac{v_{i1}}{v_{th1}}\right)^2 \left(\frac{v_{u1}}{v_{i1}}\right) \tag{7.27}$$

T_{u1}, T_{m1}, and T_{i1} are depicted in Figure 7.14. Not surprisingly, T_{m1} and T_{i1} are almost the same. However, unidirectional temperature T_{u1} is much higher because of enhanced degeneracy. There is no distinction between the three temperatures in the ND regime.

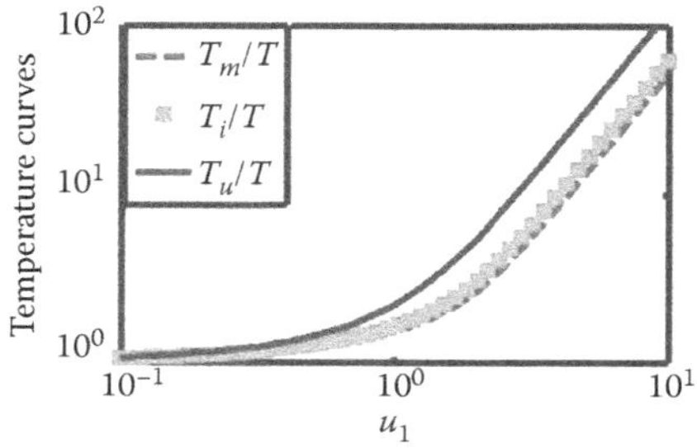

FIGURE 7.14
The degeneracy temperature $T_{i,m,u}/T$ as a function of degeneracy u_1.

This redefinition allows us to describe degenerate statistics of electron transport by a simple relationship. In the ND domain, Equation 7.21 simplifies to give

$$v_D = v_{th1} \tanh(\delta_o) \tag{7.28}$$

Here v_{th1} is the ND limit of the intrinsic velocity v_{i1}. When temperature T in $\mathscr{E}_{co} = k_B T/q\ell_{o\infty}$ is changed to unidirectional temperature T_u and quantum emission is also invoked, Equation 7.28 transforms to

$$v_D = v_{u1} \tanh\left[\delta\left(\frac{T}{T_u}\right)\right] \tag{7.29}$$

The saturation of the drift velocity then simplifies to

$$v_{sat} = v_{u1} \tanh\left[\delta_Q\left(\frac{T}{T_u}\right)\right] \tag{7.30}$$

Equation 7.30 gives insights into the nature of the limitation imposed by the quantum emission as well as by degeneracy on the ultimate saturation velocity. Saturation velocity is affected by the onset of quantum emission, but does not sensitively depend on the scattering parameters of either the phonon or impurity scattering. This shows the ballistic nature of saturation velocity.

As saturation velocity does not sensitively depend on scattering parameters, its independence from scattering-limited mobility is guaranteed. Yes, the presence of degraded ballistic mobility lowers the slope in the linear domain below $\mathscr{E} < \mathscr{E}_{co}$. However, ballistic mobility is highly unlikely to affect the saturation in drift velocity and hence the current in an NW. The current–voltage *I–V* characteristics automatically follow from Equation 7.29 as given by

$$I \approx I_{sat} \tanh\left(\frac{V}{V_c}\right) \tag{7.31}$$

with

$$I_{sat} = n_1 q v_{sat}, \quad v_{sat} = v_{u1} \tanh\left(\frac{\delta_Q T}{T_u}\right) \tag{7.32}$$

The critical voltage is $V_c = \mathcal{E}_c L = (v_{sat}/\mu_L)L$. This makes it convenient to use ballistic mobility or length-limited mfp. The detailed applications of this compact relationship are described in References 22 and 50. Since the contacts and channels are in series, the division of effective resistance is a function of driving current resulting in resistance surge. Such a surge in contact resistance results in the degradation of device behavior [51]. This contact resistance [52,53] arises from the mismatch of the number of conduction channels in the mesoscopic conductor and the macroscopic metal leads. In fact, the ohmic resistance R_o is $R_o = V_c/I_{sat}$. The effective resistance R in a high electric field is given by

$$\frac{R}{R_o} = \frac{V/V_c}{\tanh(V/V_c)} \tag{7.33}$$

The resistance is indeed ohmic (constant) in the $V < V_c$ regime in the limit $V \to 0$. However, $R > R_o$ rises linearly as the applied voltage increases beyond the critical voltage. With ohmic contacts in series with the channel, the lower-length resistor will be more resistive. Hence, the voltage division does not follow the circuit principles distinctly based on Ohm's law. This noted feature is of significant importance in device characterization. The contact resistance is normally evaluated under ohmic conditions. However, when device performance is being assessed at high voltage and high currents, the contact resistance can rise dramatically as evaluated by Greenberg and del Alamo [51]. The degradation of the conducting channel is then natural.

To conclude, the landscape of the ballistic transport with the focus on an NW is explored from the ballistic to the LC regime for arbitrary degeneracy. NEADF is endowed with a clear voyage from drift diffusion to the ballistic regime consistent with Buttiker's paradigm [54,55], with each mfp and resistors of average length ℓ_o. The ends of the mfp resistor are virtual thermalizing probes, one end being higher than the other end in the Fermi level (electrochemical potential) by $q\mathcal{E}\ell_o$. Within this scenario, carriers are removed from the device and injected into a virtual reservoir where they are thermalized and reinjected into the ballistic channel of length ℓ_o. The stochastic nature of electronic motion in equilibrium is demonstrated to transform into the unidirectional streamlined motion in a high electric field, leading to the saturation of drift velocity and current in response to the applied voltage (or electric field).

The investigated quantum electrical properties will form the strong basis of the significant interest and scholarship that exists in part due to manufacturing in advancements and innovation in fabricating nanoscale structures and synthesizing nanomaterials. Ballistic transport in the ohmic and saturation regime is particularly striking in a 1D conductor where δ-function like Van Hove singularities in the density of states plays a distinct role in the quantum character of electronic charge and the associated conduction. This singularity is not present in 2D and 3D conductors where this kind of quantum conductance is not possible. The results above bestow a different paradigm from classical ideas regarding conductance/resistance. In addition to this quantum-mechanical contact resistance, there are other sources of contact resistance, such as that produced by poor coupling between the mesoscopic conductor and the leads. The quantized resistance and conductance can be observed in clean 1D semiconductors at a very low temperature on samples that have a small number of channels [52,53]. The distinction between ballistic mfp and LC mfp is necessary as the carriers injected from the contacts traverse the channel through different quantum levels, keeping in view the Pauli Exclusion Principle. Of broad interest are the results presented above that cut across the diversity of ideas that are present in the published literature and hence useful for a variety of applications.

7.6 NW Transistor

A wide variety of materials and configurations are being tried for an NW transistor. Even the regular MOSFET with an all-around gate may become an NW transistor if the dimensions of the nondepleted part are on the nanometer scale. Figure 7.15 shows a possibility of an NW transistor made of a cylindrical NW around which a cylindrical insulator with a metal or a polysilicon gate exists. In the current state of technology, the fabrication of such a transistor is within our reach [56]. There are a number of other geometries that are fabricated. Perhaps, the most difficult part of the fabrication process is defining the photolithography of such a small structure with perfect geometry. A number of innovations on a CNT transistor have been reported. For example, in a double-walled CNT, either the inner or outer wall can become a channel while the other serves as a gate. The results reported with minor modification will apply to a channel of all cross-sections, including CNTs as well as for carriers confined to a strong magnetic field. All these geometries represent a Q1D character of the carrier gas.

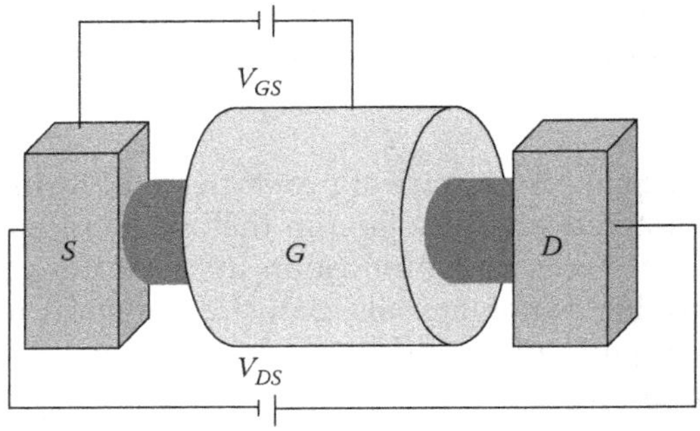

FIGURE 7.15
The schematic of an NW transistor with the gate dielectric.

The velocity saturation effect can conveniently be implemented in the modeling of an NW transistor in Figure 7.15, if the empirical relation for MOSFET used in Chapter 6 is applied. The drain current I_D as a function of the gate voltage V_{GS} and drain voltage V_D is then obtained as

$$I_D = \beta \frac{\left[2V_{GT}V_D - V_D^2\right]}{1 + (V_D/V_c)} \tag{7.34}$$

with

$$\beta = \frac{\mu_{\ell f} C_G}{2L} \tag{7.35}$$

$$V_{GT} = V_{GS} - V_T \tag{7.36}$$

$$V_c = \frac{v_{sat}}{\mu_{\ell f}} L \tag{7.37}$$

where C_G is the gate capacitance per unit length and L is the effective channel length. V_T (taken to be 0.08 V) is the threshold voltage and $\mu_{\ell f}$ is the low-field ohmic mobility that is related to the mfp ℓ_o. A value of $\mu_{\ell f} = 500$ $cm^2/V \cdot s$ is taken in the calculations. The I–V characteristics can be fine tuned by including the mobility degradation due to the quantum confinement. $v_{sat} = v_{i1}$ is the intrinsic velocity in the absence of quantum emission taken to be the thermal velocity for convenience as carrier concentration at the drain end cannot be known until the saturation voltage V_{Dsat} is identified. With the simple geometry of the NW transistor in Figure 7.15, the gate capacitance C_G is given by [13]

$$C_G = \frac{2\pi\varepsilon_{ins}}{\ln(R_{ins}/R_{ewire})} \tag{7.38}$$

where $R_{ins} = t_{ox} + R_{wire}$ is the radius of the SiO_2 insulator, t_{ox} is the thickness of the gate insulator, and $R_{ewire} = R_{wire} - (\varepsilon_{Si}/\varepsilon_{ins})Z_Q$ is the effective radius of the wire in Figure 7.15 with quantum distance Z_Q due to peaking of the wavefunction at a distance z_Q away from the interface. ε_{ins} is the permittivity of the gate insulator taken to be 3.9 that of the SiO_2. The gate capacitance C_Q may be lower than C_{ins} as electrons move away from the Si/SiO_2 interface where the wavefunction vanishes. The contribution of this quantum capacitance is also neglected in Equation 7.38. At the onset of current saturation, all carriers leave the channel with the saturation velocity where the electric field is extremely high. Therefore, the saturation current is given by

$$I_{Dsat} = C_G(V_{GT} - V_{Dsat})v_{sat} \tag{7.39}$$

When Equations 7.34 and 7.39 at the onset of current saturation are consolidated, the drain voltage at which the current saturates is given by

$$V_{Ds\ t} = V_c\left[\sqrt{1 + \frac{2V}{V_c}} - 1\right] \tag{7.40}$$

With this value of V_{Dsat} substituted in Equation 7.39, the saturation current is given by

$$I_{Dsat} = \beta V_{Dsat}^2 \tag{7.41}$$

Figure 7.16 shows the current–voltage characteristics of a transistor with $L = 300$ nm corresponding to $V_C = 0.74$ V. An analysis of the velocity distribution in the channel indicates that the velocity throughout the length of the channel is lower than that on the drain end where carriers have maximum permissible velocity. The equally spaced steps show the behavior characteristic of an SC when $V_{GS} < V_C$. However, for $V_{GS} > V_C$, the NW transistor behaves like an LC; one with steps that grow quadratically with the adjusted overdrive gate voltage V_{GT}. In fact, if Equation 7.40 is expanded to the second order in V_{GT}, one can easily notice the complete absence of the pinchoff effect. The real strength of an NW transistor lies in the elimination of the optical phonon scattering, as suggested by Arora et al. [14] that may give enhanced speed. Another factor that may give enhanced carrier concentration is the complete depletion of the NW so that any positive charge on the gate due to positive gate voltage will necessarily induce more charge in the channel. Sakaki's proposal calls for elimination of only the optical phonon

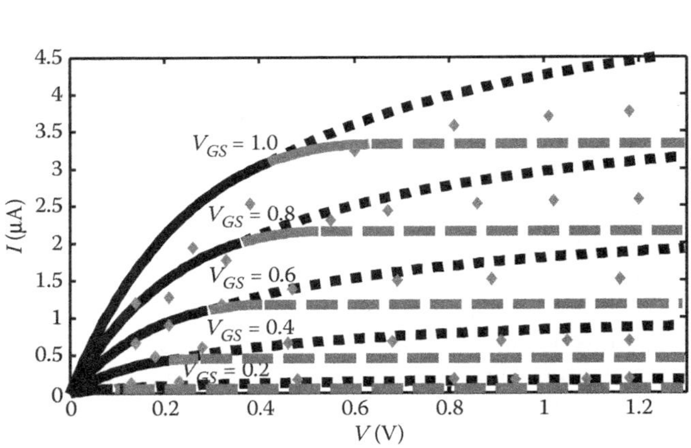

FIGURE 7.16
Current–voltage characteristics in an NW compared with experimental data (diamond) taken from Reference 4 for $V_{GS} = 0.2$, 0.4, 0.6, 0.8, and 1.0 V.

scattering. However, when an electromagnetic field is present in an NW, the possibility of absorption or emission of an electromagnetic quantum cannot be eliminated. Moreover, because of the tilted band diagram that may include quantized energy levels, the electron may easily transfer to a higher level on completing an inelastic scattering length and may emit an electromagnetic quantum. These are some of the interesting opportunities that an NW transistor presents for future investigations.

Figure 7.16 shows the comparison of theory with the experimental data. The comparison with the experimental data is not perfect due to the experimental geometry that is not either fully circular or fully rectangular. Also, there is a complete neglect of quantum capacitance that may bring experimental results closer to the experimental one. Considering some of the imperfections, the agreement of the theory and the experimental data is good. For CMOS applications for a planar transistor, normally, W/L ratios are matched for n- and p-channels for equal currents. In an NW transistor, a similar ratio will be matching the length of the n- and p-channel. However, an important factor that appears in matching the currents is the critical voltage V_C that depends on the saturation velocity. For CMOS applications, it is important not only to match dimensions proportional to the mobility, but also with the saturation velocity. That perhaps can be an important topic to be explored for circuit applications.

Figure 7.17 shows the I–V characteristics of a transistor with the channel length of $L = 10\ \mu m$. The parameters used are extracted from the experimental data to make a comparison. The mobility, μ_{ef} of 480 $cm^2/V \cdot s$ extracted from the experiment [57] is very good and is comparable with the up-to-date silicon NW FET [58]. The thickness of the NW fabricated in the experiment is 5 nm. The fabrication of this gate-all-around silicon NWs on bulk silicon is done using the top-down approach. In Figure 7.17, a correction due to the finite electric field at the drain end with α discussed for MOSFET is implemented.

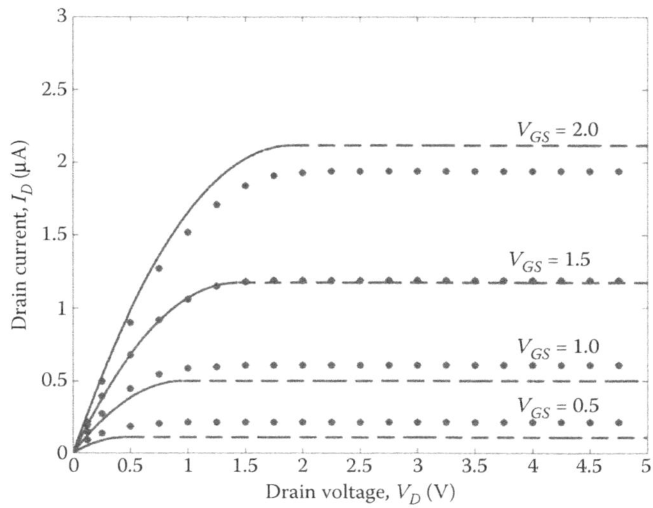

FIGURE 7.17
Current–voltage characteristics of a silicon NW transistor with experimental data.

The experimental data are found to be well fit the theoretical simulation result. There is almost a perfect match for $V_{GS} = 1.5$ V. However, as V_{GS} increases, I_{Dsat} of the experimental data is found to be lower than the simulated one ($V_{GS} = 2.0$ V). Since the quantum emission effect is ignored in this calculation, the simulated results most likely will have a higher drain current in high gate voltage compared to the experimental data. The onset of quantum emission in the experiment would lower the saturation velocity. There is a possibility where quantum emission exists in the experiment produced by Pott et al. [57] and hence lowers the saturation velocity. Lowering the saturation velocity will lead to decreasing of the saturation current as explained by the equation below:

$$I_{sat} = n_1 q v_{sat} \tag{7.42}$$

where n_1 is the concentration per unit length, v_{sat} is the saturation velocity, and I_{sat} is the saturation current.

CAD/CAE PROJECTS

C7.1 *Drift velocity in an NW*

a. Plot the normalized drift velocity v_D/v_{th1} in an NW as a function of $\mathcal{E}/\mathcal{E}_{co}$ for an arbitrary statistics for carrier concentration $u_1 = n_1/N_{c1} = 0.1$, 1, and 5.

b. On the same plot as in (a), plot an ND expression $v_D/v_{u1} = \tanh(\mathcal{E}/\mathcal{E}_c)$ as a function of $\mathcal{E}/\mathcal{E}_{co}$ with $\mathcal{E}_c = \mathcal{E}_{co}(T_{i1}/T)$, where T_{i1}/T is the normalized intrinsic temperature of an NW for $u_1 = n_1/N_{c1} = 0.1$, 1, and 5.

c. Discuss the validity of the ND and degenerate expressions.

C7.2 *Exploration of degeneracy temperature.* Considering the intrinsic, mean, unidirectional degeneracy temperatures of Chapter 3, and ohmic and electron degeneracy temperatures of Chapter 4, evaluate the relevance of each temperature with respect to the ND expressions involving tanh. Is it possible to give a single expression for all dimensionalities and all degeneracies in a unified manner?

C7.3 *Exploration of Fermi energy in a high electric field.* The concept of the Fermi energy is a complex one as most people use Fermi energy as evaluated in 3D semiconductors. In a high electric field, we have shown that the Fermi energy is a function of the electric field that gives unidirectional carrier motion. Evaluate the Fermi energy as a function of the electric field and discuss the unidirectional motion in a wire in that context.

C7.4 *Application to an NW FET.* In Chapter 6, the MOSFET is designed considering the approximation of $\tanh(x) \simeq x/(1+x)$ that give the extreme value of 0 and 1 of tanh (x) in the limit of $x \to 0$ and $x \to \infty$. Use the approximate form as well as $\tanh(x)$ to calculate I–V transistor characteristics for various gate voltages. Discuss the approximations and provide a critique.

References

1. V. K. Arora, Quantum size effect in thin-wire transport, *Physical Review B*, 23, 5611–5612, 1981.
2. M. Abul Khayer and R. K. Lake, Diameter dependent performance of high-speed, low-power InAs nanowire field-effect transistors, *Journal of Applied Physics*, 107, 014502, 2010.
3. Y. Lee, K. Kakushima, K. Shiraishi, K. Natori, and H. Iwai, Size-dependent properties of ballistic silicon nanowire field effect transistors, *Journal of Applied Physics*, 107, 113705, 2010.
4. R. Kim, S. Datta, and M. S. Lundstrom, Influence of dimensionality on thermoelectric device performance, *Journal of Applied Physics*, 105, 034506, 2009.
5. P. Yang, R. Yan, and M. Fardy, Semiconductor nanowire: What's next?, *Nano Letters*, 10, 1529–1536, 2010.
6. S. A. Dayeh, Electron transport in indium arsenide nanowires, *Semiconductor Science and Technology*, 25, 024004, 2010.

7. V. K. Arora, Theory of scattering-limited and ballistic mobility and saturation velocity in low-dimensional nanostructures, *Current Nanoscience*, 5, 227–231, May 2009.
8. E. Gnani, A. Gnudi, S. Reggiani, and G. Baccarani, Effective mobility in nanowire FETs under quasi-ballistic conditions, *IEEE Transactions on Electron Devices*, 57, 336–344, 2010.
9. R. Kim and M. S. Lundstrom, Physics of carrier backscattering in one- and two-dimensional nanotransistors, *IEEE Transactions on Electron Devices*, 56, 132–139, Jan. 2009.
10. H. Sakaki, Quantum wire superlattices and coupled quantum box arrays: A novel method to suppress optical phonon scattering in semiconductors, *Japanese Journal of Applied Physics*, 28, L314, 1989.
11. H. Sakaki, J. I. Motohisa, and K. Hirakawa, Roles of low field mobility and its carrier-concentration dependences in high electron mobility transistors and other field effect transistors, *Electron Device Letters, IEEE*, 9, 133–135, 1988.
12. H. Sakaki, Scattering suppression and high-mobility effect of size-quantized electrons in ultrafine semiconductor wire structures, *Japanese Journal of Applied Physics*, 19, L735–L738, 1980.
13. V. K. Arora, High-field electron mobility and temperature in bulk semiconductors, *Physical Review B*, 30, 7297–7298, 1984.
14. V. K. Arora, M. L. P. Tan, I. Saad, and R. Ismail, Ballistic quantum transport in a nanoscale metal–oxide–semiconductor field effect transistor, *Applied Physics Letters*, 91, 103510, 2007.
15. V. K. Arora, High-field distribution and mobility in semiconductors, *Japanese Journal of Applied Physics, Part 1: Regular Papers and Short Notes*, 24, 537–545, 1985.
16. G. Samudra, S. J. Chua, A. K. Ghatak, and V. K. Arora, High-field electron transport for ellipsoidal multivalley band structure of silicon, *Journal of Applied Physics*, 72, 4700–4704, 1992.
17. V. K. Arora, Quantum well wires: Electrical and optical properties, *Journal of Physics C: Solid State Physics*, 18, 3011, 1985.
18. V. K. Arora, M. S. Z. Abidin, M. L. P. Tan, and M. A. Riyadi, Temperature-dependent ballistic transport in a channel with length below the scattering-limited mean free path, *Journal of Applied Physics*, 111, 054301, Mar. 1, 2012.
19. M. A. Riyadi and V. K. Arora, The channel mobility degradation in a nanoscale MOSFET due to injection from the ballistic contacts, *Journal of Applied Physics*, 109, 056103, 2011.
20. V. K. Arora, M. S. Z. Abidin, S. Tembhurne, and M. A. Riyadi, Concentration dependence of drift and magnetoresistance ballistic mobility in a scaled-down metal–oxide–semiconductor field-effect transistor, *Applied Physics Letters*, 99, 063106, 2011.
21. M. Taghi Ahmadi, H. Houg Lau, R. Ismail, and V. K. Arora, Current–voltage characteristics of a silicon nanowire transistor, *Microelectronics Journal*, 40, 547–549, 2009.
22. T. Saxena, D. C. Y. Chek, M. L. P. Tan, and V. K. Arora, Microcircuit modeling and simulation beyond Ohm's law, *IEEE Transactions on Education*, 54, 34–40, Feb. 2011.
23. ITRS. http://www.itrs.net/links/2009ITRS/Home2009.htm. 2009. *International Technology Roadmap for Semiconductors*. Available: http://www.itrs.net/.

24. P. Michetti, G. Mugnaini, and G. Iannaccone, Analytical model of nanowire FETs in a partially ballistic or dissipative transport regime, *IEEE Transactions on Electron Devices*, 56, 1402–1410, Jul. 2009.
25. A. S. Mayorov, R. V. Gorbachev, S. V. Morozov, L. Britnell, R. Jalil, L. A. Ponomarenko, P. Blake et al., Micrometer-scale ballistic transport in encapsulated graphene at room temperature, *Nano Letters*, 11, 2396–2399, Jun. 2011.
26. E. V. Castro, K. S. Novoselov, S. V. Morozov, N. M. R. Peres, J. M. B. L. dos Santos, J. Nilsson, F. Guinea, A. K. Geim, and A. H. C. Neto, Electronic properties of a biased graphene bilayer, *Journal of Physics-Condensed Matter*, 175503, 2010.
27. K. S. Novoselov, Y. V. Dubrovskii, V. A. Sablikov, D. Y. Ivanov, E. E. Vdovin, Y. N. Khanin, V. A. Tulin, D. Esteve, and S. Beaumont, Nonlinear electron transport in normally pinched-off quantum wire, *Europhysics Letters*, 52, 660–666, Dec. 2000.
28. V. K. Arora, M. L. P. Tan, and C. Gupta, High-field transport in a graphene nanolayer, *Journal of Applied Physics*, 112, 114330, 2012.
29. M. D. Ganji and F. Nourozi, Density functional non-equilibrium Green's function (DFT-NEGF) study of the smallest nano-molecular switch, *Physica E-Low-Dimensional Systems and Nanostructures*, 40, 2606–2613, May 2008.
30. V. K. Arora and M. A. Al-Massári, Superoperator theory of magnetoresistance, *Physical Review B*, 21, 876–878, 1980.
31. V. K. Arora, Ballistic transport in nanoscale devices, presented at the *MIXDES 2012:19th International Conference MIXED Design of Integrated Circuits and Systems*, Wasaw, Poland, 2012.
32. V. K. Arora, M. S. Z. Abidin, M. L. P. Tan, and M. A. Riyadi, Temperature-dependent ballistic transport in a channel with length below the scattering-limited mean free path, *Journal of Applied Physics*, 111, 054301, 2012.
33. R. Qindeel, M. A. Riyadi, M. T. Ahmadi, and V. K. Arora, Low-dimensional carrier statistics in nanostructures, *Current Nanoscience*, 7, 235–239, Apr. 2011.
34. V. K. Arora, D. C. Y. Chek, M. L. P. Tan, and A. M. Hashim, Transition of equilibrium stochastic to unidirectional velocity vectors in a nanowire subjected to a towering electric field, *Journal of Applied Physics*, 108, 114314–114318, 2010.
35. V. K. Arora, High-electric-field initiated information processing in nanoelectronic devices, in *Nanotechnology for Telecommunications Handbook*, S. Anwar, ed. Oxford, UK: CRC/Taylor & Francis Group, pp. 309–334, 2010.
36. M. Buttiker, Edges, contacts and the quantized Hall-effect, *Festkorperprobleme—Advances in Solid State Physics*, 30, 41–52, 1990.
37. K. von Klitzing, Developments in the quantum Hall effect, *Philosophical Transactions of the Royal Society A—Mathematical Physical and Engineering Sciences*, 363, 2203–2219, Sep. 15, 2005.
38. K. von Klitzing, 25 years of quantum Hall effect (QHE): A personal view on the discovery, physics and applications of this quantum effect, *Quantum Hall Effect: Poincare Seminar 2004*, 45, 1–21, 2005.
39. M. S. Purewal, B. H. Hong, A. Ravi, B. Chandra, J. Hone, and P. Kim, Scaling of resistance and electron mean free path of single-walled carbon nanotubes, *Physical Review Letters*, 98, 186808, May 4, 2007.
40. H. Sakaki, Roles of quantum nanostructures on the evolution and future advances of electronic and photonic devices, in *Electron Devices Meeting, 2007. Proceedings of the IEEE International Electron Devices Meeting (IEDM)*, 9–16, 2007.

41. M. T. Ahmadi, M. L. P. Tan, R. Ismail, and V. K. Arora, The high-field drift velocity in degenerately-doped silicon nanowires, *International Journal of Nanotechnology*, 6, 601–617, 2009.
42. V. K. Arora and M. Prasad, Quantum transport in quasi-one-dimensional systems, *Physica Status Solidi (b)*, 117, 127–140, 1983
43. V. K. Arora, *Size Resonance in Quantum Well Wires*, San Francisco, CA, USA: pp. 1461–1464, 1985.
44. H. N. Spector and V. K. Arora, Mobility in a quantum well heterostructure limited by scattering from remote impurities, *Surface Science*, 159, 425–429, 1985.
45. J. Lee, H. N. Spector, and V. K. Arora, Quantum transport in a single layered structure for impurity scattering, *Applied Physics Letters*, 42, 363–365, 1983.
46. J. Lee, H. N. Spector, and V. K. Arora, Impurity scattering limited mobility in a quantum well heterojunction, *Journal of Applied Physics*, 54, 6995–7004, 1983.
47. I. Saad, M. L. P. Tan, I. H. Hii, R. Ismail, and V. K. Arora, Ballistic mobility and saturation velocity in low-dimensional nanostructures, *Microelectronics Journal*, 40, 540–542, Mar. 2009.
48. M. Lundstrom and J. Guo, *Nanoscale Transistor: Device Physics, Modeling and Simulation*. Springer, New York, 2006.
49. V. K. Arora, Hot electrons: A myth or reality? in *International Workshop on the Physics of Semiconductor Devices*, Delhi, India, pp. 563–569, 2002.
50. M. L. P. Tan, T. Saxena, and V. Arora, Resistance blow-up effect in micro-circuit engineering, *Solid-State Electronics*, 54, 1617–1624, Dec. 2010.
51. D. R. Greenberg, and J. A. d. Alamo, Velocity saturation in the extrinsic device: A fundamental limit in HFETs, *IEEE Transactions of Electron Devices*, 41, 1334–1339, 1994.
52. B. J. Vanwees, L. P. Kouwenhoven, E. M. M. Willems, C. J. P. M. Harmans, J. E. Mooij, H. Vanhouten, C. W. J. Beenakker, J. G. Williamson, and C. T. Foxon, Quantum ballistic and adiabatic electron-transport studied with quantum point contacts, *Physical Review B*, 43, 12431–12453, May 15, 1991.
53. B. J. van Wees, H. van Houten, C. W. J. Beenakker, J. G. Williamson, L. P. Kouwenhoven, D. van der Marel, and C. T. Foxon, Quantized conductance of point contacts in a two-dimensional electron gas, *Physical Review Letters*, 60, 848–850, 1988.
54. M. Buttiker and R. Landauer, The Buttiker–Landauer model generalized—Reply, *Journal of Statistical Physics*, 58, 371–373, Jan. 1990.
55. M. Büttiker, Role of quantum coherence in series resistors, *Physical Review B*, 33, 3020, 1986.
56. T. Wang, L. Lou, and C. Lee, A junctionless gate-all-around silicon nanowire FET of high linearity and its potential applications, *IEEE Electron Devices Letters*, 34, 478–480, 2013.
57. V. Pott, K. E. Moselund, D. Bouvet, L. De Michielis, and A. M. Ionescu, Fabrication and characterization of gate-all-around silicon nanowires on bulk silicon, *IEEE Transactions on Nanotechnology*, 7, 733–744, 2008.
58. Y. Cui, Z. H. Zhong, D. L. Wang, W. U. Wang, and C. M. Lieber, High performance silicon nanowire field effect transistors, *Nano Letters*, 3, 149–152, 2003.

8

Quantum Transport in Carbon-Based Devices

High fields tend to reorient carrier moments and velocity vectors under the influence of a high electric field. In this chapter, we explore the modification in the NEADF because of linear energy–momentum relationship prevalent in all carbon-based devices. Ballistic transport comes into play when the channel length is below the scattering-limited mfp. Quantum effects are natural ingredients of ballistic nonequilibrium in all carbon-based devices.

8.1 High-Field Graphene Transport

In the presence of an electric field $\mathcal{E}$, as discussed in Chapter 4, the isotropic equilibrium carrier statistics based on the FD distribution transforms into anisotropic NEADF given by [1,2]

$$f(E,\mathcal{E},\theta) = \frac{1}{e^{(E-(E_F-q\vec{\mathcal{E}}\cdot\vec{\ell}))/k_BT}+1} = \frac{1}{e^{x-H(\theta)}+1} \tag{8.1}$$

with

$$H(\theta) = \eta - \delta_o \cos\theta, \quad \eta = \frac{E_F - E_{co}}{k_BT}, \quad 0 \le \theta \le 2\pi \tag{8.2}$$

$$x = \frac{E - E_{co}}{k_BT}, \quad \delta_o = \frac{\mathcal{E}}{\mathcal{E}_{co}}, \quad \mathcal{E}_{co} = \frac{k_BT}{q\ell} = \frac{V_t}{\ell_{o\infty}}, \quad V_t = \frac{k_BT}{q} \tag{8.3}$$

NEADF of Equation 8.1 is highly asymmetric that favors electrons with velocity vectors antiparallel to the electric field as compared to those with velocity vectors parallel to the electric field. NEADF transforms randomly oriented velocity vectors or mfps or electric dipoles $q\ell$ into unidirectional ones, yielding the high-field saturation velocity equal to v_{Fo} in the absence of a quantum emission [2]. The relative drift velocity v_D/v_{Fo}, as will be shown, is a linear function of the electric field below $\delta_o = 1$ corresponding to the critical electric field $\mathcal{E} = \mathcal{E}_{co}$ and is sublinear as the electric field $\mathcal{E} > \mathcal{E}_{co}$ rises, resulting in saturation for higher values of the electric field with $\delta_o \gg 1$. The thermal voltage V_t and ohmic (o) long-channel LC (∞) mfp $\ell_{o\infty}$ in $\mathcal{E}_{co}$ play prominent roles in this transition [3].

NEADF is endowed with a clear voyage from drift diffusion to the ballistic regime consistent with Buttiker's paradigm [4,5] with each mfp and resistors of average length $\ell_{o\infty}$. The ends of the mfp resistor are virtual thermalizing probes, one end being higher than the other end in the Fermi level (electrochemical potential) by $q\mathcal{E}\ell_o$. Within this scenario, carriers are removed from the device and injected into a virtual reservoir where they are thermalized and reinjected into the ballistic channel of length ℓ_o. The description is also consistent with the Natori model [6] of ballistic transport in the ballistic domain by the Fermi level of the contacts in a MOSFET. In fact, Mugnaini and Iannaccone [7,8] validate this model by applying it to a nanoscale MOSFET. Similarly, CNT and several of its variations in the form of graphene sheets and ribbons are finding a wide variety of applications in monitoring the environment, sensing chemical elements, developing new indigenous materials, and providing interconnects for a variety of biochips. This chapter will open vista for all specialties as NEADF covers a wide range of degeneracy, phonon emission, and the electric field, consistent with experimental observations. The results can be easily converted to obtain current–voltage characteristics as indicated in References 9 and 10.

The drift response v_D to an applied electric field is the average of $v_{FO}\cos\theta$ using the NEADF and DOS, resulting in

$$\frac{v_D}{v_{Fo}} = \left(\frac{1}{2\pi u_g}\right)\int_0^{2\pi} \cos\theta\, d\theta \Im_1(H(\theta)) \tag{8.4}$$

The reduced Fermi energy η is now a function of the electric field and is evaluated from the normalization condition for carrier concentration n_g

$$u_g = \left(\frac{1}{2\pi}\right)\int_0^{2\pi} d\theta \Im_1(H(\theta)), \quad u_g = \frac{n_g}{N_g} \tag{8.5}$$

The effective density of states N_g is given by Equation 3.79. The normalized drift velocity v_D/v_{Fo} as a function of the normalized electric field $\delta_o = \mathcal{E}/\mathcal{E}_{co}$ is shown in Figure 8.1 for $u_g = 1$, 5, and 10. Three sets of curves almost overlap. The dashed lines, barely visible in the high-field domain, are from Equation 8.4. A simplified version of Equation 8.4 is obtained by substitution of cos $\theta = \pm 1/2$ in $H(\theta)$ as distribution is split into the $\pm x$-direction. $\theta = -\pi/2$ to $+\pi/2$ is for the $+x$-direction and $\theta = +\pi/2$ to $+3\pi/2$ is for the $-x$-direction. $\langle\cos\theta\rangle = \pm 1/d$ is for an arbitrary dimensionality $d = 3$, 2, and 1 [2,11]. The squares in Figure 8.1 are obtained when this approximation is applied, indicating the anisotropic character of carrier distribution. The solid lines are variations of Equation 8.5 as delineated below.

An enhanced perspective of this anisotropy is obtained when $\Delta n_g/n_g$ is also plotted in Figure 8.1, as shown by solid lines. $\Delta n_g = n_{g+} - n_{g-}$ is the

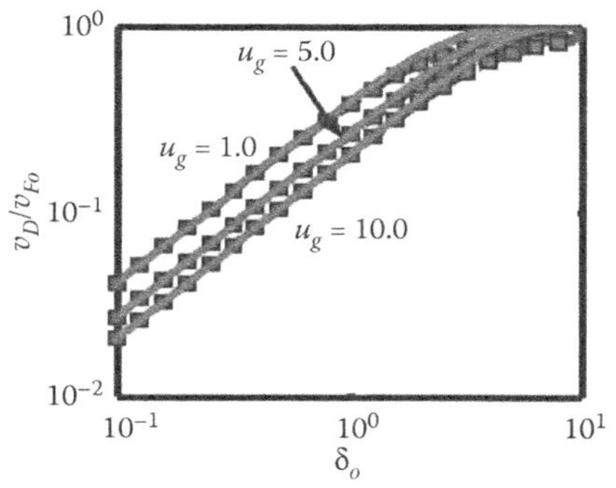

FIGURE 8.1
The dashed lines of Equation 8.4 almost overlapping the solid lines of Equation 8.8 represent the normalized drift velocity v_D/v_{Fo} as a function of the normalized electric field $\delta_o = \mathcal{E}/\mathcal{E}_c$ for $u_g = n_g/N_g = 1, 5$, and 10. The markers are obtained when $\cos\theta$ in $H(\theta)$ of Equation 8.4 is replaced with its average value of $\pm 1/2$.

excess electron concentration in the $+x$-direction (antiparallel to the electric field). $n_g = n_{g+} + n_{g-}$, where $n_{g\pm}$ is the concentration of those velocity vectors whose components are in $\pm x$-direction. $n_{g\pm}$, as obtained from Equation 8.5, is given by

$$n_{g+} = \left(\frac{1}{2\pi}\right) N_g \int_{-\pi/2}^{\pi/2} d\theta\, \Im_1(H(\theta)) \tag{8.6}$$

$$n_{g-} = \left(\frac{1}{2\pi}\right) N_g \int_{\pi/2}^{3\pi/2} d\theta\, \Im_1(H(\theta)) \tag{8.7}$$

The drift response to this orientation of net fraction $\Delta n_{g+}/n_g$ in the positive direction is given by

$$\frac{v_D}{v_{Fo}} = \frac{\Delta n_{g+}}{n_g} = \frac{n_{g+} - n_{g-}}{n_g} \tag{8.8}$$

The onset of quantum emission arises when the energy gained $q\mathcal{E}\ell_Q$ in an inelastic scattering length ℓ_Q equals Δ_Q, the expected energy of the emitted quantum. ℓ_Q is given by

$$\ell_Q = \frac{\Delta_Q}{q\mathcal{E}}, \quad \Delta_Q = \hbar\omega_q(N_o + 1) \tag{8.9}$$

where $\hbar\omega_q$ is the energy of a quantum and $(N_o + 1)$ is the probability of emission with $N_o = 1/[\exp(\hbar\omega_q/k_BT) - 1]$, the Bose–Einstein distribution for phonons/

photons. $N_o \approx 0$ when $\hbar\omega \gg k_BT$, depressing the importance of quantum emission. In the other extreme when $\hbar\omega_q \ll k_BT$, for example, for an acoustic phonon, $N_o \approx k_BT/\hbar\omega_q \gg 1$, the probability of emission is very large. In the presence of quantum emission, δ_o in Equation 8.2 is replaced by δ [2,12] given by

$$\delta = \delta_0[1 - \exp\left(\frac{-\delta_Q}{\delta_0}\right)], \quad \delta_Q = \frac{E_Q}{k_BT} \tag{8.10}$$

The replacement of δ_o with δ does not alter the linear $v - \mathcal{E}$ regime below $\mathcal{E} = \mathcal{E}_c$ and hence the mobility. The saturation velocity is lowered as δ_Q puts a damper on the rise of δ_o with the electric field. Figure 8.2 demonstrates the effect of quantum emission on the saturation velocity with $\delta_Q = 1 - 5$ for $u_g = 10$; the bottom curve is for $\delta_Q = 1$ and the topmost curve is for $\delta_Q = 5$. The quantum emission limits the saturation velocity below the Fermi velocity v_{Fo} in $\delta_o \gg 1$ regime.

A simplified version of Equation 8.8 is obtained by substituting $\cos\theta = \pm 1/2$ in $H(\theta)$ as distribution is split into $\pm x$-direction. $\theta = -\pi/2$ to $+\pi/2$ is for the $+x$-direction and $\theta = +\pi/2$ to $+3\pi/2$ is for the $-x$-direction. $\langle\cos\theta\rangle = \pm 1/d$ is for an arbitrary dimensionality $d = 3$, 2, and 1 [2,13]. The drift response to the electric field is now given by

$$v_D = v_{Fo}\frac{\Im_1(\eta + \delta_o/2) - \Im_1(\eta - \delta_o/2)}{\Im_1(\eta + \delta_o/2) + \Im_1(\eta - \delta_o/2)} \tag{8.11}$$

This expression agrees well with the numerical calculations of the exact equation as indicated in Figures 8.1 and 8.2. The mobility expression naturally follows from this equation by expansion to the low electric field using $d\Im_j(\eta)/d\eta = \Im_{j-1}(\eta)$. This expansion yields the linear drift velocity

$$v_D \approx v_{Fo}\frac{2(\delta_o/2)\Im_0(\eta)}{2\Im_1(\eta)} = \frac{q\mathcal{E}\ell}{2k_BT}v_{Fo}\frac{\Im_0(\eta)}{\Im_1(\eta)} \tag{8.12}$$

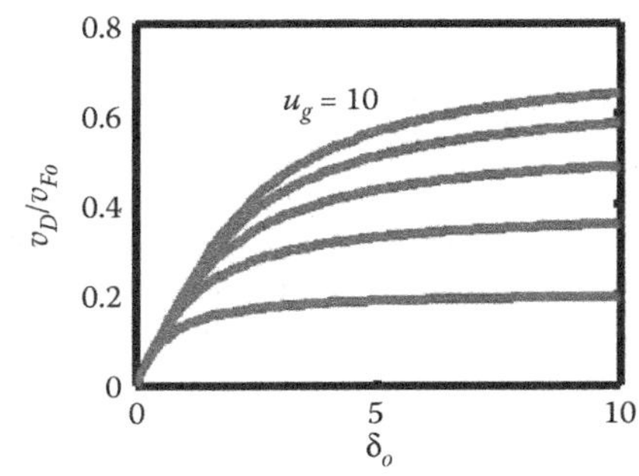

FIGURE 8.2
The effect of quantum emission for $\delta_Q = 1$ (bottom), 2, 3, 4, and 5 (top) for $u_g = 10$. Quantum emission does not affect mobility, but lowers the saturation velocity as δ_Q is reduced.

The mobility $\mu_{og} = v_D/\mathcal{E}$ is given by

$$\mu_{og} = \frac{q\ell v_{Fo}}{2k_BT}\frac{\Im_0(\eta)}{\Im_1(\eta)} \tag{8.13}$$

Graphene with zero bandgap is predominantly degenerate, $\Im_j(\eta) \approx \eta^{j+1}/\Gamma(j+2)$. The mobility expression in the degenerate regime is then obtained as

$$\mu_{og} = \frac{q\ell v_{Fo}}{E_F - E_{Fo}} \tag{8.14}$$

Even though $E_{Fo} = 0$, it is retained to differentiate from carried-induced Fermi energy $E_F = \hbar v_{Fo}\sqrt{\pi n_g}$. This degenerate limit gives the mobility expression $\mu_{og} = q\ell_{o\infty}/\hbar\sqrt{\pi n_g}$ where emphasis on the mfp is added to indicate that it is ohmic mfp for a large sample.

8.2 Application to Experimental Data for Graphene

The universal normalized curves of Figure 8.1 or 8.2 can be broken into individual curves if mfp $\ell_{o\infty}$ is extracted from the mobility values using Equation 8.14

$$\ell_{o\infty} = \frac{\mu_{og}(E_F - E_{co})}{qv_{Fo}} = \frac{\mu_{og}\hbar\sqrt{\pi n_g}}{q} \tag{8.15}$$

where μ_{og} is given by an empirical relation [14]

$$\mu_{og} = \frac{0.465\,(\mathrm{m}^2/\mathrm{V\cdot s})}{1 + (n_g/1.1\times 10^{17}\,\mathrm{m}^{-2})^{2.2}} \tag{8.16}$$

The LC mfp may become ballistic when channel length L is comparable to the ballistic mfp as discussed by Arora et al. [3,15–17]. In that case, length-limited mfp ℓ_L is given by

$$\ell_L = \ell_{o\infty}\left[1 - \exp\left(\frac{-L}{\ell_B}\right)\right], \quad \ell_B = \ell_{o\infty}\left(\frac{v_{inj}}{v_{Fo}}\right) \tag{8.17}$$

where v_{inj} is the injection velocity from the contacts that may or may not differ from the graphene Fermi velocity. The mfp obtained from the mobility is in the order of 100 nm. For a channel length of $L = 4.0$ μm, it is safe to assume that ballistic effects are negligible.

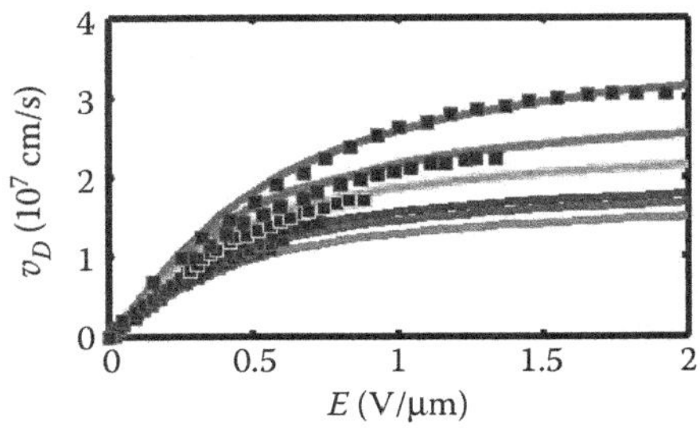

FIGURE 8.3
Drift velocity v_D as a function of the electric field $\mathscr{E}$ at room temperature $T = 280$ K for $n_g/10^{16}$ m^{-2} = 1.6 (top), 2.8, 4.2, 6.4, 8.6, and 10.3 (bottom). The solid line represents v_D from theory Equation 8.10. The markers are the experimental data of Reference 14. (*Applied Physics Letters*, 97, 082112, 2010.)

The solid lines in Figure 8.3 show the drift response to the electric field at $T = 280$ K for $n_g/10^{16}$ m^{-2} = 1.6 (top), 2.8, 4.2, 6.4, 8.6, and 10.3 (bottom). The optical phonon energy $\hbar\omega_o = 55$ meV is appropriate for the SiO_2 substrate. The markers show the data from Reference 14. The agreement is extremely good. In suspended graphene with $\hbar\omega_o = 160$ meV, the phonon emission will not affect the saturation as $\delta_Q \gg 1$. Similarly, ballistic effects with mfp limited to the length of graphene in the direction of the electric field will lower the mobility and hence the critical electric field, while saturation velocity remains the same as limited by the onset of quantum emission. The CNT shares similar characteristics as described by Purewal et al. [18].

8.3 High-Field Transport in Metallic CNT

The potential electronic applications of a CNT are particularly intriguing due to the 1D nature that parallels that of an NW [2]. The current response to applied voltage across the length of the sample results in current saturation for which the mechanism is straightforward, given the nature of intrinsic velocity discussed above. NEADF takes into account the anisotropy in the electronic motion along the length of a CNT by the energy $q\vec{\mathscr{E}} \cdot \vec{\ell}$ gained/absorbed in and opposite to the electric field $\vec{\mathscr{E}}$ in a mfp $\vec{\ell}$. $E_c = E_{Fo} = 0$ for a metallic CNT as indicated above. As NEADF favors electrons in the direction opposite to the applied electric field (say applied in the negative x-direction), the electronic concentration n_{CNT+} in the $+x$-direction increases while n_{CNT-} decreases as the electric field is raised. In a towering electric field, $n_{CNT+} \gg n_{CNT-}$ resulting in current saturation with all directed moments

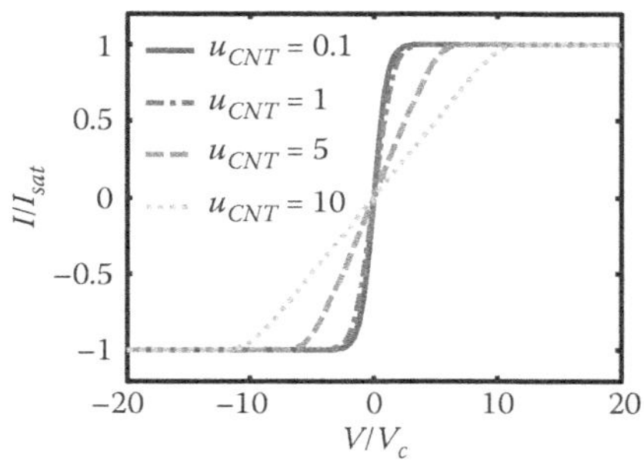

FIGURE 8.4
The normalized current I/I_{sat} as a function of the normalized electric field V/V_C for CNTs with different carrier concentrations ($u_{CNT} = 0.1, 1, 5, 10$).

streamlined. That is the origin of current saturation that has been observed [19]. The saturated current is given by

$$I_{sat} = n_{CNT} q v_{Fo} \tag{8.18}$$

The current as a function of applied voltage along the length of a metallic CNT is given by

$$I = I_{sat} \frac{\Im_o(\eta + \delta_o) - \Im_o(\eta - \delta_o)}{\Im_o(\eta + \delta_o) + \Im_o(\eta - \delta_o)}, \quad \delta_o = \frac{\mathcal{E}}{\mathcal{E}_{co}} = \frac{V}{V_c} \tag{8.19}$$

The Fermi energy η is now the function of the electric field (or applied voltage) that can be evaluated from the normalization. *I–V* characteristics for four values of $u_{CNT} = n_{CNT}/N_{CNT}$ are shown in Figure 8.4. N_{CNT} is the effective density of states for a CNT.

$n_{CNT} = 1.56 \times 10^8\ \text{m}^{-1}$ is obtained from $I_{sat} \approx 25\ \mu\text{A}$ as gathered in the experiments of Yao et al. [19]. Similarly, the critical voltage $V_c = I_{sat}\, R_o = 0.84$ V for ohmic resistance of $R_o = 40\ \text{k}\Omega$. The direct resistance $R = V/I$ surges as the slope of *I–V* curve decreases, as indicated in Figure 8.5. The signal resistance $r = dV/dI$ will rise even faster than the direct resistance.

The resistance surge can be greatly simplified for ND statistics that is not applicable for a metallic CNT. However, it may be valid for SC1(2) CNTs. The simplified relations for ND statistics are given by

$$\frac{I}{I_{sat}} = \tanh\left(\frac{V}{V_c}\right) \tag{8.20}$$

$$\frac{R}{R_o} = \frac{(V/V_c)}{\tanh(V/V_c)} \tag{8.21}$$

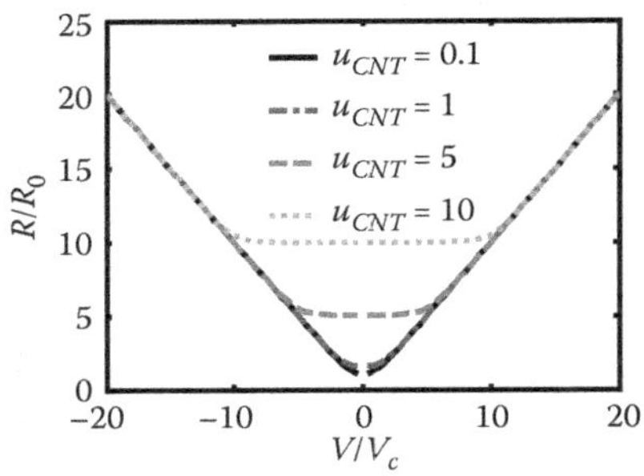

FIGURE 8.5
The plot of normalized resistance R/R_0 as a function of the normalized electric field V/V_C for CNTs with different carrier concentrations (u_{CNT} = 0.1, 1, 5, 10).

$$\frac{r}{R_o} = \frac{1}{\text{sech}^2(V/V_c)} \tag{8.22}$$

The literature is replete with distinct versions of CNTs, breaking them into zigzag, armchair, and chiral versions. However, as of now there is no cohesive paradigm that exists for CNTs of all chirality orientations. In that respect, it is the first attempt that seamlessly brings together the band structure, carrier statistics, and limiting intrinsic velocity to the forefront for applications in quantum and ballistic transport. The key feature of this work is the natural extension of graphene nanolayers to roll up into a CNT that is distinctly clear and simple in mathematical rigor. The established formalism is expected to advance the understanding of nonequilibrium carrier statistics, leading to complete velocity-field profiles similar to the work reported earlier on the graphene nanolayer [20].

8.4 High-Field GNR Transport

The DOS for a graphene nanoribbon (GNR) and carbon nanotube (CNT) are very similar [21]. Equilibrium and nonequilibrium carrier statistics [20] are identical. The DOS for a GNR follows the same pattern as that for a CNT. The differential DOS $D_{GNR}(E)$, following the same procedure [22] as that for a CNT, is given by

$$D_{GNR}(E) = D_0 \frac{|E|}{\left[E^2 - E_c^2\right]^{1/2}} \tag{8.23}$$

with

$$D_o = \frac{4}{\pi\hbar v_{Fo}} = 1.93\,\text{nm/eV}, \quad E_c = \frac{E_g}{2} \tag{8.24}$$

Band index ν from energy E is now dropped as it is redundant for a given ν. This DOS is universal for all configurations ν = 0, 1, and 2 as metallic GNR does not exist. This makes GNR more cumbersome as it is for the semiconducting CNT. In both cases, a CNT integral as discussed earlier is involved. The relative carrier concentration for a CNT $u_{CNT} = n_{CNT}/N_{CNT}$ and that of GNR $u_{GNR} = n_{GNR}/N_{GNR}$ are given by

$$u_{CNT} = \Im_{CNT}(\eta, e_g) \tag{8.25}$$

$$u_{GNR} = \Im_{CNT}(\eta, e_g) \tag{8.26}$$

$$\Im_{CNT}(\eta, e_g) = \int_0^\infty \frac{x + (e_g/2)}{\sqrt{x^2 + xe_g}} \left(\frac{1}{e^{x-\eta} + 1} \right) dx \tag{8.27}$$

with

$$N_{CNT} = D_o k_B T \tag{8.28}$$

$$e_g = \frac{E_g}{k_B T}$$

Here $\eta = (E_F - E_c)/k_B T$ is the reduced Fermi energy from the conduction band edge in terms of thermal energy.

As discussed in Section 8.3, for a metallic CNT ($E_g = 0$), $\Im_{CNT}(\eta, e_g) = \Im_0$ reduces to the FDI of order 0. However, for a semiconducting GNR or CNT, that simplification is not possible. The drift response to an electric field in a 1D CNT/GNR is given by

$$v_D = v_F \left[\frac{\Im_0(\eta + \delta) - \Im_0(\eta - \delta)}{\Im_{CNT}(\eta + \delta) + \Im_{CNT}(\eta + \delta)} \right] \tag{8.29}$$

Following the same linearization procedure as in the case of graphene, the mobility expression is obtained as

$$\mu_{oCNT} = \frac{q\ell v_{Fo}}{k_B T} \frac{\Im_{-1}(\eta)}{\Im_{CNT}(\eta, e_g)} \tag{8.30}$$

Approximate expressions in the degenerate and ND limit can be obtained by using the approximation of $\Im_{-1}$ and $\Im_{CNT}$ in those limits. The same expression equally applies to a GNR.

8.5 Ballistic Transport in Graphene, CNT, and GNR

As stated earlier, the mobility is scattering limited. However, saturation velocity is ballistic in nature in the sense that it does not depend on scattering parameters; it can be lowered by the emission of a quantum in a high electric field. Quantum emission in no way alters the scattering in the ohmic domain and hence the mobility. To understand the true nature of ballistic mobility, the same transient solution that was obtained for parabolic semiconductors can be employed for graphene, CNT, and GNR. Let us examine the response to an electric field applied in the $-x$-direction so that drift proceeds in the positive direction. The electric force is naturally proportional to the rate of change of momentum, as contained in Newton's second law. In addition, the collisions try to fetter the growth in k_x. The differential equation arising from this transient behavior is given by

$$\hbar\frac{dk_x}{dt} = q\mathcal{E} - \frac{k_x}{\tau_c} \tag{8.31}$$

The velocity response v_x related to k_x for graphene is given by

$$v_x = \frac{1}{\hbar}\frac{dE}{dk_x} = v_{Fo}\frac{k_x}{E/\hbar v_{Fo}} = \frac{\hbar v_{Fo}^2}{E}k_x \tag{8.32}$$

For graphene, assuming $E = E_F$ can lead to the transient solution of Equation 8.31 as given by

$$v_x = \mu_{og}\mathcal{E}(1 - e^{-t/\tau_c}), \quad \mu_{og} = \frac{q\ell_{o\infty}v_{Fo}}{E_F} \tag{8.33}$$

Assuming injection from the contacts with injection velocity v_{inj}, the transit time delay is $t = L/v_{inj}$, where L is the length of the ballistic channel. On the other hand, the collision time $\tau_c = \ell_{o\infty}/v_{Fo}$ is connected to mfp through the Fermi velocity. The ballistic mobility μ_{BG} for graphene is then given by

$$\mu_{BG} = \mu_{og}(1 - e^{-L/\ell_B}) \tag{8.34}$$

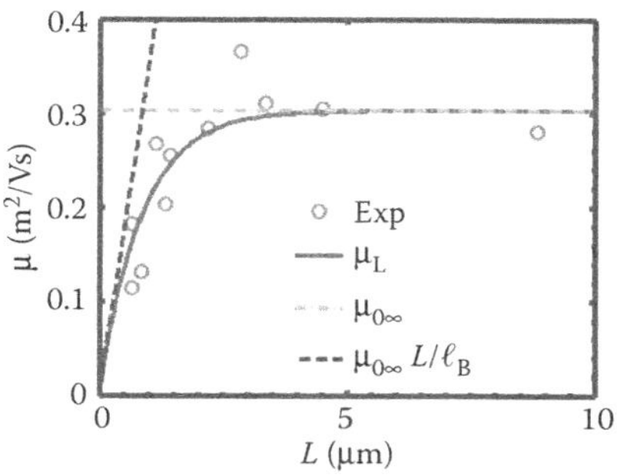

FIGURE 8.6
The mobility as a function of channel length in a graphene nanolayer. The experimental data are that of Zhihong and Appenzeller [23]. The solid line is from Equation 8.34 with ballistic mfp $\ell_B = 860$ nm.

where $\ell_B = \ell_{o\infty}(v_{inj}/v_{Fo})$ is the ballistic mfp distinct from the LC mfp $\ell_{o\infty}$ multiplied by the velocity ratio v_{inj}/v_{Fo}. As contacts are heavily doped, $\ell_B \gg \ell_{o\infty}$ is expected. Normal LC mfps encountered in practice are in the range of 10–100 nm. In the experiment of Zhihong and Appenzeller [23], $\ell_{o\infty} = 100$ nm is assessed. However, that value is inconsistent with that obtained from the $\mu_{BG}(L)$ plot that is around 860 nm. The velocity factor v_{inj}/v_{Fo} is then around 8.6, meaning that injection velocity of the contacts is $v_{inj} = 8.6v_{Fo} = 8.6 \times 10^6$ m/s. In fact, after making the correction for a series resistance of $R * W = 300\ \Omega \cdot \mu$m, the mfp of 300 ± 100 nm is extracted, which may make $v_{inj} = 2.87v_{Fo} = 2.87 \times 10^6$ m/s. The Fermi velocity of the contacts should determine the injection velocity. The uncertainty in the high carrier concentration is cited as the possible cause for wide fluctuations in the mfp. Figure 8.6 shows the length-limited mobility degrading as the channel length is reduced, and becoming linear when the length is below the ballistic mfp.

CNT, GNR, NWs, and carriers confined to 1D nanostructures fall in the category of a 1D system in the sense that propagating waves are free to move in only one direction. CNT, as shown in Figure 8.7, is an example of the propagating wave along the length of the tube when carriers are injected from left or right contacts. The carriers follow the same nonstationary nontransient solution as stated above with injection velocity v_{inj} from the contacts.

Equation 8.34 is still valid for all 1D systems except the mobility now depends on 1D parameters. The LC ohmic mobility is now given by

$$\mu_{oCNT} = \frac{v_{Fo} q \ell}{k_B T} \frac{\Im_{-1}(\eta)}{\Im_{CNT}(\eta, e_g)} \tag{8.35}$$

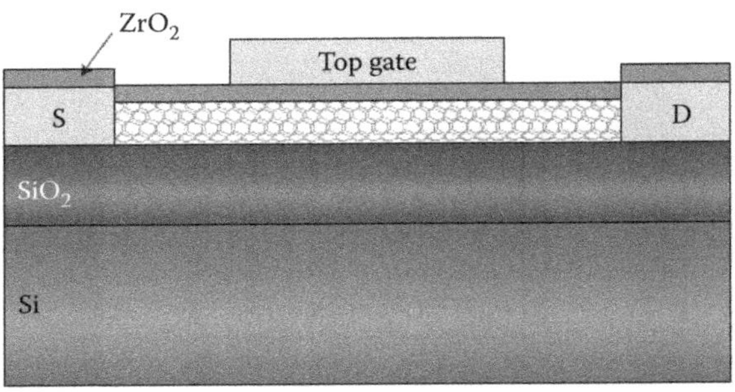

FIGURE 8.7
Configuration of a 1D CNT with the top gate put on a SiO_2 substrate.

In the low-temperature degenerate limit, $\Im_{-1}(\eta) \approx 1$ and $\Im_{CNT} = [\eta^2 + \eta e_{gv}]^{1/2}$. Only for the metallic CNT ($e_g = 0$), $\Im_{CNT} = \Im_o(\eta) = \eta$. Therefore, for a metallic CNT, the mobility expression becomes

$$\mu_{oMCNT} = \frac{v_{Fo} q \ell}{E_F} \tag{8.36}$$

which is the same as that for graphene. This mobility expression does not apply to a GNR as all GNRs are semiconducting with a finite bandgap. The expression for a GNR is

$$\mu_{oMCNT} = \frac{v_{Fo} q \ell}{\left[E_F^2 - (E_g/2)^2\right]^{1/2}} \tag{8.37}$$

As expected, for a narrow bandgap with Fermi level far above the conduction band edge, the expression approaches that of a graphene.

Figure 8.8 shows the ballistic mobility as a function of length for the extracted CNT ballistic mfp of $\ell_B = 1.0\ \mu m$ and LC mobility $\mu_{o\infty} = 1.0\ (m^2/V \cdot s)$, as extracted from experimental results. The mobility degrades when $L \le \ell_B$. The degradation is 63% to a value of $\mu_{oL} = 0.63 (m^2/V \cdot s)$ for $L = \ell_B = 1.0\ \mu m$ and to $\mu_{oL} = 0.39\ (m^2/V \cdot s)$ for $L = \ell_B/2 = 0.5\ \mu m$. This observation is also indicative of the widespread range of mobilities reported in the literature depending on the channel length vis-a-vis ballistic mfp.

Equation 8.34 for a short metallic ($L \ll \ell_{o\infty}$) CNT $E_{FM} = n_{CNT} v_{Fo} h/8$ becomes $\ell_B = \ell_{o\infty}$ as injection is limited to the Fermi velocity for a CNT. In the limit of $L \ll \ell_{o\infty}$, ballistic mobility for a CNT degenerates to

$$\mu_{oL} = \frac{8qL}{n_{CNT} h} \tag{8.38}$$

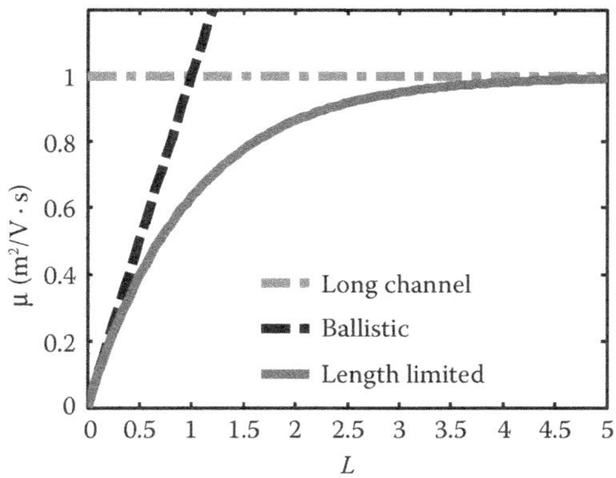

FIGURE 8.8
Length-limited mobility as a function of channel length with long and SC (ballistic) limits indicated.

The resistivity for a 1D metallic CNT is given by

$$\rho_{CNT} = \frac{1}{(n_{CNT}/2)q\mu_{oL}} = \frac{h}{4q^2}\frac{1}{L} \tag{8.39}$$

Here, only half the electrons are propagating along the tube as Pauli Exclusion Principle does not permit back flow to the already-occupied states. The quantum resistance R_{QMCNT} now follows naturally

$$R_Q = \rho_{CNT}L = \frac{h}{4q^2} = 6.453\,\text{k}\Omega \tag{8.40}$$

Using scaling arguments, we can write the resistance of a channel of arbitrary length equal to

$$R_L = R_Q\frac{L/\ell_B}{1-e^{-L/\ell_B}} \tag{8.41}$$

Figure 8.9 shows the length-limited resistance as a function of channel length. The expression converges to R_Q in the low-length limit ($L \ll \ell_B$) and rises linearly $R_L = R_Q(L/\ell_B)$ with the length of the sample in the long-length limit ($L \ll \ell_B$). With $R_L/L = R_Q/\ell_B = 6$ kΩ/μm reported by Reference 24 on a 0.4-cm SWCNT, the ballistic mfp is evaluated as $\ell_B = R_Q/(6$ kΩ/μm$) = 1.076$ μm, which is compatible with experimental values reported in the literature. It is compatible with $\ell_B = \sigma R_Q/2 = 1.0$ μm of what is extracted from measured conductivity $\sigma = 1.4 \times 10^{-8} (\Omega\text{-cm})^{-1}$ on a 0.4-cm CNT.

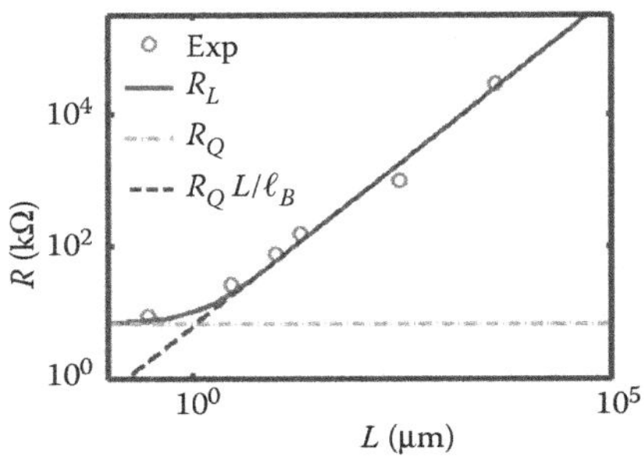

FIGURE 8.9
The length-limited resistance of a ballistic conductor reaching ballistic value $R_Q = h/4q^2$.

Equation 8.41, the length-dependent resistance of a channel of arbitrary length, can be rewritten by employment of the transmission coefficient $\mathfrak{T}$

$$R_L = \frac{R_Q}{\mathfrak{T}} \tag{8.42}$$

with the ballistic transmission factor $\mathfrak{T}$ given by

$$\mathfrak{T} = \frac{1 - e^{-L/\ell_B}}{L/\ell_B} \tag{8.43}$$

Figure 8.10 shows $\mathfrak{T}$ as a function of channel length with $\ell_B = 1.0$ μm and $\ell_{0\infty} = 0.073$ μm. The degradation is much more drastic when ℓ_B is replaced with $\ell_{0\infty}$. In the low-length limit, the transmission is perfect because of the absence of collisions. However, as the channel becomes longer, the transmission goes to zero in an LC limit as scattering degrades the ballisticity of the

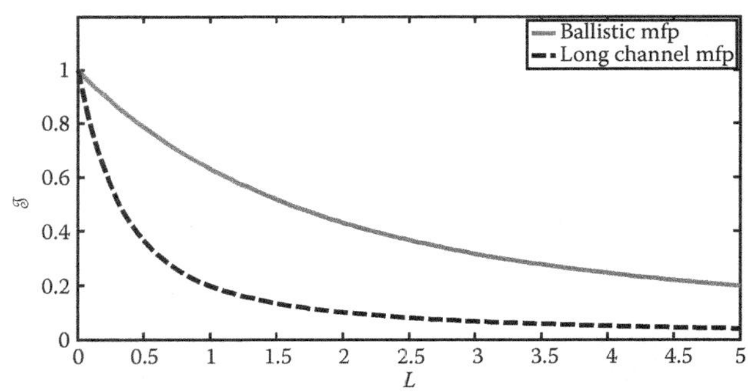

FIGURE 8.10
Transmission through the channel with ballistic mean free path (mfp) and long-channel mean free path.

channel. The transmission $\mathfrak{T}$ should not be confused with the tunneling coefficient, as the channel length is much larger than the De Broglie wavelength. As expected, the transmission is perfect ($\mathfrak{T} = 1$) in the limit of $L \to 0$. Equation 8.42 can be written as

$$R_L = R_Q\left[1 + \frac{1 - \mathfrak{T}}{\mathfrak{T}}\right] = R_Q\left[1 + \frac{\mathfrak{R}}{\mathfrak{T}}\right] \tag{8.44}$$

where $\mathfrak{R} = 1 - \mathfrak{T}$ is the reflection from the other contacts. Perfect transmission is possible when there is no backward transmission ($\mathfrak{R} = 0$).

CAD/CAE PROJECTS

C8.1 *CAD/CAE for a CNT.*

a. In this project, you will create and use the function file for the CNT integral to plot $\eta = (E_F - E_c)/k_BT$ as a function of the relative carrier concentration $u_{CNT} = n_{CNT}/N_{CNT}$. Take one example each of a metallic, SC1, and SC2 on the same graph. On the same graph, plot the degenerate and ND approximations to assess in what range the ND approximation is good and in what range the degenerate approximation is good.

b. Extend the analysis to nonequilibrium statistics by inclusion of the electric field factor $\delta = \delta_o[1 - \exp(E_Q/\delta_o)]$. At this stage, several options open up for your exploration. The first thing you will find is that $\eta = (E_F - E_c)/k_BT$ is now a declining function of the electric field. While using this field-dependent η, you should be able to determine the fraction of electrons $\Delta n_{CNT+}/n_{CNT}$ driven opposite to an electric field applied, say in the $-x$-direction. You may multiply with the intrinsic velocity to get the drift velocity $v_D = v_{iCNT}(\Delta n_{CNT+}/n_{CNT})$ for all chirality directions you have chosen for the rollover into CNT. You may compare this with the exact calculation using the average of velocity from NEADF.

c. You can assess the role of quantum emission E_Q and how it lowers the saturation velocity, but not the ohmic mobility. In fact, you should be able to determine an expression for ohmic mobility. You may be able to compare the results if you find the relevant experimental data from the literature.

C8.2 *CAD/CAE project for a GNR*. Repeat C3.1 for a GNR keeping in view that there is no metallic state. Hence, $\Im_{CNT}$ that applies equally well for a GNR cannot be approximated to FDI $\Im_0(\eta)$ making determination of the Fermi energy, both in equilibrium and in extreme nonequilibrium, much more challenging. Again, you need to carefully choose the zigzag or armchair GNR of a given configuration.

C8.3 *Approximation by a tanh function for a CNT.* In this project, you will assess the validity of a tanh expression for a CNT in the metallic and semiconducting states by replacing ambient temperature with degeneracy temperature using one of the definitions appearing in the book. As tanh function is easy to apply, the key is to carefully assess the critical value of the electric field as well as the critical value of voltage and apply them to the current–voltage relationship. Assess whether the saturation velocity matches with $\tanh(E_Q/k_BT)$ which for a large quantum is unity and for small quanta, lower the saturation velocity to tanh(1), that is, $v_{sat} = v_i \tanh(1)$. The saturation velocity can be anywhere between v_i and $v_{sat} = v_i \tanh(1)$. In fact, you can redefine the critical electric field to $\mathcal{E}_c = (V_t/\ell_{o\infty})$ modified by the degeneracy temperature and the onset of a quantum emission that affects only the saturation velocity, but not the ohmic mobility.

C8.4 *Approximation by a tanh function for a GNR.* In this project, you will repeat what has been said in C3.3 for a GNR.

C8.5 *Design of a CNT transistor.* Taking leads from Chapter 6, design a CNT transistor for signal processing with relevant parameters evaluated.

C8.6 *Design of a GNR transistor.* Taking leads from Chapter 6, design a GNR transistor for signal processing with relevant parameters evaluated.

C8.7 *I–V characteristics of a graphene nanolayer.* Extend the velocity-field characteristics to obtain current–voltage characteristics for a graphene nanolayer, similar to what has been done for a metallic CNT except the carrier statistics applicable in this case is that of a graphene.

C8.8 *Bilayer graphene.* Bilayer graphene, as indicated in the following figure is known to follow the zero-bandgap structure with a parabolic *E–k* relationship with effective mass *m**. Search the literature to identify the essential ingredients and mix with the NEADF defined in this chapter to describe quantum transport in bilayer graphene.

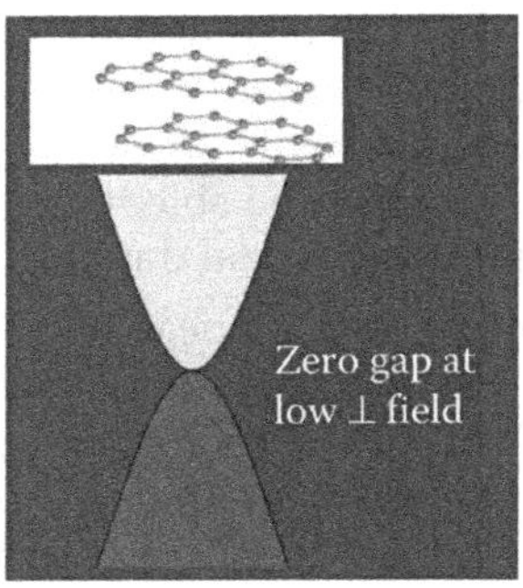

Measurement of the scattering rate and minimum conductivity in graphene.

C8.9 *Trilayer graphene.* Trilayer graphene, as indicated in the following figure, is known to follow bandgap overlap structure with a parabolic E–k relationship with effective mass m^*. Search the literature to identify the essential ingredients and mix with the NEADF defined in this chapter to describe quantum transport in trilayer (or multilayer) graphene. Remember graphite is a multilayered graphene.

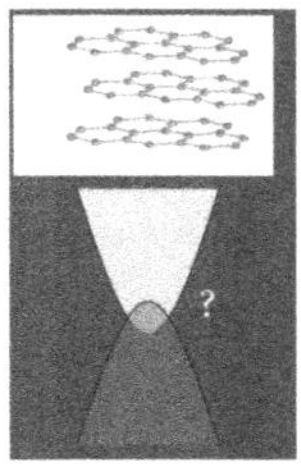

Band structure of trilayer graphene with overlapping bands posing questions to be explored.

References

1. V. K. Arora, High-field distribution and mobility in semiconductors, *Japanese Journal of Applied Physics, Part 1: Regular Papers and Short Notes*, 24, 537–545, 1985.
2. V. K. Arora, D. C. Y. Chek, M. L. P. Tan, and A. M. Hashim, Transition of equilibrium stochastic to unidirectional velocity vectors in a nanowire subjected to a towering electric field, *Journal of Applied Physics*, 108, 114314–114318, 2010.
3. M. A. Riyadi and V. K. Arora, The channel mobility degradation in a nanoscale MOSFET due to injection from the ballistic contacts, *Journal of Applied Physics*, 109, 056103, 2011.
4. M. Buttiker and R. Landauer, The Buttiker–Landauer model generalized—Reply, *Journal of Statistical Physics*, 58, 371–373, Jan. 1990.
5. M. Büttiker, Role of quantum coherence in series resistors, *Physical Review B*, 33, 3020, 1986.
6. K. Natori, Compact modeling of quasi-ballistic silicon nanowire MOSFETs, *IEEE Transactions on Electron Devices*, 59, 79–86, Jan. 2012.
7. G. Mugnaini and G. Iannaccone, Physics-based compact model of nanoscale MOSFETs—Part II: Effects of degeneracy on transport, *IEEE Transactions on Electron Devices*, 52, 1802–1806, Aug. 2005.
8. G. Mugnaini and G. Iannaccone, Physics-based compact model of nanoscale MOSFETs—Part I, *IEEE Transactions on Electron Devices*, 52, 1795–1801, 2005.
9. V. Arora, M. L. P. Tan, and T. Saxena, Resistance blow-up effect in micro-circuit engineering, *Solid-State Electronics*, 54, 1617–1624, Dec. 2010.
10. T. Saxena, D. C. Y. Chek, M. L. P. Tan, and V. K. Arora, Microcircuit modeling and simulation beyond Ohm's law, *IEEE Transactions on Education*, 54, 34–40, Feb. 2011.

11. V. K. Arora, *Quantum Nanoengineering*. Wilkes-Barre, PA: Wilkes University, 2012.
12. G. Samudra, S. J. Chua, A. K. Ghatak, and V. K. Arora, High-field electron transport for ellipsoidal multivalley band structure of silicon, *Journal of Applied Physics*, 72, 4700–4704, 1992.
13. V. K. Arora, *Nanoelectronics: Quantum Engineering of Low-Dimensional Nanoensemble*. Wilkes-Barre, PA: Wilkes University, 2013.
14. V. E. Dorgan, M. H. Bae, and E. Pop, Mobility and saturation velocity in graphene on SiO_2, *Applied Physics Letters*, 97, 082112, Aug. 2010.
15. V. K. Arora, M. S. Z. Abidin, M. L. P. Tan, and M. A. Riyadi, Temperature-dependent ballistic transport in a channel with length below the scattering-limited mean free path, *Journal of Applied Physics*, 111, 054301, Mar. 1, 2012.
16. V. K. Arora, D. C. Y. Chek, and A. M. Hashim, Digital signal propagation delay in a nano-circuit containing reactive and resistive elements, *Solid-State Electronics*, 61, 87–92, Jul. 2011.
17. V. K. Arora, M. S. Z. Abidin, S. Tembhurne, and M. A. Riyadi, Concentration dependence of drift and magnetoresistance ballistic mobility in a scaled-down metal–oxide–semiconductor field-effect transistor, *Applied Physics Letters*, 99, 063106, 2011.
18. M. S. Purewal, B. H. Hong, A. Ravi, B. Chandra, J. Hone, and P. Kim, Scaling of resistance and electron mean free path of single-walled carbon nanotubes, *Physical Review Letters*, 98, 186808, May 4, 2007.
19. Z. Yao, C. L. Kane, and C. Dekker, High-field electrical transport in single-wall carbon nanotubes, *Physical Review Letters*, 84, 2941–2944, 2000.
20. V. K. Arora, M. L. P. Tan, and C. Gupta, High-field transport in a graphene nanolayer, *Journal of Applied Physics*, 112, 114330, 2012.
21. R. Qindeel, M. A. Riyadi, M. T. Ahmadi, and V. K. Arora, Low-dimensional carrier statistics in nanostructures, *Current Nanoscience*, 7, 235–239, Apr. 2011.
22. V. K. Arora and A. Bhattacharyya, Cohesive band structure of carbon nanotubes for applications in quantum transport, *Nanoscale*, 5, 10927–10935, 2013.
23. C. Zhihong and J. Appenzeller, Mobility extraction and quantum capacitance impact in high performance graphene field-effect transistor devices, in *Proceedings of the International Electron Devices Meeting (IEDM), 2008*, 1–4, 2008.
24. S. Li, Z. Yu, C. Rutherglen, and P. J. Burke, Electrical properties of 0.4 cm long single-walled carbon nanotubes, *Nano Letters*, 4, 2003–2007, 2004.

9

Magneto- and Quantum-Confined Transport

The magnetic field is indispensable in the characterization and performance evaluation of semiconductor devices. Magnetoresistance (MR) is the relative change $\Delta\rho/\rho_o$ in resistivity ρ from zero-field resistivity ρ_o when an external magnetic field is applied to them. In ordinary magnetoresistivity (OMR), the effect is rather small and can be classically described as a result of the force acting on conducting electrons in the magnetic and electric field. A 3D sample can be converted into a 1D configuration when the magnetic field is in the quantum domain. Unfortunately, for transverse configuration with the electric field perpendicular to the applied magnetic field, the traditional framework based on FD statistics and Boltzmann transport equation (BTE) is of limited use and density matrix takes on an increasing importance [1]. A quantum-mechanical theory using a density matrix for electrical conductivity in the presence of a magnetic field is described. The conventional methods using the BTE are not satisfactory because the magnetic field introduces a curvature in the free path of an electron. The expectation value of the current density and the components of the conductivity tensor in a magnetic field are obtained. The advantages of the density matrix in electric transport problems with a magnetic field are discussed.

9.1 Classical Theory of MR

As discussed earlier, the conductivity of a 3D material is the reciprocal of resistivity:

$$\sigma_3 = \frac{1}{\rho_3} = \frac{n_3 q^2 \tau}{m^*} = \frac{n_3 q^2 \ell}{m^* \, v_{i3}} \tag{9.1}$$

In the following equations, σ_0 is used to mean the 3D zero-magnetic-field conductivity in place of σ_3, and ρ_o is used for ρ_3.

In the presence of the electric field $\mathcal{E}$ and magnetic field B, the force on the electron is

$$\vec{F} = -q(\vec{\mathcal{E}} + \vec{v} \times \vec{B}) \tag{9.2}$$

The transient effect of the change in velocity in response to this force can now be described by the differential equation

$$m^* \frac{d\vec{v}}{dt} = -q(\vec{\mathcal{E}} + \vec{v} \times \vec{B}) - \frac{\vec{v}}{\tau} \tag{9.3}$$

where τ is used to mean the collision time τ_c. The steady-state solution $(d\vec{v}/dt = 0)$ of this equation is obtained as

$$\vec{v} = -\frac{\mu_0}{1+(\mu_0 B)^2}\left(\vec{\mathcal{E}} + \mu_0 \vec{\mathcal{E}} \times \vec{B}\right) \quad \text{with} \quad \mu_o = \frac{q\tau}{m^*} \tag{9.4}$$

For a magnetic field in the z-direction, the components of velocity vector $\vec{v}$ are obtained as

$$v_x = -\frac{\mu_0}{1+(\mu_0 B)^2}\left(\mathcal{E}_x - \mu_0 \mathcal{E}_y B\right) \tag{9.5}$$

$$v_x = -\frac{\mu_0}{1+(\mu_0 B)^2}\left(\mathcal{E}_y + \mu_0 \mathcal{E}_x B\right) \tag{9.6}$$

$$v_z = \mu_o \mathcal{E}_z \tag{9.7}$$

The current density $\vec{j} = -n_3 q \vec{v}$ is now obtained as

$$j_x = \sigma_{xx}\mathcal{E}_x + \sigma_{xy}\mathcal{E}_y \tag{9.8}$$

$$j_y = \sigma_{yx}\mathcal{E}_x + \sigma_{yy}\mathcal{E}_y \tag{9.9}$$

$$j_z = \sigma_{zz}\mathcal{E}_z = \sigma_o \mathcal{E}_z \tag{9.10}$$

with

$$\sigma_{xx} = \sigma_{yy} = \frac{\sigma_0}{1+(\mu_0 B)^2} \tag{9.11}$$

$$\sigma_{zz} = \sigma_o \tag{9.12}$$

$$\sigma_{xy} = -\sigma_{yx} = -\frac{\sigma_0}{1+(\mu_0 B)^2}\mu_0 B \tag{9.13}$$

The z-component of the current is not affected by a magnetic field in the z-direction. The diagonal components σ_{xx} and σ_{yy} of the magnetoconductivity decrease monotonically as the magnetic field is increased. The magnitude of the off-diagonal components σ_{xy} and σ_{yx} first increases and then decreases as B is increased. The relative MR in the classical model vanishes as

$$\frac{\Delta\rho_{xx}}{\rho_0} = \frac{\sigma_{xx}\sigma_0}{\sigma_{xx}^2 + \sigma_{xy}^2} - 1 = 0 \tag{9.14}$$

when the conductivity tensor is inverted into the resistivity tensor. Since conductivity is a tensor, the inversion of $\rho_{xx} = 1/\sigma_{xx}$ is not appropriate, but is used in the published literature to extract mobility from MR experiments. This incorrect form of the relative incremental MR is given by

$$\frac{\Delta\rho_{xx}}{\rho_0} = \frac{\sigma_0}{\sigma_{xx}} - 1 = (\mu_0 B)^2 \tag{9.15}$$

Comparison of Equation 9.15 with Equation 9.14 clearly indicates that Equation 9.15 is not correct for a constant collision time. Arora et al. [2] resolved the apparent contradiction between the rise of MR mobility and fall of drift mobility with increasing channel concentration. They attribute this to the scattering-dependent MR factor that is evaluated by $\mu_{MR} = MR\mu_o$. Inversion of tensors cannot be term by term as is done in Equation 9.15. However, as shown later, the expression can be reproduced when the energy dependence of the relaxation time is taken into account with a proportionality factor A_o so that $\Delta\rho_{xx}/\rho_0 = A_o(\mu_0 B)^2$ with $A_o = 0$ for the energy-independent mean-free time τ and $A_o = MR^2$ when the energy dependence of collision time is taken into account.

9.2 Rationale for Density Matrix

There have been many studies of electronic transport in solids by using the BTE [3]. BTE has been very successful for problems involving zero or magnetic field in the order of a fraction of a tesla (T) that is a few kilogauss (kG). For high magnetic fields, a density-matrix approach is required. Unfortunately, a density-matrix approach used in the literature is quite difficult to understand. A simple quantum-mechanical derivation of the field terms of the BTE based on the gage-independent density-matrix formalism was given by Weisenthal and de Graaf [4]. With the magnetic field present, extreme care is required in the correct interpretation of the physical quantities involved for the conduction process. The scalar conductivity σ_3 in the absence of a magnetic field for a simplified free-electron model is given by

$$\sigma_3 = n_3 q \mu_o = \frac{n_3 q^2 \tau}{m^*} \tag{9.16}$$

where n_3 is the electron density, $-e$ is the electronic charge, m^* is the effective mass, and τ_c is the average collision time of an electron. The binding energy of an electron to host atoms will increase in the presence of a magnetic field. This decreases the probability of the electron to be found in a free state, and hence decreases the number of conduction electrons. This "freeze-out effect" certainly could be neglected for an ND semiconductor, and hence the number of conduction electrons remains essentially unchanged in a magnetic field:

$$n_3(B) \approx n_3(0) \tag{9.17}$$

The effective mass m^* of an electron that is strongly related to the band structure can be assumed to be constant for a semiconductor if the radius of the cyclotron orbit in a magnetic field B, $R_1 = (\hbar/qB)^{1/2} \gg a$, where a is the interatomic spacing. The fluctuation in the wave function over the lattice cell is then small, justifying the use of the "effective-mass theorem." Any change in conductivity then should be through the effect of the magnetic field on τ. This can easily be seen by examining the curvature in the free path of an electron. In the longitudinal configuration, where electric and magnetic fields are parallel, the component of the free path in the direction of the electric field is not affected, making the conductivity field independent in the classical model. The quantized motion of the electron in a magnetic field makes τ field-dependent [5,6]. But, the absence of curvature still allows the problem to be treated by the BTE. However, in the transverse configuration where the magnetic field is perpendicular to the electric field, τ is strongly affected by the change (resulting from the curvature of electron paths in the magnetic field) in the electron mfp; in addition, the conductivity of Equation 9.16 becomes a tensor. For small magnetic fields, this curvature could be treated as a perturbation with field-independent relaxation time. Theories based on this ansatz predicted a saturation [7] in relative change in MR, the relative resistance change in a magnetic field, which is now known to not be present in experimental results [8]; rather, MR varies almost linearly with the magnetic field in the high-field domain.

A review paper by Dresden [9] gives an excellent explanation of the limitations of the BTE. Implicit in the BTE are many assumptions, including the existence of a scalar relaxation time, which are not always satisfied. The semiclassical BTE has been fairly successful because the de Broglie wavelength λ_D of the electron has been smaller than the mfp ℓ and radius of the cyclotron orbit R_1 ($\lambda_D < \ell < R_1$), $R_1 = \infty$ for zero-magnetic field. For such situations, an electron can be treated like a classical particle and the effect of the magnetic field on its motion can be treated as a perturbation. But, for stronger magnetic fields, this semiclassical situation may change when $R_1 \sim \ell$, or equivalently $\omega_c\tau \sim 1$, where $\omega_c = qB/m^*$ is the cyclotron frequency of the electron. In this case, the effect of the magnetic field cannot be treated as a perturbation. The curvature in the free path introduces nondiagonal matrix elements in a quantum-mechanical representation and gives zero expectation value of the current if its averaging is attempted with the BTE in transverse configurations in conflict with the experimental observation on transverse current. When $R_1 \sim \lambda_D$ or $\hbar\omega_c \sim k_BT$, T being the temperature of the crystal ($\hbar\omega_c \sim E_F$, the Fermi energy for degenerate electrons), the quantization of energy levels of an electron plays a prominent role.

9.3 Density Matrix

The concept of a Hamiltonian is crucial to the density-matrix formalism. In a Hamiltonian, the energy is raised to an operator. For example, the momentum p is raised to $-j\hbar\nabla$ that operates on the wavefunction ψ and converts it into a differential equation known as Schrodinger's equation as discussed in Chapter 2. The Hamiltonian of an electron, in the simultaneous presence of a magnetic field with magnetic potential $A = (0, Bx, 0)$, and electric field $\mathcal{E} = (\mathcal{E}_x, \mathcal{E}_y, \mathcal{E}_z)$ could be written as a sum of an unperturbed part $\mathcal{H}_0$ and a perturbation $\mathcal{H}'$ (neglecting spin effects):

$$\mathcal{H} = \mathcal{H}_0 + \mathcal{H}' \tag{9.18}$$

with

$$\mathcal{H}_0 = \mathcal{H}_e + \mathcal{H}_L \tag{9.19}$$

$$\mathcal{H}_e = \frac{\left[p + (q/c)A\right]^2}{2m^*} \tag{9.20}$$

$$\mathcal{H}' = V + F = V + e\vec{\mathcal{E}} \cdot \vec{r} \tag{9.21}$$

where $\mathcal{H}_L$ is the lattice Hamiltonian, V is the electron–phonon interaction, and $F = e\mathcal{E} \cdot r$ is the electron–electric field interaction. The electric field interaction is treated as a perturbation in the same spirit as in the solution of the BTE for ohmic transport with the electric field less than its critical value $\mathcal{E}_c$. This provides the convenience of making the electronic distribution uniform in the absence of a perturbation. The distribution function is found to contain

$\exp(-e\mathcal{E}r/k_BT)$ where position is restricted to the mfp ℓ as discussed in the previous chapters. It is then not consistent to expand to the first-order for small $\mathcal{E}$. Thus, if one wishes to investigate ohmic currents linear in $\mathcal{E}$, this field must ordinarily be treated as small from the beginning.

The eigenvalue solution of $\mathcal{H}_e$ is well known [10] with eigenfunctions (normalized in a unit volume)

$$|\alpha\rangle = |nk\rangle = \exp\left[j\left(k_y y + k_z z\right)\right]\phi_n\left(\frac{x - x_k}{R_1}\right) \tag{9.22}$$

where ϕ_n is the harmonic-oscillator wave function centered at

$$x_k = -R_1^2 k_y \tag{9.23}$$

The eigenvalues of $\mathcal{H}_e$ of Equation 9.20 are Landau levels:

$$\varepsilon_\alpha = \varepsilon_{nk} = \left(n + \frac{1}{2}\right)\hbar\omega_c + \frac{\hbar^2 k_z^2}{2m^*} \tag{9.24}$$

where k stands for k_y and k_z. Equations 9.22 and 9.24 show that the motion transverse to the magnetic field **B** applied parallel to the z-direction is quantized with harmonic-oscillator-type energy levels and eigenfunctions, whereas motion parallel to the magnetic field is unquantized with plane-wave eigenfunctions. This makes the electron in a magnetic field quasi-1D, which in the quantum limit ($n = 0$ for $\hbar\omega_c \gg k_BT$) is strictly 1D. We will use the quantum-mechanical representation of Equation 9.22 as the basis to specify all matrix elements. In treating quantum-mechanical systems, it is necessary to deal with two types of uncertainties. The first type is due to the probabilistic nature of the wave function, as illustrated by the uncertainty principle. The second uncertainty occurs when we do not have sufficient information about the quantum-mechanical state of the system. This can be handled by the density matrix. In the absence of perturbation ($\mathcal{H}' = 0$), the state of an electron is well described by the wave function of Equation 9.22. When perturbation is present, we do not have sufficient information about the state of the system, necessitating an expansion of wave function ψ^i of the ith electron in terms of an orthonormal set of Equation 9.22:

$$\psi^i(t) = \sum a_\alpha^i(t)|\alpha\rangle \tag{9.25}$$

In general, $|\alpha\rangle$ should be an eigenfunction of $\mathcal{H}_0$, the unperturbed Hamiltonian. But, in this work, we assume the phonon distribution to be undisturbed by the weak mutual electron–phonon interaction. Hence, in the description of the matrix elements of electronic motion, only eigenfunctions of $\mathcal{H}_e$ are required. The expectation value of the electric current j^i of the ith electron then is

$$j^i = \langle \psi^i | j | \psi^i \rangle = \sum_{\alpha\alpha'} a^i_{\alpha'} a^i_{\alpha} \langle \alpha' | j | \alpha \rangle \tag{9.26}$$

An ensemble average for N electrons in the system can then be described by

$$\langle j \rangle = \frac{1}{N} \sum_i j^i = \sum_{\alpha\alpha'} \langle \alpha | \rho | \alpha' \rangle \langle \alpha' | j | \alpha \rangle = Tr(\rho j) \tag{9.27}$$

with

$$\langle \alpha | \rho | \alpha' \rangle \equiv \frac{1}{N} \sum_i a^i_{\alpha'}(t) a^{i*}_{\alpha}(t) \tag{9.28}$$

The time dependence of a_α's can be obtained from the time-dependent Schrödinger equation for wave function ψ^i by using the orthonormality of $|\alpha\rangle$

$$i\hbar \frac{da^i_\alpha(t)}{dt} = \sum_{\alpha'} \langle \alpha | \mathcal{H} | \alpha' \rangle a^i_{\alpha'}(t) \tag{9.29}$$

which leads to the time dependence of the density matrix given by

$$i\hbar \frac{d\rho}{dt} = [\mathcal{H}, \rho] \tag{9.30}$$

To solve Equation 9.30, we assume the perturbation to be absent at $t = -\infty$, when it is turned on slowly:

$$\mathcal{H}(t) = \mathcal{H}_0 + \mathcal{H}' e^{st} \tag{9.31}$$

where $s \to 0^+$ is a small positive number. In fact, it is not always necessary to turn on the electron–phonon interaction in the same way as the electric field interaction. The final expression for the matrix elements of ρ' is linearized in the electric field interaction, whereas the electron–phonon interaction provides the damping effect. In the steady-state behavior of the system, when limit $s \to 0$ is taken, the damping effect is unaffected whether or not the electron–phonon interaction is turned on in the same way. In calculations of magneto-microwave conductivity, it has been found convenient not to include the turning on of the electron–phonon interaction the same way as the electric field [11]. A similar time dependence of the density matrix follows:

$$\rho(t) = \rho_0 + \rho' e^{st} \tag{9.32}$$

with ρ_0 being the uniform density matrix independent of the external electric field, which is diagonal in the Landau representation:

$$\langle \alpha' | \rho_0 | \alpha \rangle = f_\alpha \delta_{\alpha'\alpha} = \left[\exp\left(\frac{\varepsilon_{nk} - E_F}{k_B T} \right) + 1 \right]^{-1} \delta_{\alpha'\alpha} \tag{9.33}$$

where E_F is the Fermi energy obtained from the normalization condition

$$Tr(\rho) = \sum_\alpha \langle \alpha | \rho_0 | \alpha \rangle = N \tag{9.34}$$

with the result for ND electrons:

$$\exp\left(\frac{E_F}{k_B T} \right) = N \left(\frac{2\pi\hbar^2}{m^* \, k_B T} \right)^{3/2} \frac{\sinh\left(\hbar\omega_c/2k_B T\right)}{\hbar\omega_c/k_B T} \tag{9.35}$$

The substitution of the time dependence of $\mathcal{H}(t)$ and $\rho(t)$ in Equation 9.30 gives a coupled equation for ρ':

$$\begin{aligned} &(\varepsilon_{\alpha'\alpha} - i\hbar s)\langle \alpha' | \rho' | \alpha \rangle \\ &= \langle \alpha' | [\rho, \mathrm{F}] | \alpha \rangle + f_{\alpha'\alpha} \langle \alpha' | V | \alpha \rangle + \langle \alpha' | [\rho, \mathcal{H}'] | \alpha \rangle e^{st} \end{aligned} \tag{9.36}$$

$$\varepsilon_{\alpha'\alpha} = \varepsilon_{\alpha'} - \varepsilon_\alpha \tag{9.37}$$

$$f_{\alpha'\alpha} = f_{\alpha'} - f_\alpha \tag{9.38}$$

A method for solving these coupled equations was suggested by Arora and Peterson [5,10]. Following their approach, the steady-state solution of Equation 9.36 at $t = 0$ in the ohmic limit is shown to be

$$\langle \alpha' | \rho' | \alpha \rangle = \frac{\langle \alpha' | [\rho_0, F] | \alpha \rangle}{\varepsilon_{\alpha'\alpha} - i\hbar/\tau_{\alpha'\alpha}} \tag{9.39}$$

$$\langle\alpha'|[\rho_0,F]|\alpha\rangle = \left(\frac{qR_1}{\sqrt{2}}\right) f_{\alpha'\alpha}\mathcal{E}_x\left[(n+1)^{1/2}\delta_{n',n+1} + (n)^{1/2}\delta_{n',n-1}\right]\delta_{k',k}$$
$$-\left(\frac{jqR_1}{\sqrt{2}}\right) f_{\alpha'\alpha}\mathcal{E}_y\left[(n+1)^{1/2}\delta_{n',n+1} + (n)^{1/2}\delta_{n',n-1}\right]\delta_{k',k}$$
$$-\left(\frac{jq\hbar^2 k_z}{m^*}\right)\left(\frac{df}{d\varepsilon_\alpha}\right)\mathcal{E}_z\delta_{\alpha',\alpha} \tag{9.40}$$

$$\frac{1}{\tau_{\alpha'\alpha}} = \frac{1}{2\tau_{\alpha'}} + \frac{1}{2\tau_\alpha} \tag{9.41}$$

$$\frac{1}{\tau_\alpha} = \frac{2\pi}{\hbar}\sum_{\alpha''}\left|\langle\alpha|V|\alpha''\rangle\right|^2\delta(\varepsilon_{\alpha\alpha''}) \tag{9.42}$$

Equation 9.39 contains a Breit–Wigner type of collision broadening. The neglect of this term would be in violation of the uncertainty principle, leading to divergence difficulty encountered by others. This built-in broadening is thus quite important in magnetotransport work. Owing to the curvature effect, the relaxation rate is now an arithmetic average of the two states. In Equation 9.41, we have assumed electron scattering to be elastic and have neglected energy shift by electron–phonon interaction.

9.4 Magnetoresistance

The matrix elements of one-electron current operator $\hat{j}$ obtained from the Heisenberg equation of motion

$$\hat{j} = -qv = -\left(\frac{jq}{\hbar}\right)[\mathcal{H}, r] \tag{9.43}$$

can be shown to be

$$\langle\alpha'|\hat{j}_x|\alpha\rangle = \left(\frac{-j\hbar q}{\sqrt{2}R_1 m^*}\right)\left[(n+1)^{1/2}\delta_{n',n+1} - (n)^{1/2}\delta_{n',n-1}\right]\delta_{k'k} \tag{9.44}$$

$$\langle\alpha'|\hat{j}_y|\alpha\rangle = \left(\frac{\hbar q}{\sqrt{2}R_1 m^*}\right)\left[(n+1)^{1/2}\delta_{n',n+1} + (n)^{1/2}\delta_{n',n-1}\right]\delta_{k'k} \tag{9.45}$$

$$\langle \alpha' | \hat{j}_z | \alpha \rangle = -\left(\frac{\hbar q k_z}{m^*} \right) \delta_{\alpha'\alpha} \tag{9.46}$$

The structure of these matrix elements clearly suggests the importance of the density matrix. The diagonal matrix elements of $\hat{j}_x$ and $\hat{j}_y$ vanish in the eigenfunction scheme of Equation 9.22. The expectation values of $\hat{j}_x$ and $\hat{j}_y$ thus cannot be described by the diagonal Boltzmann transport function, making use of the density matrix of Equation 9.39 that is more effective in such problems. On the other hand, the longitudinal current operator is diagonal, justifying the use of the BTE. This extension of the zero-field method is reasonable since the factor $\exp(jk_z z)$ associated with the electronic wave function remains unchanged, with or without a magnetic field. The density matrix of Equation 9.39 can now be used in conjunction with Equations 9.44 through 9.46 to find the expectation values of the current by using Equation 9.27. The components of the conductivity tensor defined by

$$\langle j \rangle = \sigma \cdot \mathcal{E} \tag{9.47}$$

are then given by

$$\sigma = \begin{bmatrix} \sigma_1 & -\sigma_2 & 0 \\ \sigma_2 & \sigma_1 & 0 \\ 0 & 0 & \sigma_3 \end{bmatrix} \tag{9.48}$$

with

$$\sigma_1 = \frac{q^2}{m^*} \sum_{nk\pm} f_{nk,(n+1)k} (n+1) \frac{\tau^{-1}_{nk,(n+1)k}}{\omega_c^2 + \tau^{-2}_{nk,(n+1)k}} \tag{9.49}$$

$$\sigma_2 = \frac{q^2}{m^*} \sum_{nk\pm} f_{nk,(n+1)k} (n+1) \frac{\omega_c}{\omega_c^2 + \tau^{-2}_{nk,(n+1)k}} \tag{9.50}$$

$$\sigma_3 = q^2 \sum_{nk\pm} \frac{df}{d\varepsilon_{nk}} \left(\frac{\hbar k_z}{m^*} \right)^2 \tau_{nk} \tag{9.51}$$

where ± in the summation above stands for two-spin states. $f_{nk,(n+1)k}$ is an abbreviation for $f_{nk} - f_{(n+1)k}$. These expressions reduce to those obtained from the transport equation when the assumption of a low magnetic field is made. As $B \to 0$, the conductivity tensor σ of Equation 9.48 becomes diagonal with all components equal to

$$\sigma_0 = \frac{1}{3} q^2 \sum_k \frac{df}{d\varepsilon_k} \left(\frac{\hbar k}{m^*} \right)^2 \tau(\varepsilon_k) \tag{9.52}$$

where $\varepsilon_k = \hbar^2k^2/2m^*$ is the unquantized energy of an electron in the absence of a magnetic field, when $n \to \infty$. This reduction is in accordance with Bohr's correspondence principle. For the constant τ model, independent of energy, we can use the identity

$$\sum_{nk\pm} f_{nk,(n+1)k}(n+1) = n_3 \tag{9.53}$$

to obtain the textbook expressions for conductivity

$$\sigma_1 = \frac{\left(n_3q^2/m^*\right)\tau^{-1}}{\omega_c^2 + \tau^{-2}} \tag{9.54}$$

$$\sigma_2 = \frac{\left(n_3q^2/m^*\right)\omega_c}{\omega_c^2 + \tau^{-2}} \tag{9.55}$$

$$\sigma_3 = \left(\frac{n_3q^2}{m^*}\right)\tau \tag{9.56}$$

The experimentally measurable parameters, the relative change in resistivity $\Delta\rho_{xx}/\rho_0$ (transverse MR), $\Delta\rho_{zz}/\rho_0$ (longitudinal MR), and the Hall coefficient R_H can be expressed in terms of σ_1, σ_2, σ_3, and σ_0 by inverting σ ($\rho = \sigma^{-1}$):

$$\frac{\Delta\rho_{xx}}{\rho_0} = \frac{\sigma_1\sigma_0}{\left(\sigma_1^2 + \sigma_2^2\right)} - 1 \tag{9.57}$$

$$\frac{\Delta\rho_{zz}}{\rho_0} = \frac{\sigma_0}{\sigma_3} - 1 \tag{9.58}$$

$$R_H = \frac{\rho_{yx}}{B} = \frac{-\sigma_2}{\left(\sigma_1^2 + \sigma_2^2\right)B} \tag{9.59}$$

When Equations 9.54 through 9.56 are inserted for constant τ, the excess MR vanishes as indicated before:

$$\frac{\Delta\rho_{xx}}{\rho_0} = \frac{\Delta\rho_{zz}}{\rho_0} = 0 \tag{9.60}$$

$$R_H = \frac{1}{n_3qB} \tag{9.61}$$

This shows that the experimental MR is due to quantum effects.

In general, if $\tau = \tau_o x^r$, where $x = (E - E_c)/k_BT$, the average value of $\tau = \tau_o x^r$ for 3D is evaluated as follows:

$$\langle \tau \rangle = \frac{2}{3}\tau_o \frac{\Gamma(r + (5/2))}{\Gamma(3/2)} \frac{\Im_{r+(1/2)}}{\Im_{1/2}} \tag{9.62}$$

τ_o is the collision time appropriate for $(E - E_c) = k_BT$ or $x = 1$. This relation can be generalized to all dimensionalities by rewriting

$$\langle \tau \rangle = \frac{2}{d}\tau_o \frac{\Gamma(r + (d/2))}{\Gamma(d/2)} \frac{\Im_{r+(d-2/2)}}{\Im_{(d-2)/2}} \tag{9.63}$$

The above formalism shows the complexity of magnetotransport in the simultaneous presence of a magnetic and electric field. When the radius of the cyclotron orbit becomes comparable to other parameters of interest, the semiclassical character of the BTE breaks down. Also, the energy of the electron is quantized in a strong magnetic field. The density-matrix technique as used above can be successfully applied to describe electronic transport. Device engineers have not yet become familiar with the advantages of the density matrix as a tool to interpret a practical situation. Calculations are done without reference to density matrices even when the maximum information is not available and averaging is done over unknown parameters. With the density matrix, calculations can be made without introducing unnecessary variables. In an era of strong magnetic fields available from superconducting magnets, a more meaningful interpretation of experimental data is possible only through the use of density matrix.

In the limit when only $n = 0$ level is appreciably occupied, the expression simplifies considerably for the ND limit [6]:

$$\sigma_{xx} = \frac{2n_3q^2\hbar}{3\pi m^*\, k_BT\omega_c\tau_o}\left[\ln\left(\frac{3\sqrt{\pi}k_BT}{2\hbar/\tau_o}\right)^2 - 0.577\right] \tag{9.64}$$

$$\sigma_{xy} = \frac{n_3q}{B} \tag{9.65}$$

where $\tau_o = m^*/n_3q^2\rho_o$, derived from zero-field 3D resistivity ρ_o, is the collision time appropriate for acoustic phonon scattering. The relative transverse MR is now given by

$$\frac{\Delta\rho_{xx}}{\rho_0} = \frac{2}{3\pi}\frac{\hbar\omega_c}{k_BT}\left[\ln\left(\frac{3\sqrt{\pi}k_BT}{2\hbar/\tau_o}\right)^2 - 0.577\right] \tag{9.66}$$

And the longitudinal MR is given by

$$\frac{\Delta\rho_{zz}}{\rho_0} = \frac{1}{3}\frac{\hbar\omega_c}{k_B T} \tag{9.67}$$

Both $\Delta\rho_{xx}$ and $\Delta\rho_{zz}$ are now linear functions of the magnetic field. The ratio $\Delta\rho_{xx}/\Delta\rho_{zz} \approx 10.9$ for indium antimonide (InSb) at $T = 77$ K.

In the other extreme when the magnetic field is small, the summation over n can be converted into an integral as the spacing between quantum levels is miniscule. This is known as the classical limit. In this classical limit, longitudinal MR vanishes ($\Delta\rho_{zz}/\rho_o = 0$). The expressions for transverse components of conductivity are given by

$$\sigma_{xx}^{C} = \frac{n_3 q^2}{m^*}\frac{4}{3\sqrt{\pi}}\int \frac{\tau(x)}{1+\{\omega_c\tau(x)\}^2} e^{-x} x^{3/2}\,dx \tag{9.68}$$

$$\sigma_{xy}^{C} = \frac{n_3 q^2}{m^*}\frac{4}{3\sqrt{\pi}}\int \frac{\omega_c\tau(x)^2}{1+\{\omega_c\tau(x)\}^2} e^{-x} x^{3/2}\,dx \tag{9.69}$$

In the limit of $\omega_c\tau \ll 1$ (low magnetic field) at which most experiments are performed, the classical transverse MR is a quadratic function of the magnetic field with a coefficient that depends on the scattering interaction

$$\frac{\Delta\rho_{xx}}{\rho_0} = A_o\left(\omega_c\tau_o\right)^2 = A_o\left(\mu_o B\right)^2 \tag{9.70}$$

with $A_o = 0.38$ for acoustic phonon scattering and $A_o = 2.15$ for impurity scattering. $\mu_o = q\tau_o/m^*$ is the zero-field mobility. It is good to define MR mobility μ_{MR} that differs from conductivity mobility by a scattering-dependent factor [2]

$$\mu_{MR} = \sqrt{A_o}\,\mu_o \tag{9.71}$$

In general, $\sqrt{A_o} = \langle\tau^2\rangle/\langle\tau\rangle^2$, also known as the Hall factor, $r_H = \sqrt{A_o}$. For a 3D bulk semiconductor, the average value of $\langle\tau^s x^t\rangle$ can be calculated to

$$\langle\tau^s x^t\rangle_{3D} = \frac{4}{3\sqrt{\pi}}\Gamma\left(sr + t + \frac{5}{2}\right)\tau_o^s \frac{\Im_{sr+t+(1/2)}(\eta)}{\Im_{1/2}(\eta)} \tag{9.72}$$

The Hall factor for a ND semiconductor is given by

$$r_H = \sqrt{A_o} = \frac{3\sqrt{\pi}}{4}\frac{(2r + (3/2))!}{\left[(r + (3/2))!\right]^2} \tag{9.73}$$

Similar relationships can be determined for 2D statistics and strongly degenerate statistics.

In a strong degenerate configuration for which the Fermi energy $E_F \gg k_BT$, the MR expression can be simplified. The FD distribution $f_{nk} = f_o(\varepsilon_{nk})$ decreases from value 1 to 0 within a range of thermal energy k_BT around the point $\varepsilon_{nk} = E_F$. Therefore, the factor $f_{nk,(n+1)k} = f_o(\varepsilon_{nk}) - f_o(\varepsilon_{(n+1)k})$ is approximately –1 in the range of $E_F - \hbar\omega_c$ and E_F and zero elsewhere. In the quantum limit ($n = 0$) considering $k_BT < E_F < (3/2)\hbar\omega_c$ in an extremely high magnetic field, the Fermi level is

$$E_F - \frac{1}{2}\hbar\omega_o = \frac{4}{9}\left(\frac{E_{Fo}}{\hbar\omega_o}\right)^2 E_{Fo}, \quad E_{Fo} = 3^{2/3}\pi^{4/3}\left(\frac{\hbar^2}{2m^*}\right)n_3^{2/3} \tag{9.74}$$

In this extreme, the Hall coefficient $R = \rho_{xy}/B$ and MR $\Delta\rho_{xx}/\rho_o$ is given by [12]

$$R = \frac{1}{n_3 q}\left(1 + \frac{3\pi}{8\tau_o\omega_f}\frac{\hbar\omega_c}{E_{Fo}}\right) \tag{9.75}$$

$$\frac{\Delta\rho_{xx}}{\rho_o} = \left(\frac{\hbar\omega_c}{E_{Fo}}\right)^2 \ln\left(\frac{4\tau_o\omega_f}{3}\frac{E_{Fo}}{\hbar\omega_c}\right) \tag{9.76}$$

where zero-field resistivity $\rho_o = 1/\sigma_o$ and $\tau_o = \tau_o(\varepsilon = E_{F0})$ is the relaxation time at $\varepsilon = E_{F0}$ and $\omega_f = E_{Fo}/\hbar$. These equations state that for sufficiently high magnetic fields, the Hall coefficient is a linear function of the magnetic field, whereas the resistivity is essentially proportional to B^2.

9.5 An Application

In this section, an application to the parabolic model with parameters appropriate for InSb is made [10].

What is normally measured in an experiment is the relative change in MR $\Delta R_{xx}/R(0) = \Delta\rho_{xx}/\rho_o$ (transverse), $\Delta R_{zz}/R(0) = \Delta\rho_{zz}/\rho_o$ (longitudinal), and the Hall coefficient $R_H = R_{yx}/B$. The high-field value of the Hall coefficient is $R_H^0 = 1/n_3 q$. ρ_o is zero-field resistivity given by [13]

$$\rho_o = 3(2\pi m^* k_BT)^{1/2}\frac{m^{*2}\ E_1^2 k_BT}{4n_3 q^2\pi\hbar^4\rho_d u_s^2} \tag{9.77}$$

where $E_1 = 30$ eV is the electron–acoustic phonon coupling, $m^* = 0.016m_o$ is the effective mass for InSb, $n_3 = 10^{14}$ cm^{-3}, $\rho_d = 5.77$ g/cm^3, $u_s = 3.7 \times 10^5$ cm/s, and

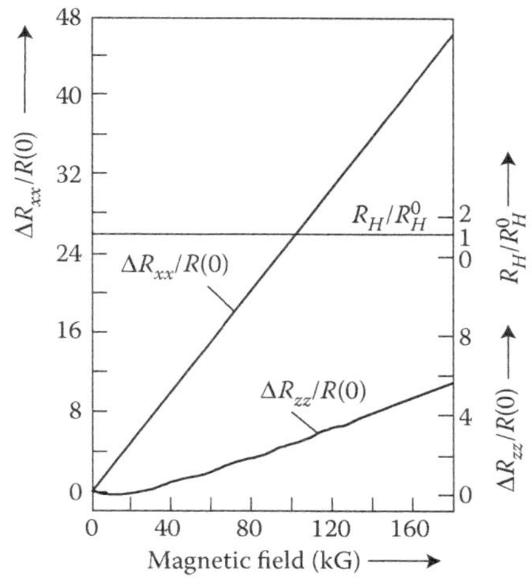

FIGURE 9.1
MR ratio and the normalized Hall coefficient versus the magnetic field for the parabolic band model of n-type InSb at temperature 77 K assuming electron-acoustic scattering that is proportional to the DOS to be the dominant mechanism of scattering.

$T = 77$ K. The numerical results for the MR and Hall coefficient are shown in Figure 9.1. The general trend is in agreement with experimental data [14] considering the uncertainty in the value of E_1. Although $\Delta R_{xx}/R(0)$ is sensitive to the value of, E_1, $\Delta R_{zz}/R(0)$ is independent of E_1.

9.6 Other Types of MR

There are other types of MR that are commonly used. Colossal MR (CMR) is a property of some materials, mostly manganese-based peroxides (they could be normal or superconductors) that enable them to dramatically change their electrical resistivity in the presence of a magnetic field. The OMR of conventional materials enables changes in resistivity of up to 5%, but materials featuring CMR may demonstrate resistivity changes by orders of magnitude. This property is not explainable by any classical theories, unless quantum processes indicated above are included. The tunneling MR (TMR) occurs when two ferromagnets are separated by a thin (about 1 nm) layer of insulation. Resistivity of the tunneling current changes with the relative orientation of the two magnetic layers. The resistivity is normally higher in the antiparallel case. There is a renewed interest in this field fueled by the discovery of the giant magnetoresistive (GMR) effect. GMR is a quantum-mechanical effect observed in thin-film structures composed of an alternating ferromagnetic and nonmagnetic metal multilayer. The difference between a tunnel and a giant MR is that in TMR, we compose a multilayer structure of alternating ferromagnetic insulators, and in GMR, the structure is composed of alternating ferromagnetic and nonmagnetic layers. The GMR effect is a large decrease in electrical resistivity that occurs when the magnetization of the layered sample is aligned by an external magnetic field. It is similar to a polarization experiment in optics, where an aligned polarizer allows light to pass through, but a crossed polarizer does not. The first magnetic layer allows electrons in only one-spin state to pass through easily. If the second magnetic layer is aligned, then those spins can easily pass through the structure, and the resistivity is low. If the second magnetic layer is misaligned, then neither spin can get through the structure easily so that the electrical resistivity is high. The GMR effectively measures the difference in angle between the two magnetizations in the magnetic layers. Small angles (parallel alignment) give a low resistivity, and large angles (antiparallel alignment) give a higher resistivity. It is easy to produce the state where the two magnetic layers are parallel—simply apply a field large enough to magnetically saturate both layers.

The advantage of a detailed study of 3D semiconductors in a magnetic field can help us understand the 1D NWs whose energy spectrum resembles that of a 3D semiconductor in a high magnetic field. When linear density $n_1 = n_3\pi R_1^2$ in a magnetic field is related to an NW of the circular cross-section of radius R, many features of an NW can be studied by conducting an experiment on 3D semiconductors in a high magnetic field, thereby saving time in fabricating and testing the properties of NWs.

9.7 NW Effect in High Electric and Magnetic Fields

NWs having a cylindrical geometry with 1D character can be effectively studied to get predictive outcomes when compared to a similar geometry in a high magnetic field. A number of problems can be explored during the comparative study of NWs and 3D bulk in a high magnetic field exhibiting 1D character. The energy of an NW of radius R_{NW} is given by [15]

$$E_{nmk_z} = \alpha_{nm}^2\varepsilon_o + \frac{\hbar^2 k_z^2}{2m^*}, \quad \varepsilon_o = \frac{\hbar^2}{2m^* R_{NW}^2} \tag{9.78}$$

where α_{nm} is the mth zero of Bessel function $J_n(x)$ of order n ($J(\alpha_{nm}) = 0$). A few values are given by $\alpha_{01} = 2.405$, $\alpha_{11} = 3.832$, $\alpha_{21} = 5.136$, $\alpha_{021} = 5.520$, and so on. For higher roots, the asymptotic formula $\alpha_{nm} \approx m\pi + (n - (1/2)\pi/2)$ can be used.

Electrons in a magnetic field in a 3D material follow the same 1D pattern

$$E_{nk} = (2n+1)\varepsilon_o + \frac{\hbar^2 k_z^2}{2m^*}, \quad \varepsilon_o = \frac{\hbar^2}{2m^* R_{MF}^2}, \quad R_{MF} = \left(\frac{\hbar}{m^* \omega_c}\right)^{1/2} = \left(\frac{\hbar}{qB}\right)^{1/2} \tag{9.79}$$

where R_{MF} is the radius of the orbit in a magnetic field **B**. The area of confinement is $A = \pi R_{NW(MF)}^2$. Therefore, a comparative study of an NW and 3D material in a strong magnetic field can be done with similar results. From an experimental point of view, it is more difficult to fabricate an NW, but it is easy to create an NW effect in a 3D material subjected to the magnetic field. The number of such applications is unlimited depending on the imagination of the researcher.

An NW and an electron in a magnetic field share 1D configuration, and it is easy to make predictions about the NW by changing the magnetic field in the quantum domain in the 3D material, provided mfp ℓ is determined either from mobility experiments or from zero-field resistance. 1D density $n_1 = n_3 \pi R_{MF}^2$. As the magnetic field changes, the radius $R_{MF} = (\hbar/qB)^{1/2}$ will change and with that the 1D density also changes. This will be a useful exercise to determine whether the NW will provide any special advantage in the application of high electric and magnetic fields, following the paradigm discussed in Chapter 7. This itself will provide a large set of problems to explore.

Arora [16] has done a comparative quantum theoretical study on the electrical and optical properties of NWs under strong quantum confinement conditions when the acoustic phonon, point defect, or alloy scattering is enhanced, and ionized-impurity scattering is suppressed. A quantum size resonance is expected to take place when the energy of a photon polarized perpendicular to the wire is equal to the spacing between the quantized energy levels. The linewidth is shown to be proportional to $(\lambda_D^2/A)\tau_{3D}$, where λ_D is the de Broglie wavelength of an electron, A is the confinement area, and τ_{3D} is the bulk relaxation time. A quantum freeze-out of confined carriers is expected to induce a semimetal–semiconductor transition in semimetallic thin wires.

9.8 Quantum-Confined Transport

A lot has been said of transport in bulk, nanolayers, NWs, and carbon-based nanostructures predominantly using NEADF as opposed to NEGF that is too complicated to implement in practical situations. In all the situations considered before, it has been in transporting electric field in an analog-type direction where electron waves are propagating waves. However, a number of problems arise especially with optical excitations when the electric field is applied in the confinement direction perpendicular to a nanolayer or NW. Naturally, in such cases, the conductivity is modified by assuming a nonequilibrium part of the distribution by $\exp(\pm j\omega t)$ with angular frequency $\omega = 2\pi f$ of the electromagnetic wave of frequency f. A linearly polarized wave can be broken down into two circularly polarized waves, one clockwise ($+\omega$) and the other in a counterclockwise direction ($-\omega$) with phase factors $\exp(\pm j\omega t)$. In such circumstances, Equation 9.3 can still be used except now with $v = v_o \exp(\pm j\omega t)$, $dv/dt = \pm j\omega v$. When combined with the collision term for $B = 0$, $1/\tau$ is replaced by $(1/\tau) \pm j\omega$. All expressions based on this transient model then make mobility μ or conductivity $\sigma_d = n_d q \mu_d$ (with $d = 3$ (bulk 3D), 2 (layered 2D), and 1 (line 1D)). $\sigma_d(\omega)$ in terms of $\sigma_{do} = \sigma_d(0)$ (dc or zero-frequency conductivity) becomes

$$\sigma_d(\omega) = \frac{\sigma_{do}}{1 \pm j\omega\tau} \tag{9.80}$$

This complex conductivity can be used for a variety of optical problems on free carrier absorption. A detailed theory using the density matrix is discussed in Reference 17.

When the cyclotron resonance frequency in a magnetic field ($\omega_c = qB/m^*$) is also included in Equation 9.3, a resonant component arises at $\omega = \omega_c = qB/m^*$ that allows us to determine the effective mass and give various resonances. This cyclotron resonance for collision time varying with energy is evaluated in References 11 and 18.

Another set of problems come into focus when the electromagnetic field with electric field polarization in the confinement direction (say the z-direction) is applied. In that case, standing waves give $\langle v \rangle = 0$ that can give a finite result only when the nondiagonal element of the density matrix is considered, and the conductivity is finite. This is discussed in Reference 19 and is called quantum size resonance, as it depends on the size of confinement. It becomes resonant at frequency $\hbar\omega = \varepsilon_2 - \varepsilon_1 + \hbar\Delta$, where $\varepsilon_2 - \varepsilon_1$ is the spacing between two quantized levels and $\hbar\Delta$ is the shift in the frequency due to the principal part of the scattering interaction, whereas the width is dictated by scattering $\hbar/\tau$ depending on the dominant mechanism of scattering. Reference 19 shows the formalism that can open vista for a number of opportunities on optical properties that can be designed for ongoing research. At resonance, the conductivity σ_2 becomes real and degrades with decreasing thickness of the confinement width d when compared to bulk conductivity σ_3. For any isotropic scattering proportional to the DOS, the relative conductivity σ_2^{res}/σ_3 at resonance is proportional to d/λ_D, where d is the thickness of the quantum well and λ_D is the de Broglie wavelength. The linewidth of resonance decreases with d as d^{-1} in the quantum limit only when the first level is appreciably populated.

Another set of problems related to NWs are discussed in Reference 16. In this reference, quantum theoretical results on the electrical and optical properties of quantum well wires (QWWs) are described under strong quantum confinement conditions when the acoustic phonon, point defect, or alloy scattering is enhanced, and ionized-impurity scattering is suppressed. A quantum size resonance is expected to take place when the energy of a photon polarized perpendicular to the wire is equal to the spacing between the quantized energy levels. The linewidth is shown to be proportional to $(\lambda_D^2/A)\tau_3$, where A is the confinement area of the cross-section of the NW and τ_3 is the bulk relaxation time. Reference 16 also compares the NWs with the bulk semiconductors in a magnetic field with a comparison of the size and cyclotron resonance.

CAD/CAE PROJECTS

C9.1 *NW-like effect in a 3D semiconductor in a magnetic field.* Cylindrical NWs offer the same quantum description as for an electron in a 3D semiconductor in a longitudinal high magnetic field. Considering this equivalence of quantized energies, and taking $n_1 = n_3(\pi R_{MF}^2)$, transform the high-field description, extensively discussed for an NW in the earlier chapters, into an electron confined to a 3D semiconductor in the presence of a magnetic field. Make sure that the magnetic field is in the quantum domain before proceeding with the comparison. Take appropriate values of n_1 and n_3 for ND statistics. Assume constant mfp so that $\tau = \ell/v = \tau_o x^{-1/2}$, where $\tau_o = \ell/v_{th}$ and $x = (E - E_c)/k_B T$.

C9.2 *NW-like effect in a 3D semiconductor in a magnetic field-degenerate domain.* Repeat C8.1 for degenerate statistics.

C9.3 *Ballistic transport.* Considering the equivalence of an NW and electron in a 3D magnetic field with length that is below the scattering-limited mfp, assess if resistance quantum is observable in 3D semiconductors in a magnetic field.

C9.4 *Ballistic transport in 2D MOSFET.* The MOSFET is well known to be a 2D semiconductor. The quantum Hall effect (QHE) has been observed in a magnetic field applied perpendicular to the nanolayer where, naturally, the thickness is much smaller than the mfp. Using the results from a transverse magnetic field, is it possible to observe QHE as a ballistic phenomenon? In the ND domain? In the degenerate domain?

C9.5 *Ballistic transport in 2D MOSFET.* Repeat C9.4 for a graphene nanolayer to observe QHE and compare with what has been experimentally observed in the literature.

C9.6 *Classical MR.* The classical and quantum limit of transverse MR is discussed by Arora and Spector (*Physics Review B*, Vol. 24(6), September 1981). In that paper, they argue that in the low-field limit, the incremental relative MR, is given by $\Delta\rho_{xx}/\rho_o = A_o(\beta)(\omega_c\tau_o)^2 = \mu_{MR}^2 B^2$. Obviously, μ_{MR} differs from drift mobility $\mu_o = q\tau_o/m^*$ by an MR factor $\sqrt{A_o(\beta)}$, where β is a scattering-dependent factor. Considering the scattering by acoustic phonons and impurity scattering combined, assess this MR factor for a bulk semiconductor.

C9.7 *Classical MR for the MOSFET nanolayer.* Repeat C9.6 for a 2D MOSFET nanolayer. Does it provide a useful experimental method to extract mobility?

References

1. V. Arora, Ohmic electrical conductivity in a magnetic field, *American Journal of Physics*, 44, 643–646, 1976.
2. V. K. Arora, M. S. Z. Abidin, S. Tembhurne, and M. A. Riyadi, Concentration dependence of drift and magnetoresistance ballistic mobility in a scaled-down metal–oxide–semiconductor field-effect transistor, *Applied Physics Letters*, 99, 063106–063106-3, 2011.
3. C. M. Wolfe, J. Nick Holonyak, and G. E. Stillman, *Physical Properties of Semiconductors*. Upper Saddle River, NJ: Prentice-Hall, 1989.
4. L. Weisenthal and A. M. de Graaf, On the quantum theory of electrical transport in a uniform magnetic field, *American Journal of Physics*, 40, 1469–1473, 1972.
5. V. K. Arora and R. L. Peterson, Quantum theory of ohmic galvano- and thermomagnetic effects in semiconductors, *Physical Review B*, 12, 2285–2296, 1975.
6. V. K. Arora, D. R. Cassiday, and H. N. Spector, Quantum-limit magnetoresistance for acoustic-phonon scattering, *Physical Review B*, 15, 5996–5998, 1977.
7. S. Wang, *Fundamentals of Semiconductor Theory and Device Physics*. Upper Saddle River, NJ: Prentice-Hall, 1989.
8. V. K. Arora and S. C. Miller, Effect of electron–phonon drag on the magnetoconductivity tensor of n-germanium, *Physical Review B*, 10, 688–697, 1974.
9. M. Dresden, Recent developments in the quantum theory of transport and galvanomagnetic phenomena, *Reviews of Modern Physics Journal* 33, 265–342, 1961.
10. V. K. Arora, Linear magnetoresistance in parabolic semiconductors, *Physica Status Solidi (b)*, 71, 293–303, 1975.
11. V. K. Arora, Quantum theory of magnetomicrowave conductivity in semiconductors, *Physical Review B*, 14, 679–684, 1976.
12. P. N. Argyres, Quantum theory of galvanomagnetic effects, *Physical Review*, 109, 1115–1128, 1958.
13. E. Conwell, *High Field Transport in Semiconductors*. Vol. Supplement 9. New York: Academic Press, 1967.
14. P. I. Baranski and O. P. Gorodnic, Longitudinal magnetoresistance of n-type InSb, *Soviet Physics Semiconductors—USSR*, 2, 708, 1968.
15. V. K. Arora and M. Prasad, Quantum transport in quasi-one-dimensional systems, *Physica Status Solidi (b)*, 117, 127–140, 1983.

16. V. K. Arora, Quantum well wires: Electrical and optical properties, *Journal of Physics C: Solid State Physics*, 18, 3011, 1985.
17. V. Arora and E. Qureshi, Quantum theory of microwave conductivity, *Physica Status Solidi (b)*, 77, 77–84, 1976.
18. V. Arora and H. Spector, Quantum-limit cyclotron resonance linewidth in semiconductors, *Physica Status Solidi (b)*, 94, 701–709, 1979.
19. V. Arora, Quantum size resonance in semiconducting thin films, *Journal of Vacuum Science and Technology*, 20, 94–95, 1982.

10

Drift-Diffusion and Multivalley Transport

A theory that makes an explicit connection between scattering-limited ohmic mobility and quantum-emission-limited saturation velocity is presented. The theory is applied to electrons in bulk silicon by taking a quantum equal to the energy of an optical phonon. Because this quantum emission is indicated, a modification in the mfp appears only in the high-field regime. This modification is shown to lead to electric-field-induced degradation of the diffusion coefficient. The theory presented agrees with the drift-diffusion experimental data and empirical relations utilized in modeling devices. The theory makes connections with an alternate description in terms of electron temperature under ac and dc conditions. As drift-diffusion processes are central in the performance evaluation of submicron-scale devices where high fields are necessarily present, these results contribute significantly in reshaping thinking processes in the high-field regime. Transfer to higher valleys in multivalley band structure is also discussed.

10.1 Primer

Considerable interest in the high-field transport will continue to exist because of scaled-down dimensions of devices that are now of submicron scale in the quasi-free direction of the carrier flow. As we have seen, with a high-electric field present, the familiar linear velocity-field characteristics become sublinear and the velocity eventually saturates at an applied high-electric field. This nonlinear response of the carrier velocity to a high electric field has been extensively explored, both theoretically and experimentally (for a review, see Reference 1). However, not much work has been reported on diffusive transport in the presence of a high electric field. An acceptable procedure for studying such processes is that of the Monte Carlo simulation with input parameters varied to get a desired output. A cumbersome numerical procedure makes this approach impractical for device applications. A simplified framework that extrapolates the well-established ohmic behavior to the saturation regime is of more practical value to device designers and researchers. Moreover, it provides an insight into processes that control the high-field transport behavior.

The Einstein relation with ratio of diffusion coefficient D_o to ohmic mobility μ_o for macro-devices has been around for many years. In the ohmic domain, it can be described as

$$\frac{D_o}{\mu_o} = V_t = \frac{k_B T}{q} \tag{10.1}$$

where $V_t = k_B T/q$ is the thermal voltage at temperature T with a value at room temperature ($T = 300$ K) 25.9 mV. In the high-field domain, $D(\mathcal{E})$, $\mu(\mathcal{E})$, and electron temperatures $T(\mathcal{E})$ in the published litrature are sporadically indicated to depend upon the electric field $\mathcal{E}$. Considerable confusion exists while evaluating each of these parameters in isolation. A unified expression for each of these parameters allows us to define hot-electron temperature and also obtain its anlaytical expression.

10.2 Simplified Drift-Diffusion

In a free segment indicated by the free path, the electrons are traveling with intrinsic velocity v_i as discussed earlier. In a simplified model [1,2], this intrinsic velocity for one dimension is the rms thermal velocity that strictly applies to ND statistics. The kinetic energy related to the random motion of an electron, $(1/2)\, m_n^* v_{th}^2$, is the thermal energy $(1/2)k_B T$ for one-degree of freedom in the direction of the carrier flow. The thermal velocity v_{th} is given by

$$v_{th} = \sqrt{\frac{k_B T}{m_n^*}} \tag{10.2}$$

where m_n^* is the electron effective mass. For silicon at room temperature, this thermal velocity is in the order of 10^7 cm/s when a carrier mobility effective mass $m_n^* = 0.26 m_o$ is considered. This random thermal velocity does not contribute to any current (charge flow) in equilibrium as random velocity vectors cancel each other out, resulting in zero ensemble (or time) average of the velocity in a given direction.

When an electric field, $\mathcal{E}$, is applied, the random network drifts in a direction that is opposite to that of an applied electric field, in this case $-x$-direction from the right to left. As an electron is accelerated by the electric field, collisions impede their motion. An electron after suffering a collision makes a fresh start of its drift velocity $v(t)$ whose transient response, as discussed in Chapter 4, is given by

$$v(t) = -\frac{q\tau_c \mathcal{E}}{m_n^*}\left(1 - e^{-t/\tau_c}\right) \tag{10.3}$$

Here, τ_C is the mean collision time and q is electronic charge. $v(t)$ is a constant that results in a steady-state drift velocity in the absence of transients for $t \gg \tau_C$ (steady state). The ohmic mobility (the slope of linear v–$\mathcal{E}$ curve), μ_{no}, is then given by

$$\mu_{no} = \frac{q\tau_C}{m_n^*} = \frac{q\ell_o}{m_n^* v_{th}} \tag{10.4}$$

Here, ℓ_o is the mfp under low-field (ohmic) conditions. The mfp extracted from experimental mobilities $\mu_o = 1322$, 2975, and 20,000 $\text{cm}^2/\text{V}\cdot\text{s}$ is $\ell_o = 0.099$, 0.18, and 0.764 μm at $T = 300$, 200, and 77 K, respectively, for silicon. The mobility varies with temperature approximately as T^{-2}, while the mfp varies as $T^{-3/2}$. This linear paradigm ignores the fact that during its approach toward a steady state, an electron may accumulate enough energy for it to be able to emit a quantum, for example, an optical phonon. In a quantum well, this is the energy corresponding to the difference between the two lowest quantized levels. In the presence of quantum emission, the time t is constrained by τ_Q during which the energy gained by an electron in an electric field is enough to emit a quantum of energy $\hbar\omega_o$ with probability of emission given by $N_o + 1$. The average energy of the emitted quantum is $E_Q = (N_o + 1)\hbar\omega_o$. The effective collision time τ for modified mobility $\mu_n = q\tau/m_n^*$ is then given by

$$\tau = \tau_C\left(1 - e^{-\tau_Q/\tau_C}\right) \tag{10.5}$$

The time, τ_Q, at the onset of an emission of a quantum $\hbar\omega_o$ of energy is given by

$$\tau_Q = \frac{\ell_Q}{v_{th}} = \frac{E_Q}{v_{th} q\mathcal{E}} \tag{10.6}$$

where $\ell_Q = E_Q/q\mathcal{E}$ is the inelastic scattering length for the onset of a quantum emission. When Equation 10.5 is multiplied by the thermal velocity v_{th} of Equation 10.2, the mfp ℓ also degrades similarly with the applied electric field:

$$\ell = \ell_o\left(1 - e^{-\ell_Q/\ell_o}\right) \tag{10.7}$$

where ℓ_o is the momentum randomizing mfp under ohmic conditions. The composite mfp, ℓ, of Equation 10.7 is in agreement with the one obtained by the probabilistic arguments of Schwarz and Russek [3]. Equation 10.7 indicates that $\ell = \ell_o$ in the low-electric-field ohmic regime and approaches ℓ_Q in the high-electric-field saturation regime. This paradigm transforms the way we view scattering events. The velocity of carriers in a low electric field is, therefore, limited by traditional momentum randomizing scattering events. On the other hand, the velocity in a high electric field is virtually independent of these scattering events and can only be limited by the onset of a quantum emission that limits the saturation velocity. In this model, the electric field may be viewed as an organizer of a completely random motion to a streamlined one where all thermal velocity vectors have become unidirectional and, hence, limit the drift velocity to a streamlined thermal velocity in the nondegenerate regime. Although not considered here, one can easily conjecture that in the degenerate regime this orientation will limit the drift velocity to the Fermi velocity.

Based on the distribution function presented for a tilted band diagram in an electric field [4], it is easy to show that there is a left–right asymmetry in the otherwise random motion of the carriers. The fraction of carriers $n_+(x)/n(x)$ of electrons at a location x traveling to the right and fraction $n_-(x)/n(x)$ of those traveling to the left, with an electric field applied in the $-x$-direction, is given by

$$\frac{n_{\pm}(x)}{n(x)}=\frac{e^{\pm\delta}}{e^{+\delta}+e^{-\delta}}=\frac{e^{\pm\delta}}{2\cosh(\delta)} \tag{10.8}$$

with

$$\delta=\delta_o\left(1-e^{-\delta_Q/\delta_o}\right) \tag{10.9}$$

$$\delta_o=\frac{q\,\mathcal{E}\,\ell_o}{k_BT}=\frac{\mathcal{E}}{\mathcal{E}_{co}}=\frac{V}{V_{co}} \tag{10.10}$$

$$\delta_Q=\frac{E_Q}{k_BT} \tag{10.11}$$

Here, $E_{co}=k_BT/q\ell_o=V_t/\ell_o$ is the critical electric field and $V_{co}=\mathcal{E}_{co}\,L$ is the associated critical voltage that defines the onset of nonohmic behavior for the drifting electrons. $\mathcal{E}_{co}$ is an intrinsic property of the material at a given temperature. $\mathcal{E}_{co}=260.0$, 94.4, and 8.68 kV/m for $T=300$, 200, and 77, respectively, as obtained from the mfp ℓ_o extracted from the ohmic mobility in silicon; $\hbar\omega_o$ = 63 meV for silicon.

Following the procedure outlined in Reference [4], and Chapter 4, the current density, $J(x)$, at a position x is evaluated as

$$J(x)=n(x)\,q\,v_{th}\,\tanh(\delta)+q\,v_{th}\,\ell\,\frac{dn}{dx} \tag{10.12}$$

The first term in Equation 10.12 is the drift term giving drift velocity v_D:

$$v_D = v_{th} \tanh(\delta) \tag{10.13}$$

The second term in Equation 10.12 is the diffusion term. The coefficient of dn/dx is qD_n, where D_n is the diffusion coefficient given by

$$D = v_{th}\ell = D_o\left(1 - e^{-E_Q/q\mathcal{E}\ell_o}\right) \tag{10.14}$$

The normalized diffusion coefficient (D/D_o) differs from its unity ohmic value and degrades in the presence of a quantum emission. If quantum emission is absent, the diffusion coefficient remains constant. The ratio D/D_o is plotted as a function of the electric field in Figure 10.1 for electrons in silicon at temperatures $T = 77$, 200, and 300 K. The scattered points are from the empirical relation given by

$$\frac{D}{D_o} = \frac{1}{\left[1 + \left(\mathcal{E}/\mathcal{E}_{co}\delta_Q\right)^{2.0}\right]^{1/2.0}} \tag{10.15}$$

The equation provides a good fit for the scattered data in Reference [5]. This equation also explicitly shows that the critical field for the onset of deviation

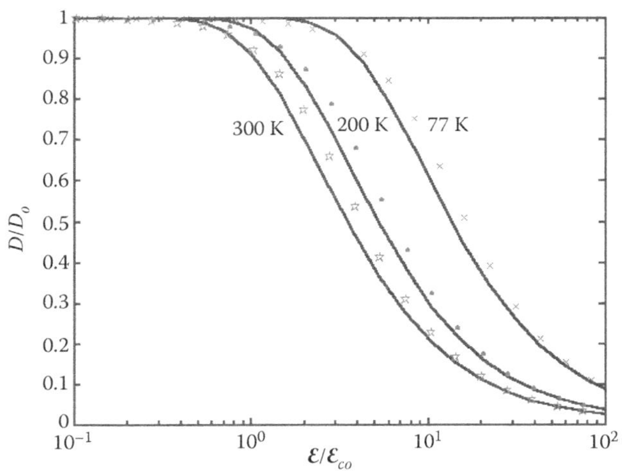

FIGURE 10.1
Normalized diffusion coefficient versus normalized electric field $\delta_o = \mathcal{E}/\mathcal{E}_{co}$. Solid lines are from Equation 10.13 with $\hbar\omega_o = 63$ meV and experimental points are obtained from the empirical Equation 10.14.

from the unity of this ratio is given by $\mathcal{E}_{cD} = \mathcal{E}_{co}\delta_Q$, which is not an empirical number, but extracted from the theory above.

10.3 Einstein Ratio

To evaluate the Einstein ratio, we need to obtain the mobility expression in a high electric field. As expected in a low electric field, mobility is a function of temperature, but not of an electric field. The deviation from a constant value is noticeable when the electric field exceeds $\mathcal{E}_{c\mu} = \mathcal{E}_{co}\tanh(\delta_Q)$. As is well known, the velocity saturation is a direct result of mobility degradation. The saturation velocity is $v_{th}\tanh(\delta_Q)$. As velocity-field curves are nonlinear, we need to distinguish between dc and small-signal ac mobilities.

The relative mobility, $\mu_n(\mathcal{E})/\mu_{no} = v_d/\mathcal{E}\,\mu_{no}$, under dc conditions, is given by an expression

$$\frac{\mu_n}{\mu_{no}} = \frac{\tanh(\delta)}{\delta_o} \tag{10.16}$$

The relative mobility is plotted in Figure 10.2. The points are the data points obtained from the empirical relation

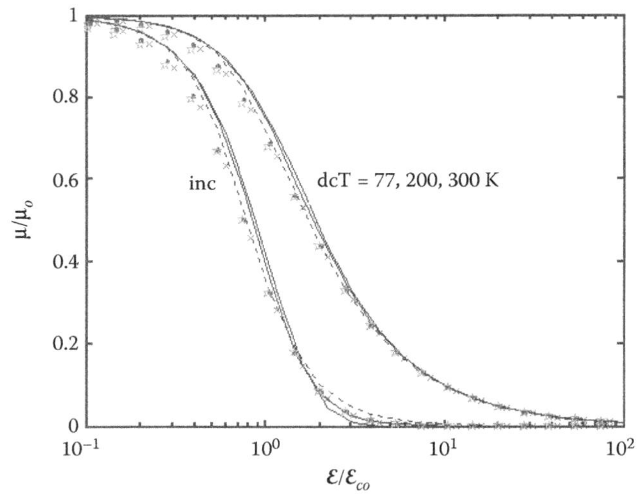

FIGURE 10.2
The normalized mobility, direct (dc) and incremental (inc) as a function of normalized electric field.

$$\frac{\mu_n}{\mu_{no}} = \frac{1}{\left[1 + \left(\mathcal{E}/\mathcal{E}_{co}\tanh(\delta_Q)\right)^{2.0}\right]^{1/2.0}} \tag{10.17}$$

This degradation enhances the Einstein ratio D_n/μ_n as is given by

$$\frac{D_n}{\mu_n} = V_t \frac{\delta}{\tanh(\delta)} \tag{10.18}$$

This enhanced ratio may give a misleading impression of electrons becoming hot as effective V_t is now multiplied by $\delta/\tanh(\delta)$. In this viewpoint, the apparent hot electron temperature T_e in terms of normal temperature is given by

$$\frac{T_e}{T} = \frac{D_n/D_o}{\mu_n/\mu_o} = \frac{\delta}{\tanh(\delta)} \tag{10.19}$$

The electrons appear hot due to the mobility degradation effect. However, no new additional randomness in electronic motion is generated by the electric field. Rather, electronic motion is streamlined. The relative temperature as a function of the electric field is plotted in Figure 10.3. The ratio is unity in low values of the electric field and rises at intermediate electric fields and finally saturates to a value of $\delta_Q/\tanh(\delta_Q)$. The saturated value of T_e/T can give useful information about the energy of the quantum involved.

The incremental differential mobility $\mu_{n,inc} = dv_d/d\mathcal{E}$, as obtained from Equation 10.13, when compared to the ohmic mobility, gives for $\mu_{n,inc}/\mu_{no}$, an expression

$$\frac{\mu_{n,inc}}{\mu_{no}} = \frac{\delta - \delta_Q e^{-\delta_Q/\delta_o}}{\delta_o} \sec h^2(\delta) \tag{10.20}$$

This degraded mobility is shown in Figure 10.2. It may be noticed that the degradation is much stronger under incremental conditions appropriate for signal propagation than under direct dc conditions. Also shown are the data points obtained from the empirical equation:

$$\frac{\mu_{n,inc}}{\mu_{no}} = \frac{1}{\left[1 + \left(\mathcal{E}/\mathcal{E}_{co}\tanh(\delta_Q)\right)^{2.0}\right]^{1.5}} \tag{10.21}$$

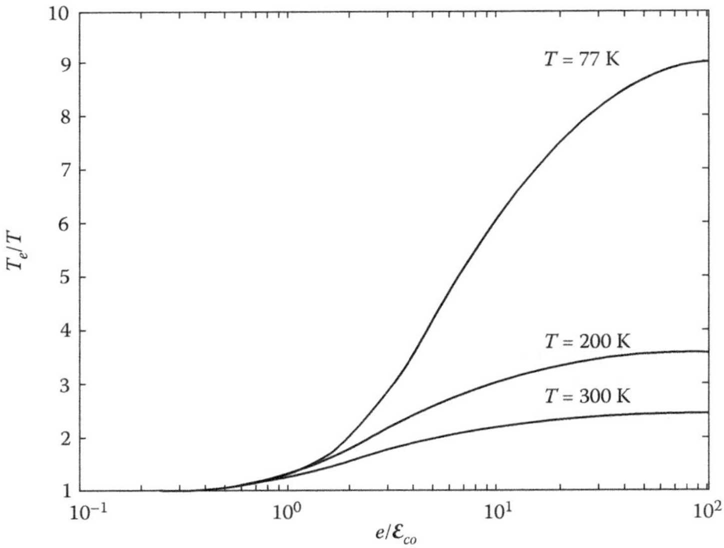

FIGURE 10.3
The normalized electron temperature as a function of normalized electric field under dc conditions.

The Einstein relation now gets modified to

$$\frac{D_n}{\mu_{n,inc}} = V_t \frac{\delta}{\delta - \delta_Q e^{-\delta_Q/\delta_o}} \cosh^2(\delta) \tag{10.22}$$

The associated relative small-signal incremental noise temperature T_n/T is now given by

$$\frac{T_n}{T} = \frac{D_n/D_o}{\mu_{n,ac}/\mu_{no}} = \frac{\delta}{\delta - \delta_Q e^{-\delta_Q/\delta_o}} \cosh^2(\delta) \tag{10.23}$$

This noise temperature is plotted in Figure 10.4.

These closed expressions for diffusion coefficient and mobility in the presence of a high electric field are obtained in a simplified model. The deviations from the well-known ohmic behavior are apparent when the energy gained by an electron in an mfp is comparable to the thermal energy. The high-field distribution function presented transforms randomness into an ordered streamlined motion, resulting in thermal velocity as the limiting velocity. A higher mobility does not necessarily leads to a higher saturation velocity. The above simplistic paradigm clearly indicates

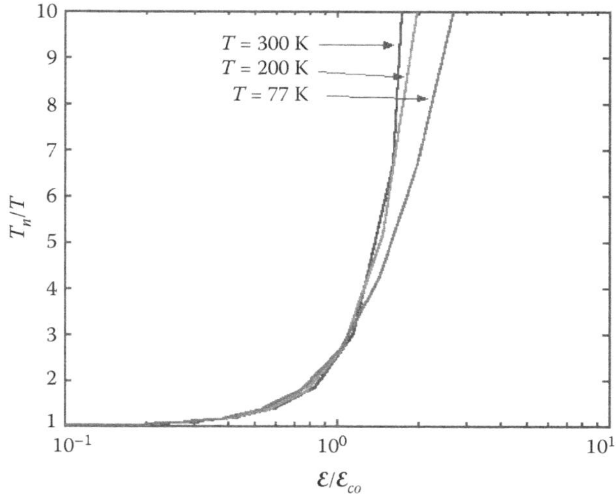

FIGURE 10.4
The normalized noise temperature as a function of normalized electric field.

that the velocity-limiting mechanisms in the ohmic and nonohmic regimes are quite different. Hence, saturation velocity and ohmic mobility should be considered independent parameters. In a high electric field, the quantum emission plays a prominent role in limiting the mobility. If somehow this emission process is curtailed or eliminated, higher electron velocities are possible. The diffusion coefficient in response to the concentration gradient remains unaffected if quantum emission is ignored. However, a change in the mfp due to the quantum emission process degrades the diffusion coefficient. It may give important clues on quantum emission processes in diffusion experiments. Similarly, the enhanced ratio of the diffusion coefficient with high-field-limited mobility can be understood in terms of mobility degradation. The apparent increase in an electron temperature is also attributed to this mobility degradation, not to an increase in the randomness of the electron distribution. These results show that the hot-electron concept is an elusive one depending on the context in which it is being applied. Thermodynamically, higher temperature is attributed to a higher entropy or higher randomness that is not present. The effective electron temperature may be defined for convenience to indicate the degradation of mobility in varying low dimensions. But such a description comes at a price to the reader who has to interpret different versions of hot-electron temperatures. The other version of electric-field-induced bandgap temperature appears in Figure 4.16. It is directly connected to bandgap narrowing and enhanced intrinsic carrier concentration in an applied electric field.

10.4 A Refined Model

The above model reproduces the salient features of drift-diffusion with reasonable fit to the empirical data from experiments. Certainly, one may expect that diffusion is driven by the concentration gradient and not by the presence of an electric field. However, it could be criticized on the ground that the mfp used above is for low-field mobility and should only reproduce the low-field mobility, not the degraded mfp relevant to degraded mobility in a high electric field. A more refined model relies on the traditional equation for drift-diffusion

$$I = n_3 q v_{sat} \tanh\left(\frac{\mathcal{E}}{\mathcal{E}_c}\right) A_c + q D_3 \frac{dn_3}{dx} A_c \tag{10.24}$$

with

$$n_3(x) = N_{c3} \Im_{1/2}(\eta(x)) \tag{10.25}$$

The first term in Equation 10.24 is the drift term in tanh model discussed in Chapter 4 and the second term is the diffusion term.

In equilibrium ($I = 0$), the Fermi energy is constant, the concentration gradient by using Equation 10.25 is obtained as

$$\frac{dn_3}{dx} = N_{c3} \frac{d\Im_{1/2}}{d\eta} \frac{d\eta}{dx} = N_{c3} \Im_{-3/2} \left(-\frac{1}{k_B T}\right) \frac{dE_c}{dx} \tag{10.26}$$

Here, identity $d\Im_j / d\eta = \Im_{j-1}$ is used assuming E_F constant in equilibrium. The electric field is $\mathcal{E} = (1/q)(dE_c/dx)$. Now with $I = 0$ (in equilibrium) and using Equation 10.24 gives

$$D_3 = V_t \frac{\Im_{-1/2}}{\Im_{1/2}} v_{sat} \tanh \frac{(\mathcal{E}_b / \mathcal{E}_c)}{\mathcal{E}_b} \tag{10.27}$$

where we have added the subscript "b" as a reminder that this is a built-in electric field. Using $\mathcal{E}_c = v_{sat}/\mu_o$, we obtain

$$\frac{D_3}{\mu_o} = V_t \frac{\Im_{-1/2}}{\Im_{1/2}} \frac{\tanh(\mathcal{E}_b / \mathcal{E}_c)}{\mathcal{E}_b / \mathcal{E}_c} \tag{10.28}$$

The Einstein relation is reproduced in the ND regime where $\Im_j(\eta) = \exp(\eta)$, independent of j, and with a low electric field ($\tanh(x)/x \to 1$) gives

$$\frac{D_3}{\mu_o} = V_t \tag{10.29}$$

However, the mobility μ in a high electric field is

$$\frac{\mu}{\mu_o} = \frac{\tanh(\mathcal{E}_b/\mathcal{E}_c)}{\mathcal{E}_b/\mathcal{E}_c} \tag{10.30}$$

Giving us the generalized Einstein ratio

$$\frac{D_3}{\mu} = V_t \frac{\Im_{-1/2}}{\Im_{1/2}} \tag{10.31}$$

The same procedure can be applied to 2D and 1D, giving the generalized Einstein ratio as

$$\frac{D_d}{\mu_d} = V_t \frac{\Im_{(d/2)-2}}{\Im_{(d-2)/2}}, \quad d = 3,2,1 \tag{10.32}$$

A common expression to calculate the built-in electric field is

$$\mathcal{E}_b = -V_t \frac{1}{n_d}\frac{dn_d}{dx} \tag{10.33}$$

Equation 10.33 is built on ohmic drift ($v_d = \mu_o\mathcal{E}$). For an example of the exponential profile of $n_3 = n_3(0)\exp(-x/x_o)$ with $x_o = 0.042$ μm extracted for the base of a bipolar transistor, this relation gives $\mathcal{E}_b = -V_t/x_o = -0.62$ V/μm = −6.2 kV/cm. This field is much larger than its critical value where $v_d = \mu_o\mathcal{E}$ is not applicable. Hence, a revisit of this paradigm is needed. If indeed the electric field is that high, $\tanh(x) \approx 1$, both diffusion coefficient and mobility will degrade with the electric field. A careful thought is required to study this problem whether the applied field is in the same direction or opposite direction to the built-in electric field. Maybe diffusion is eliminated with the built-in electric field and only the applied field over and above the built-in one plays an active role for current, either to get saturated or to be in the ohmic domain. The distinction between a built-in and applied electric field is of paramount importance to delineate the drift diffusion and the role it plays in nonhomogenous nanostructures. This dilemma remains unresolved even today. Clear distinction is needed between the built-in electric field due to inhomogeneous carrier concentration and an external field.

10.5 Multivalley Transport

As shown in previous chapters, the effective mfp and, hence, the effective mobility, degrades as the electric field is increased. As an applied electric field increases, electrons tend to follow velocity-field profiles of those valleys with a higher Ohmic mfp [6]. If we consider two valleys with energy spacing, with the upper one having a larger mfp, carrier transfer to a higher valley takes place, giving negative differential mobility or conductivity. GaAs in a bulk form is experimentally known to show this negative differential mobility as a carrier transfer to the higher valley where degraded mobility due to the high electric field is lower. The major attraction for GaAs is its high low-field electron mobility (~8000 $cm^2/V \cdot s$) at room temperature, which exceeds 10^6 $cm^2/V \cdot s$ at low temperatures. If these high mobilities are preserved in a high electric field, substantial improvement in the speed of a given device is possible. Unfortunately, these high mobilities degrade when a semiconductor sample is subjected to a high electric field. The mobility degradation arises due to velocity-field profiles becoming sublinear as the electric field is increased beyond the critical field. This sublinear character, as shown in Chapter 4, degrades dc mobilities (slopes from the origin) and differential mobilities (tangential slopes at various points). A high electric field is always present in short-channel devices. These short channels are designed to cut down the transit time delay as electrons transit through a given device. Due to the ellipsoidal energy surface of higher valleys in GaAs, the DOS effective mass is distinctly different from the anisotropic conductivity effective mass.

The electronic energy of anisotropic valleys is given by

$$E_k = E_c + \frac{\hbar^2 k_x^2}{m_1^*} + \frac{\hbar^2 k_y^2}{m_2^*} + \frac{\hbar^2 k_z^2}{m_3^*} \tag{10.34}$$

where $m_{1,2,3}^*$ is the effective mass in the x-, y-, or z-directions, that can be extracted from the longitudinal and transverse effective mass of a given valley. E_c is the energy of the conduction band edge.

Using the NEADF extensively discussed in Chapter 4, the velocity response in the ND approximation to the applied electric field in z-direction is obtained as [7]

$$v_{3D} = \langle v_z \rangle = \sqrt{\frac{8k_BT}{\pi m_o}} \sum_{\mathrm{v}} \sqrt{\frac{m_o}{m_{n3}^*}} \frac{\left((\cosh\delta_\upsilon/\delta_\upsilon) - (\sinh\delta_\upsilon/\delta_\upsilon^2)\right)}{\sum_{\mathrm{v}} (\sinh\delta_{\mathrm{v}}/\delta_{\mathrm{v}})} \tag{10.35}$$

with

$$\delta_{\mathrm{v}} = \frac{\mathcal{E}}{\mathcal{E}_{c\mathrm{v}}} = \frac{\mathcal{E}\ell_{\mathrm{v}}}{V_t} \tag{10.36}$$

Here, index v is for a valley with an mfp ℓ_{v} in a given multivalley conduction band (Γ, L, or X). m_o is the free electron mass. For an isotropic semiconductor, the effective mass is the same in all directions, as is the case for single

Γ-valley of GaAs. In such circumstances, $m_1^* = m_2^* = m_3^* = m_n^*$ and Equation 10.35 is simplified to give

$$v_{3D} = \sqrt{\frac{8k_BT}{\pi m_n^*}}\; L(\delta) \tag{10.37}$$

where

$$L(\delta) = \coth(\delta) - \frac{1}{\delta} \tag{10.38}$$

is the Langevin function. With single valley is considered in Equation 10.38, the index ν is avoided.

As expected, in the limit of a low electric field ($\delta \to 0$), Equation 10.37 gives $v = \mu_o E$. The Ohmic mobility, μ_o, in this limit is obtained as

$$\mu_o = \frac{4}{3}\cdot\frac{q\ell}{\sqrt{2\pi m_n^* k_BT}} \tag{10.39}$$

In the other extreme, $\delta \to \infty$, $v = v_{sat}$ that is given by

$$v_{sat} = \sqrt{\frac{8k_BT}{\pi m_n^*}} \tag{10.40}$$

Although not considered here, if the distribution function in the strongly degenerate limit is considered (by preserving 1 in the denominator of the distribution function), it is not difficult to conjecture that this saturation velocity is comparable to the Fermi velocity, which is a function of the carrier concentration. The emission of a quantum in the high electric field will further lower the saturation velocity. However, Γ–*L*–*X* spacing in GaAs is too large for this quantum to be emitted. The drift velocity obtained in Equation 10.35 is a direct result of finding the expected value of the velocity for multiple valleys in a given conduction band. The fraction f_i of electrons in each conduction band in Γ–*L*–*XP*–*XL* hierarchy is given by

$$f_\nu = \frac{n_\nu}{n_\Gamma + n_L + n_{XP} + n_{XL}} \tag{10.41}$$

with

$$n_\nu = N_{c3\nu}\, e^{\eta_{ci}} \frac{\sinh(\delta_i)}{\delta_i}, \quad \text{with } N_{c3\nu} = 2\left(\frac{m_{ds\nu}^* k_BT}{2\pi\hbar^2}\right)^{3/2}, \quad \eta_{c\nu} = \frac{\varepsilon_F - \varepsilon_{c\nu}}{k_BT} \tag{10.42}$$

$$\nu = \Gamma, L, XP, XL$$

TABLE 10.1

Parameters and Other Properties for Ellipsoidal Valleys of GaAs

Valley (Number)	Conductivity Effective Mass (m_c^*/m_o)	Thermal Velocity v_{th3} (10^5 m/s)	Mean Free Path (nm)	Ohmic Mobility μ_o ($m^2/V \cdot s$)	Degraded dc Mobility $E = 3.3$ kV/cm	Degraded Diff Mobility $E = 3.3$ kV/cm	Effective Density of States (N_C)
Γ(1)	0.067	4.16	150	0.80	0.66	0.439	4.21×10^{17} cm^{-3}
L(4)	0.11	3.24	850	3.55	0.89	0.091	22.4 $N_{c\Gamma}$
XP(2)	0.19	2.47	1380	4.38	0.71	0.042	15.6 $N_{c\Gamma}$
XL(1)	1.90	0.78	1435	1.44	0.22	0.013	31.3 $N_{c\Gamma}$

Equation 10.41, along with Equation 10.42, is an indication of the condition under which intervalley transfer is possible. The intervalley transfer is possible when the ohmic mfp in the higher valley is higher than that of the lower valley. The relevant parameters used in the multivalley transfer are shown in Table 10.1. It is clear from Table 10.1 that the effective mfp degrades at the onset of intervalley transfer. The higher the value of the mfp, the higher the degradation at a given electric field. Even in the same conduction band with multivalleys (e.g., the three valleys in the *X*-band), the mfps can differ depending on the orientation of the valleys. The mfp is different in those valleys (*XL* valleys) with longitudinal axis in the direction of the electric field when compared to those (*XP*) with the longitudinal axis perpendicular to the electric field.

The fraction of electrons as a function of the electric field in each valley is depicted in Figure 10.5. At a low electric field (applied in the [001]-direction), all electrons are confined in the Γ-valley. At higher electric fields, the electrons first transfer to *L* valley, then on to *XP* valley, and ultimately to *XL* valley. In a very high electric field, most of the electrons are in *XL* valleys. *XP* is a set of four half valleys whose longitudinal direction is perpendicular to the applied electric field and *XL* is a set of two valleys with the longitudinal axis parallel to the electric field. The mfp of those valleys with a longitudinal axis perpendicular to the applied electric field is higher than those with the longitudinal axis parallel to the electric field.

The combined velocity response is obtained as

$$v(E) = v_{th3o} \sum_{\nu} f_{\nu} \left(\frac{m_o^*}{m_{ci}^*} \right) L(\delta_{\nu}) \tag{10.43}$$

with

$$v_{th3o} = \sqrt{\frac{8k_B T}{\pi m_o}} \tag{10.44}$$

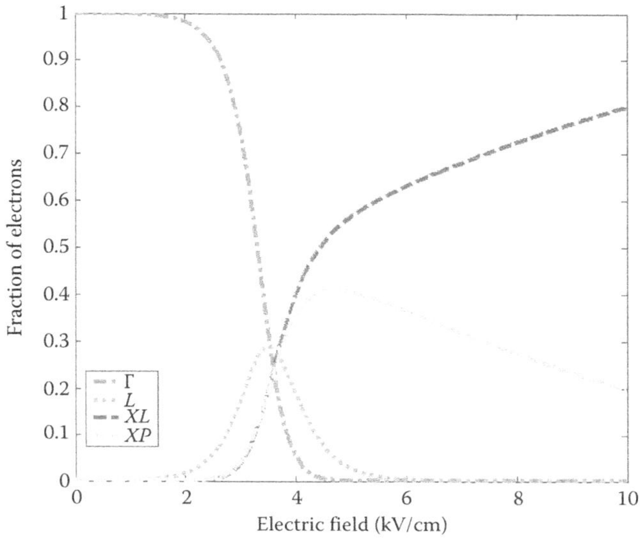

FIGURE 10.5
The fraction of electrons in each valley in the Γ–*L*–*XP*–*XL* hierarchy of GaAs.

Here, m^*_{ci} is the conductivity effective mass in the *i*th conduction band. Figure 10.6 shows the combined *v*–*E* characteristics along with those of each individual valley. As one can see, the velocity-field characteristics tend to follow the velocity-field curves of Γ-valley. As electrons transfer to *L*-valleys, the curve starts following the velocity-field graphs of that valley. Finally, the electrons tend to settle in the *XL*-valleys where the combined drift velocity at 10 kV/cm is 1.1×10^5 m/s. The combined differential mobility is maximum with a value 0.74 m²/V · s at an electric field of $\mathcal{E} = 3.75$ kV/cm. We have included in the graph the experimental points and the empirical curves that are found to agree with the Monte Carlo simulations. Some Monte Carlo simulations predict saturation of velocity. However, if the curve is continued to extend beyond 10 kV/cm, it is sloping toward a saturation value of 0.8×10^5 m/s. At this point, it is difficult to predict the role of X_7 valleys. That may change the saturation velocity if the conditions are appropriate for transfer to X_7 valleys. The parameters used in the calculations and those extracted from the velocity–field profiles obtained are listed in Table 10.1. The extraction of the Ohmic mfps for Γ-valley can be easily obtained from the Ohmic mobility using Equation 10.39. The other mfps are extracted from the experimental data at a peak value of the velocity-field curves and negative-differential region.

The multivalley model of GaAs is considered to investigate the velocity-field characteristics and clearly shows that transfer to higher valleys is possible if mfp of the higher valley is larger than that of lower valley, as shown

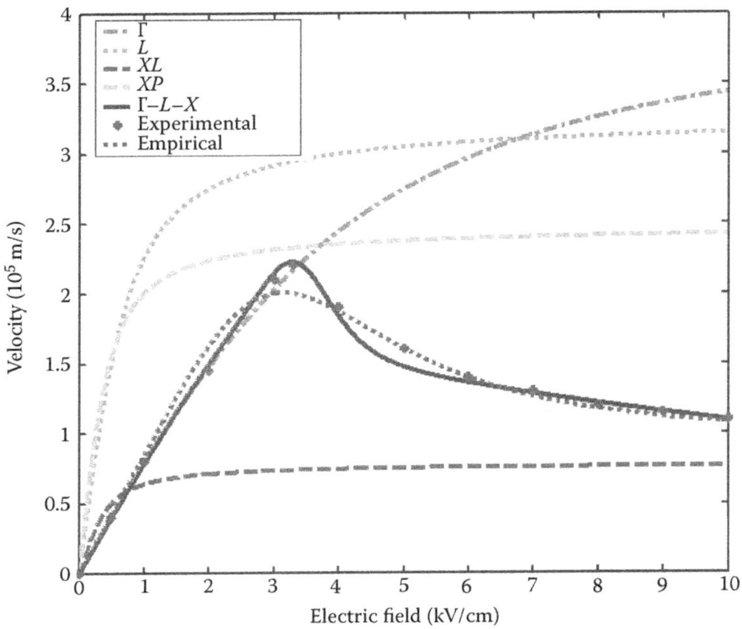

FIGURE 10.6
The velocity-field characteristics in the Γ–*L*–*XP*–*XL* hierarchy of GaAs. Solid curve is the results of all valleys combined. The dashed curve is from empirical relation derived from Monte Carlo simulations. The points represent the experimental data.

in Figure 10.6. The results are in reasonable agreement with the published experimental data found in the literature. The ultimate saturation velocity is comparable to the thermal velocity of the *XL*-valleys. The higher conductivity effective mass will give lower saturation velocity. Contrary to the paradigm that considers saturation of velocity arising from the enhanced scattering in an electric field, the paradigm presented here gives the saturation velocity as arising from the realignment of velocity vectors.

CAD/CAE PROJECTS

C10.1 *Built-in electric field + applied electric field.* As established in the chapter, built-in electric field and applied electric field must be differentiated. The incremental current above the equilibrium value of $I = 0$ may be ascribed to only high-field effects. Considering this viewpoint, revisit drift and diffusion currents to assess the correct form of current with an applied electric field in and opposite to the direction of a built-in electric field.

C10.2 *Diffusion in a MOSFET.* Diffusion is normally neglected in a channel of MOSFET even if we know that the channel concentration varies with the distance along the channel with an extreme profile at the onset of current saturation. Evaluate this assumption to see if, indeed, diffusion is negligible in a MOSFET and only the drift current dominates.

C10.3 *Revisiting I–V Characteristics of a Diode.* Junctions are present in every bulk or nanostructure. The *I–V* characteristics are modeled after diffusion of minority carrier in neutral layers with the assumption that the current at the edge of neutral layer remains the same as it traverses the depletion layer. What about the drift through the depletion layer? It is well known that the electric field in a junction is very high. Evaluate the p–n junction from a drift point-of-view with injected carriers from a neutral layer. *Physics of Semiconductor Devices* by S. M. Sze and Kwok K. Ng by inclusion of drift-diffusion have established

$$I = \frac{qN_cD_nA_c}{\int_0^{W_D} \exp[E_c(x)/k_BT]}\left[\exp\left(\frac{qV}{k_BT}\right) - 1\right] \tag{C10.1}$$

In the light of the discussion in the chapter and knowing well that the depletion layers are ND, evaluate the integral in the expression above by considering a linear approximation of $E_c(x) = qV(x)$ from $V(0)$ to $V(W_D) = V_{bi}$ (built-in potential). Compare with the diffusion theory of a diode. The diffusion theory model in the textbooks incorrectly displays the experimental data. The introduction of fudge factors, generation–regeneration current, series resistance, and so on are needed to display $I = I_o[\exp(V/\eta V_t) - 1]$. Considering the ballistic models and drift-diffusion processes in a high electric field, assess the values of I_o and η. Check if your understanding establishes a twist to traditional diode theory.

C10.4 *High electric field expression in drift-diffusion.* Drift and diffusion are now well established to be field-dependent with drift velocity saturating to the thermal velocity in a ND domain, especially for depletion layers. As in the case of MOSFET, $\tanh(x) \approx x/(1 + x)$ is established to evaluate the effect of a high electric field in a MOSFET channel. Can this be interjected into C10.3 to see if it reduces the current by a factor of $(1 + V/V_c)$ for a diode as well? Compare with the thermionic emission–diffusion theory cited on p. 159 of *Physics of Semiconductor Devices* by S. M. Sze and Kwok K. Ng. Argue in favor or against the critical electric field and/or critical voltage as far as its contribution to diode theory is expected.

References

1. V. K. Arora, Drift diffusion and Einstein relation for electrons in silicon subjected to a high electric field, *Applied Physics Letters*, 80, 3763–3765, May 2002.
2. V. K. Arora, Quantum engineering of nanoelectronic devices: The role of quantum emission in limiting drift velocity and diffusion coefficient, *Microelectronics Journal*, 31, 853–859, 2000.
3. S. A. Schwarz and S. E. Russek, Semi-empirical equations for electron velocity in silicon: Part II. MOS inversion layer, *IEEE Transactions on Electron Devices*, 30, 1634–1639, 1983.
4. S. M. Sze and K. K. Ng, *Physics of Semiconductor Devices*. Wiley, New York, 2006.
5. M. A. Omar and L. Reggiani, Drift and diffusion of charge carriers in silicon and their empirical relation to the electric field, *Solid-State Electronics*, 30, 693–697, 1987.
6. V. K. Arora, D. S. Mui, and H. Morkoc, Mobility degradation and transferred electron effect in gallium arsenide and indium gallium arsenide, *IEEE Transactions on Electron Devices*, 34, 1231–1238, 1987.
7. A. Sharma and V. K. Arora, Velocity-field characteristics in the multivalley model of gallium arsenide, *Journal of Applied Physics*, 97, 093704, 2005.

Appendix A: Physical Constants and Scales

TABLE A.1

Physical Constants

Name	Symbol	Value	Units
Boltzmann's constant	k_B	1.38×10^{-23}	J/K
		8.62×10^{-5}	eV/K
Electron rest mass	m_0	9.1×10^{-31}	kg
Electron volt	eV	1.6×10^{-19}	J
Electronic charge	q	1.6×10^{-19}	C
Planck's constant	h	6.625×10^{-34}	J s
		4.135×10^{-15}	eV s
Planck's constant divided by 2 pi (h-bar)	$\hbar$	1.055×10^{-34}	J s
		6.58×10^{-16}	eV s
Permittivity of free space	ε_o	8.85×10^{-14}	F/cm
		8.85×10^{-12}	F/m
Proton rest mass	m_p	1.673×10^{-27}	kg
Speed of light (vaccum)	c	3×10^{8}	m/s
Thermal energy ($T = 300$ K)	k_BT	0.0259	eV
Thermal voltage ($T = 300$ K)	$V_t = k_BT/q$	0.0259	V

TABLE A.2

Prefixes

Name	Abbreviation	Value
Peta	P	10^{15}
Tera	T	10^{12}
Giga	G	10^{9}
Mega	M	10^{6}
Kilo	k	10^{3}
Hecto	h	10^{2}
Deka	da	10^{1}
Deci	d	10^{-1}
Centi	c	10^{-2}
Milli	m	10^{-3}
Micro	μ	10^{-6}
Nano	n	10^{-9}
Pico	p	10^{-12}
Femto	f	10^{-15}

Appendix B: Silicon Parameters

Name	Symbol	Value	Units
Density of states (conduction band)	N_{c3}	2.86×10^{19}	cm^{-3}
		2.86×10^{25}	m^{-3}
Density of states (valence band)	N_{v3}	3.10×10^{19}	cm^{-3}
		3.10×10^{25}	m^{-3}
Dielectric constant (relative permittivity)	κ or ε_r	11.8	
Dielectric constant (relative permittivity) of SiO_2	$\kappa_{0x}(\varepsilon_{rox})$	3.9	
Effective mass (density of states) of an electron	m^*_{dsn}	$1.09m_o$	
Effective mass (density of states) of a hole	m^*_{dsp}	$1.15m_o$	
Effective mass (conductivity) of an electron	m^*_{cn}	$0.26m_o$	
Effective mass (conductivity) of a hole	m^*_{cp}	$0.36m_o$	
Effective mass—longitudinal of an electron	m^*_{ℓ}	$0.92m_o$	
Effective mass—transverse of an electron	m^*_t	$0.198m_o$	
Effective mass—heavy holes	m^*_{hh}	$0.48m_o$	
Effective mass—light holes	$m^*_{\ell h}$	$0.16m_o$	
Effective mass—split-off band	m^*_{sh}	$0.24m_o$	
Electron mobility	μ_n	1400	$cm^2/V \cdot s$
		0.14	$m^2/V \cdot s$
Electron affinity	χ	4.05	eV
Energy bandgap	E_g	1.1242	eV
Energy of an optical phonon	E_{ph}	0.063	eV
Hole mobility	μ_p	400	$cm^2/V \cdot s$
		0.04	$m^2/V \cdot s$
Intrinsic carrier concentration	n_{i3}	1.08×10^{10}	cm^{-3}
		1.08×10^{16}	m^{-3}
Split-off band separation	Δ	0.044	eV

Appendix C: Gallium Arsenide Parameters

Name	Symbol	Value	Units
Density of states (conduction band)	N_{c3}	4.4×10^{17}	cm^{-3}
		4.4×10^{23}	m^{-3}
Density of states (valence band)	N_{v3}	8.4×10^{18}	cm^{-3}
		8.4×10^{24}	m^{-3}
Dielectric constant (relative permittivity)	κ or ε_r	13.2	
Effective mass (density of states) of an electron	m_{dsn}^*	$0.067m_o$	
Effective mass (density of states) of a hole	m_{dsp}^*	$0.48m_o$	
Effective mass (conductivity) of an electron	m_{cn}^*	$0.067m_o$	
Effective mass (conductivity) of a hole	m_{cp}^*	$0.34m_o$	
Effective mass—longitudinal of an electron	m_ℓ^*	$0.067m_o$	
Effective mass—transverse of an electron	m_t^*	$0.067m_o$	
Effective mass—heavy holes	m_{hh}^*	$0.45m_o$	
Effective mass—light holes	$m_{\ell h}^*$	$0.082m_o$	
Effective mass—split-off band	m_{sh}^*	$0.15m_o$	
Electron mobility	μ_n	1400	$cm^2/V \cdot s$
		0.14	$m^2/V \cdot s$
Electron affinity	χ	4.07	eV
Energy bandgap	E_g	1.42	eV
Energy of an optical phonon	E_{ph}	0.034	eV
Hole mobility	μ_p	400	$cm^2/V \cdot s$
		0.04	$m^2/V \cdot s$
Intrinsic carrier concentration	n_{i3}	2.2×10^6	cm^{-3}
		2.2×10^{12}	m^{-3}
Split-off band separation	Δ	0.34	eV

Appendix D: Semiconductor Properties

	Crystal Structure	Lattice Constant (Å)		Energy Gap (eV)	Electron Effective Mass		Hole Effective Mass		Static Dielectric Constant	Refractive Index	Electron Mobility (cm²/V · s)	Hole Mobility (cm²/V · s)	
C	Dia												C
Si	Dia		5.4310	1.11*X*	m_1	0.98	m_l	0.16	11.7	3.44	1350	480	Si
					m_t	0.19	m_h	0.5					
Ge	Dia		5.6461	0.67*L*	m_1	1.58	m_l	0.04	16.3	4.00	3900	1900	Ge
					m_t	0.08	m_h	0.3			2000	1000	
α-Sn	Dia		6.4892	0.08Γ		0.02			10.2	2.65	500		Sn
α-SiC	Wur	a	3.0865	2.8									α-SiC
		c	15.117										
AIN	Wur	a	3.111	5.9*i*									AIN
		c	4.978										
AIP	Sph		5.4625	2.43*X*		0.13			9.8	3.0	80		AIP
AIAs	Sph		5.6605	2.16*X*		0.5	m_l	0.49	12.0		1000	180	
							m_h	1.06					
AISb	Sph		6.1355	1.52*X*		0.11		0.39	11	3.4	200	300	AISb
GaN	Wur	a	3.189	3.4Γ		0.2		0.8	12	2.4	300		GaN
		c	5.185										
GaP	Sph		5.4506	2.26*X*		0.13		0.67	10	3.37	300	150	GaP
GaAs	Sph		5.6535	1.43Γ		0.067		0.12	12.5	3.4	8500	400	GaAs
								0.5					
GaSb	Sph		6.0954	0.72Γ		0.045		0.39	15	3.9	5000	1000	GaSb
InN	Wur	a	3.533	2.4Γ									InN
		c	5.693										

InP	Sph		5.8688	1.35Γ		0.07		0.40	12.1	3.37	4000	600	InP
InAs	Sph		6.0584	0.36Γ		0.028		0.33	12.5	3.42	22,600	200	InAs
InSb	Sph		6.4788	0.18Γ		0.013		0.18	18	3.75	100,000	1700	InSb
ZnS	Sph		5.4109	3.6Γ		0.39			8.3	2.4	110		ZnS
ZnSe	Sph		5.6686	2.58Γ		0.17			8.1	2.89	600		ZnSe
ZnTe	Sph		6.1037	2.25Γ		0.15			9.7	3.56			ZnTe
CdS	Wur	a		2.42Γ		0.20		0.7	8.9	2.5	250		CdS
		c											
CdSe	Wur	a		1.73Γ		0.13		0.4	10.6		650		CdSe
		c											
CdTe	Sph		6.4816	1.50Γ		0.11		0.35	10.9	2.75	1050	100	CdTe
HgS	Sph		5.852	2.0Γ		0.045			25		50		HgS
HgSe	Sph		6.084	–0.15Γ		0.029		0.3	20	3.7	18,500		HgSe
HgTe	Sph		6.4616	–0.15Γ							22,000	100	HgTe
GeS	Orth			1.8									GeS
GeSe	Orth			1.16								70	GeSe
GeTe	NaCl		5.986									100	GeTe
SnS	Orth			1.08								90	SnS
SnSe	Orth			0.9								110	SnSe
SnTe	NaCl		6.325	0.18L								400	SnTe
PbS	NaCl		5.936	0.37L		0.1		0.1	170	3.7	500	600	PbS
PbSe	NaCl		6.147	0.26L	m_l	0.07	m_l	0.06	250		1,800	930	PbSe
					m_h	0.039	m_h	0.03					

Continued

	Crystal Structure	Lattice Constant (Å)		Energy Gap (eV)	Electron Effective Mass		Hole Effective Mass		Static Dielectric Constant	Refractive Index	Electron Mobility ($cm^2/V \cdot s$)	Hole Mobility ($cm^2/V \cdot s$)	
PbTe	NaCl		6.45	0.29*L*	m_l	0.24	m_l	0.3	412		1400	1100	PbTe
					m_h	0.02	m_h	0.02					
$ZnSiP_2$	Chal	a	5.400	2.96Γ		0.07					100		$ZnSiP_2$
		c	10.441									140	
$ZnSiAs_2$	Chal	a	5.606	2.12Γ				0.07					$ZnSiAs_2$
		c	10.890									20	
$ZnGeP_2$	Chal	a	5.465	2.34Γ				0.5					$ZnGeP_2$
		c	10.771									23	
$ZnGeAs_2$	Chal	a	5.670	1.15Γ				0.4					$ZnGeAs_2$
		c	11.153									55	
$ZnSnP_2$	Chal	a	5.651	1.66Γ									$ZnSnP_2$
		c	11.302									200	
$ZnSnAs_2$	Chal	a	5.852	0.73Γ				0.35					$ZnSnAs_2$
		c	11.705										
$CdSiP_2$	Chal	a	5.678	2.45Γ		0.09		0.7					$CdSiP_2$
		c	10.431								150		
$CdSiAs_2$	Chal	a	5.884	1.55Γ								500	$CdSiAs_2$
		c	10.882										

Appendix E: Metal Properties

Element	E_F (eV)	v_F (10^6 m/s)	λ_F (nm)	n_3 ($10^{28}/\text{m}^3$)	σ ($10^7/\Omega$ m)
Cesium (Cs)	1.58	0.75	0.976	0.91	0.50
Rubidium (Rb)	1.85	0.81	0.902	1.15	0.80
Potassium (K)	2.12	0.86	0.842	1.40	1.39
Sodium (Na)	3.23	1.07	0.682	2.65	2.11
Barium (Ba)	3.65	1.13	0.642	3.20	0.26
Strontium (Sr)	3.95	1.18	0.617	3.56	0.47
Calcium (Ca)	4.68	1.28	0.567	4.60	2.78
Lithium (Li)	4.72	1.29	0.565	4.70	1.07
Silver (Ag)	5.48	1.39	0.524	5.85	6.21
Gold (Au)	5.51	1.39	0.522	5.90	4.55
Copper (Cu)	7.00	1.57	0.464	8.45	5.88
Magnesium (Mg)	7.13	1.58	0.459	8.60	2.33
Cadmium (Cd)	7.46	1.62	0.449	9.28	1.38
Indium (In)	8.60	1.74	0.418	11.49	1.14
Zinc (Zn)	9.39	1.82	0.400	13.10	1.69
Lead (Pb)	9.37	1.82	0.401	13.20	0.48
Gallium (Ga)	10.35	1.91	0.381	15.30	0.67
Aluminum (Al)	11.63	2.02	0.360	18.06	3.65
Beryllium (Be)	14.14	2.23	0.326	24.20	3.08

Appendix F: Fermi–Dirac Integral

The FD integral of order j is given by

$$\Im_j(\eta) = \frac{1}{\Gamma(j+1)} \int_0^\infty dx \, \frac{x^j}{1 + e^{x-\eta}} \quad \text{(F.1)}$$

where $\Gamma(j+1)$ is the Gamma function. $\Gamma(j+1) = j!$ for an integer j. Gamma function follows the recursion relation $\Gamma(j+1) = j\Gamma(j)$. $\Gamma(1/2) = \sqrt{\pi}$ and $\Gamma(3/2) = \sqrt{\pi}/2$. In general,

$$\Gamma(j+1) = 1.3.5\ldots(2j-3)(2j-1)\frac{\sqrt{\pi}}{2^n} \quad \text{(F.2)}$$

$\Im_j(\eta)$ follows the relation

$$\frac{d\Im_j(\eta)}{d\eta} = \Im_{j-1}(\eta) \quad \text{(F.3)}$$

$\Im_0(\eta)$ can be evaluated in a closed form

$$\Im_0(\eta) = \ln(e^\eta + 1) \quad \text{(F.4)}$$

Using Equation F.3, $\Im_{-1}(\eta)$ can be evaluated as

$$\Im_{-1}(\eta) = \frac{e^\eta}{e^\eta + 1} \quad \text{(F.5)}$$

Two extremes of $\Im_j(\eta)$ are particularly noticeable. One is when $\eta < -2$. In this ND regime,

$$\Im_j(\eta) \approx e^\eta \quad \text{for } \eta < -2 \quad \text{(F.6)}$$

The other extreme is the degenerate regime when $\eta > 2$. In this extreme,

$$\Im_j(\eta) \approx \frac{1}{\Gamma(j+1)} \frac{\eta^{j+1}}{j+1} = \frac{\eta^{j+1}}{\Gamma(j+2)} \quad \text{for } \eta > 2 \quad \text{(F.7)}$$

As $\Im_{1/2}(\eta)$ is extensively used in the context of 3D semiconductors, a number of approximations are available in the published literature. Here is one possibility

$$\Im_{1/2}(\eta) \approx \frac{e^{\eta}}{1+0.27e^{\eta}} \quad \text{for } \eta \geq 1.3 \tag{F.8}$$

$$\Im_{1/2}(\eta) \approx \frac{4}{3\sqrt{\pi}}\eta^{3/2}\left(1+\frac{1.15}{\eta^2}\right) \quad \text{for } \eta \leq 1 \tag{F.9}$$

In general, there is an analytic expression for any FD integral of integer order j, for $j \leq -2$, given by

$$\Im_j(\eta) = \frac{e^{\eta}}{(1+e^{\eta})^{-j}} P_{-j-2}(e^{\eta}) \tag{F.10}$$

where $P_k(\eta)$ is a polynomial of degree k, and the coefficients $p_{k,i}$ are generated from a recurrence relation

$$p_{k,0} = 1 \tag{F.11}$$

$$p_{k,i} = (1+i)p_{k-1,j} - (k+1-i)p_{k-1,i-1}, \quad i = 1,\ldots,k \tag{F.12}$$

Appendix G: Table of Fermi–Dirac Integrals

η	$F_{-3/2}$	F_{-1}	$F_{-1/2}$	F_0	$F_{1/2}$	F_1	$F_{3/2}$	F_2	$F_{5/2}$	F_3	$F_{7/2}$	F_4	η
−4.0	1.78 (−2)	1.799 (−2)	1.808 (−2)	1.815 (−2)	1.8199 (−2)	1.8232 (−2)	1.8256 (−2)	1.8274 (−2)	1.8287 (−2)	1.8295 (−2)	1.8301 (−2)	1.8305 (−2)	−4.0
−3.9	1.96 (−2)	1.984 (−2)	1.995 (−2)	2.003 (−2)	2.0099 (−2)	2.0140 (−2)	2.0170 (−2)	2.0191 (−2)	2.0206 (−2)	2.0216 (−2)	2.0224 (−2)	2.0229 (−2)	−3.9
−3.8	2.17 (−2)	2.188 (−2)	2.203 (−2)	2.213 (−2)	2.2195 (−2)	2.2247 (−2)	2.2283 (−2)	2.2309 (−2)	2.2327 (−2)	2.2340 (−2)	2.2349 (−2)	2.2355 (−2)	−3.8
−3.7	2.38 (−2)	2.413 (−2)	2.429 (−2)	2.442 (−2)	2.4510 (−2)	2.4572 (−2)	2.4617 (−2)	2.4648 (−2)	2.4670 (−2)	2.4686 (−2)	2.4697 (−2)	2.4705 (−2)	−3.7
−3.6	2.63 (−2)	2.660 (−2)	2.681 (−2)	2.6% (−2)	2.7063 (−2)	2.7139 (−2)	2.7193 (−2)	2.7231 (−2)	2.7259 (−2)	2.7277 (−2)	2.7291 (−2)	2.7301 (−2)	−3.6
−3.5	2.89 (−2)	2.931 (−2)	2.956 (−2)	2.975 (−2)	2.9880 (−2)	2.9972 (−2)	3.0037 (−2)	3.0084 (−2)	3.0118 (−2)	3.0141 (−2)	3.0158 (−2)	3.0169 (−2)	−3.5
−3.4	3.18 (−2)	3.230 (−2)	3.260 (−2)	3.283 (−2)	3.2986 (−2)	3.3099 (−2)	3.3179 (−2)	3.3235 (−2)	3.3276 (−2)	3.3304 (−2)	3.3325 (−2)	3.3339 (−2)	−3.4
−3.3	3.50 (−2)	3.557 (−2)	3.595 (−2)	3.625 (−2)	3.6412 (−2)	3.6549 (−2)	3.6645 (−2)	3.6715 (−2)	3.6764 (−2)	3.6799 (−2)	3.6824 (−2)	3.6841 (−2)	−3.3
−3.2	3.85 (−2)	3.917 (−2)	3.962 (−2)	3.995 (−2)	4.0187 (−2)	4.0354 (−2)	4.0473 (−2)	4.0557 (−2)	4.0617 (−2)	4.0659 (−2)	4.0690 (−2)	4.0711 (−2)	−3.2
−3.1	4.23 (−2)	4.311 (−2)	4.367 (−2)	4.407 (−2)	4.4349 (−2)	4.4552 (−2)	4.4696 (−2)	4.4800 (−2)	4.4872 (−2)	4.4924 (−2)	4.4961 (−2)	4.4986 (−2)	−3.1
−3.0	4.65 (−2)	4.743 (−2)	4.810 (−2)	4.858 (−2)	4.8933 (−2)	4.9181 (−2)	4.9356 (−2)	4.9482 (−2)	4.9571 (−2)	4.9634 (−2)	4.9679 (−2)	4.9710 (−2)	−3.0
−2.9	5.10 (−2)	5.215 (−2)	5.298 (−2)	5.356 (−2)	5.3984 (−2)	5.4284 (−2)	5.4498 (−2)	5.4651 (−2)	5.4759 (−2)	5.4836 (−2)	5.4891 (−2)	5.4929 (−2)	−2.9
−2.8	5.60 (−2)	5.732 (−2)	5.831 (−2)	5.904 (−2)	5.9545 (−2)	5.9910 (−2)	6.0170 (−2)	6.0356 (−2)	6.0488 (−2)	6.0582 (−2)	6.0649 (−2)	6.0695 (−2)	−2.8
−2.7	6.13 (−2)	6.297 (−2)	6.417 (−2)	6.504 (−2)	6.5665 (−2)	6.6109 (−2)	6.6425 (−2)	6.6652 (−2)	6.6813 (−2)	6.6927 (−2)	6.7009 (−2)	6.7066 (−2)	−2.7
−2.6	6.71 (−2)	6.914 (−2)	7.059 (−2)	7.164 (−2)	7.2398 (−2)	7.2938 (−2)	7.3323 (−2)	7.3599 (−2)	7.3795 (−2)	7.3934 (−2)	7.4033 (−2)	7.4103 (−2)	−2.6
−2.5	7.35 (−2)	7.586 (−2)	7.762 (−2)	7.889 (−2)	7.9804 (−2)	8.0459 (−2)	8.0927 (−2)	8.1263 (−2)	8.1501 (−2)	8.1671 (−2)	8.1791 (−2)	8.1877 (−2)	−2.5
−2.4	8.02 (−2)	8.317 (−2)	8.529 (−2)	8.684 (−2)	8.7944 (−2)	8.8740 (−2)	8.9309 (−2)	8.9716 (−2)	9.0006 (−2)	9.0213 (−2)	9.0360 (−2)	9.0464 (−2)′	−2.4
−2.3	8.76 (−2)	9.112 (−2)	9.369 (1)	9.555 (−2)	9.6887 (−2)	9.7852 (−2)	9.8544 (−2)	9.9038 (−2)	9.9391 (−2)	9.9643 (−2)	9.9822 (−2)	9.9949 (−2)	−2.3
−2.2	9.55 (−2)	9.975 (−2)	1.0284 (−1)	1.051 (−1)	1.0671 (−1)	1.0788 (−1)	1.0872 (−1)	1.0932 (−1)	1.0975 (−1)	1.1005 (−1)	1.1027 (−1)	1.1042 (−1)	−2.2
−2.1	1.040 (−1)	1.091 (−1)	1.1280 (−1)	1.155 (−1)	1.1748 (−1)	1.1890 (−1)	1.1992 (−1)	1.2065 (−1)	1.2117 (−1)	1.2154 (−1)	1.2181 (−1)	1.2200 (−1)	−2.1
−2.0	1.132 (−1)	1.192 (−1)	1.2366 (−1)	1.269 (−1)	1.2930 (−1)	1.3101 (−1)	1.3225 (−1)	1.3313 (−1)	1.3377 (−1)	1.3422 (−1)	1.3454 (−1)	1.3477 (−1)	−2.0
−1.9	1.229 (−1)	1.301 (−1)	1.3546 (−1)	1.394 (−1)	1.4225 (−1)	1.4432 (−1)	1.4581 (−1)	1.4689 (−1)	1.4766 (−1)	1.4821 (−1)	1.4860 (−1)	1.4888 (−1)	−1.9
−1.8	1.331 (−1)	1.419 (−1)	1.4826 (−1)	1.530 (−1)	1.5642 (−1)	1.5893 (−1)	1.6074 (−1)	1.6204 (−1)	1.6297 (−1)	1.6364 (−1)	1.6412 (−1)	1.6446 (−1)	−1.8
−1.7	1.442 (−1)	1.545 (−1)	1.6213 (−1)	1.678 (−1)	1.7193 (−1)	1.7496 (−1)	1.7714 (−1)	1.7872 (−1)	1.7986 (−1)	1.8067 (−1)	1.8125 (−1)	1.8166 (−1)	−1.7
−1.6	1.558 (−1)	1.680 (−1)	1.7712 (−1)	1.839 (−1)	1.8889 (−1)	1.9253 (−1)	1.9517 (−1)	1.9708 (−1)	1.9846 (−1)	1.9944 (−1)	2.0015 (−1)	2.0066 (−1)	−1.6
−1.5	1.680 (−1)	1.824 (−1)	1.9330 (−1)	2.014 (−1)	2.0740 (−1)	2.1178 (−1)	2.1497 (−1)	2.1728 (−1)	2.1895 (−1)	2.2015 (−1)	2.2099 (−1)	2.2162 (−1)	−1.5

−1.4	1.808 (−1)	1.978 (−1)	2.1074 (−1)	2.204 (−1)	2.2759 (−1)	2.3286 (−1)	2.3671 (−1)	2.3950 (−1)	2.4152 (−1)	2.4297 (−1)	2.4401 (−1)	2.4476 (−1)	−1.4
−1.3	1.941 (−1)	2.142 (−1)	2.2948 (−1)	2.410 (−1)	2.4959 (−1)	2.5592 (−1)	2.6055 (−1)	2.6392 (−1)	2.6636 (−1)	2.6812 (−1)	2.6938 (−1)	2.7029 (−1)	−1.3
−1.2	2.080 (−1)	2.315 (−1)	2.4958 (−1)	2.633 (−1)	2.7353 (−1)	2.8112 (−1)	2.8669 (−1)	2.9075 (−1)	2.9370 (−1)	2.9583 (−1)	2.9736 (−1)	2.9846 (−1)	−1.2
−1.1	2.222 (−1)	2.497 (−1)	2.7108 (−1)	2.873 (−1)	2.9955 (−1)	3.0863 (−1)	3.1533 (−1)	3.2022 (−1)	3.2378 (−1)	3.2636 (−1)	3.2822 (−1)	3.2955 (−1)	−1.1
−1.0	2.367 (−1)	2.689 (−1)	2.9402 (−1)	3.133 (−1)	3.2780 (−1)	3.3865 (−1)	3.4667 (−1)	3.5256 (−1)	3.5686 (−1)	3.5997 (−1)	3.6222 (−1)	3.6384 (−1)	−1.0
−0.9	2.517 (−1)	2.891 (−1)	3.1845 (−1)	3.412 (−1)	3.5841 (−1)	3.7135 (−1)	3.8096 (−1)	3.8804 (−1)	3.9321 (−1)	3.9698 (−1)	3.9970 (−1)	4.0166 (−1)	−0.9
−0.8	2.667 (−1)	3.100 (−1)	3.4438 (−1)	3.711 (−1)	3.9154 (−1)	4.0695 (−1)	4.1844 (−1)	4.2693 (−1)	4.3316 (−1)	4.3770 (−1)	4.4098 (−1)	4.4336 (−1)	−0.8
−0.7	2.820 (−1)	3.318 (−1)	3.7181 (−1)	4.032 (−1)	4.2733 (−1)	4.4564 (−1)	4.5936 (−1)	4.6953 (−1)	4.7702 (−1)	4.8249 (−1)	4.8646 (−1)	4.8933 (−1)	−0.7
−0.6	2.971 (−1)	3.543 (−1)	4.0077 (−1)	4.375 (−1)	4.6595 (−1)	4.8766 (−1)	5.0400 (−1)	5.1617 (−1)	5.2515 (−1)	5.3174 (−1)	5.3653 (−1)	5.4000 (−1)	−0.6
−0.5	3.121 (−1)	3.775 (−1)	4.3123 (−1)	4.741 (−1)	5.0754 (−1)	5.3322 (−1)	5.5265 (−1)	5.6718 (−1)	5.7795 (−1)	5.8587 (−1)	5.9164 (−1)	5.9584 (−1)	−0.5
−0.4	3.268 (−1)	4.013 (−1)	4.6318 (−1)	5.130 (−1)	5.5224 (−1)	5.8255 (−1)	6.0561 (−1)	6.2294 (−1)	6.3583 (−1)	6.4533 (−1)	6.5229 (−1)	6.5736 (−1)	−0.4
−0.3	3.410 (−1)	4.256 (−1)	4.9657 (−1)	5.544 (−1)	6.0022 (−1)	6.3590 (−1)	6.6321 (−1)	6.8382 (−1)	6.9923 (−1)	7.1063 (−1)	7.1899 (−1)	7.2510 (−1)	−0.3
−0.2	3.548 (−1)	4.502 (−1)	5.3137 (−1)	5.981 (−1)	6.5161 (−1)	6.9350 (−1)	7.2577 (−1)	7.5026 (−1)	7.6863 (−1)	7.8228 (−1)	7.9234 (−1)	7.9969 (−1)	−0.2
−0.1	3.677 (−1)	4.750 (−1)	5.6750 (−1)	6.444 (−1)	7.0654 (−1)	7.5561 (−1)	7.9365 (*t* − 1)	8.2267 (−1)	8.4455 (−1)	8.6088 (−1)	8.7294 (−1)	8.8179 (−1)	−0.1
0.0	3.800 (−1)	5.000 (−1)	6.0490 (−1)	6.932 (−1)	7.6515 (−1)	8.2247 (−1)	8.6720 (−1)	9.0154 (−1)	9.2755 (−1)	9.4703 (−1)	9.6148 (−1)	9.7212 (−1)	0.0
0.1	3.915 (−1)	5.250 (−1)	6.4348 (−1)	7.444 (−1)	8.2756 (−1)	8.9430 (−1)	9.4680 (−1)	9.8730 (−1)	1.0182 (0)	1.0414 (0)	1.0587 (0)	1.0715 (0)	0.1
0.2	4.019 (−1)	5.498 (−1)	6.8317 (−1)	7.981 (−1)	8.9388 (−1)	9.7150 (−1)	1.0328 (0)	1.0806 (0)	1.1171 (0)	1.1448 (0)	1.1654 (0)	1.1807 (0)	0.2
0.3	4.114 (−1)	5.744 (−1)	7.2384 (−1)	8.544 (−1)	9.6422 (−1)	1.0541 (0)	1.1257 (0)	1.1818 (0)	1.2250 (0)	1.2578 (0)	1.2824 (0)	1.3008 (0)	0.3
0.4	4.196 (−1)	5.987 (−1)	7.6540 (−1)	9.130 (−1)	1.0387 (0)	1.1424 (0)	1.2258 (0)	1.2916 (0)	1.3425 (0)	1.3814 (0)	1.4107 (0)	1.4326 (0)	0.4
0.5	4.269 (−1)	6.225 (−1)	8.0774 (−1)	9.7410 (−1)	1.1173 (0)	1.2367 (0)	1.3336 (0)	1.4105 (0)	1.4704 (0)	1.5164 (0)	1.5513 (0)	1.5774 (0)	0.5
0.6	4.328 (−1)	6.457 (−1)	8.5074 (−1)	1.0375 (0)	1.2003 (0)	1.3373 (0)	1.4494 (0)	1.5391 (0)	1.6095 (0)	1.6638 (0)	1.7052 (0)	1.7363 (0)	0.6
0.7	4.378 (−1)	6.682 (−1)	8.9429 (−1)	1.1032 (0)	1.2875 (0)	1.4443 (0)	1.5738 (0)	1.6782 (0)	1.7606 (0)	1.8246 (0)	1.8736 (0)	1.9106 (0)	0.7
0.8	4.415 (−1)	6.900 (−1)	9.3826 (−1)	1.1711 (0)	1.3791 (0)	1.5580 (0)	1.7071 (0)	1.8282 (0)	1.9246 (0)	1.9998 (0)	2.0577 (0)	2.1017 (0)	0.8
0.9	4.441 (−1)	7.110 (−1)	9.8255 (−1)	1.2412 (0)	1.4752 (0)	1.6786 (0)	1.8497 (0)	1.9900 (0)	2.1023 (0)	2.1906 (0)	2.2589 (0)	2.3111 (0)	0.9
1.0	4.457 (−1)	7.311 (−1)	1.0271 (0)	1.3133 (0)	1.5756 (0)	1.8063 (0)	2.0023 (0)	2.1642 (0)	2.2948 (0)	2.3982 (0)	2.4787 (0)	2.5404 (0)	1.0
1.1	4.463 (−1)	7.503 (−1)	1.0717 (0)	1.3873 (0)	1.6806 (0)	1.9413 (0)	2.1650 (0)	2.3515 (0)	2.5031 (0)	2.6239 (0)	2.7184 (0)	2.7913 (0)	1.1
1.2	4.459 (−1)	7.685 (−1)	1.1163 (0)	1.4633 (0)	1.7900 (0)	2.0838 (0)	2.3385 (0)	2.5527 (0)	2.7282 (0)	2.8690 (0)	2.9799 (0)	3.0658 (0)	1.2
1.3	4.447 (−1)	7.858 (−1)	1.1608 (0)	1.5410 (0)	1.9038 (0)	2.2340 (0)	2.5232 (0)	2.7685 (0)	2.9712 (0)	3.1349 (0)	3.2647 (0)	3.3658 (0)	1.3
1.4	4.427 (−1)	8.022 (−1)	1.2052 (0)	1.6204 (0)	2.0221 (0)	2.3921 (0)	2.7194 (0)	2.9997 (0)	3.2332 (0)	3.4232 (0)	3.5747 (0)	3.6936 (0)	1.4

Continued

η	$F_{-3/2}$	F_{-1}	$F_{-1/2}$	F_0	$F_{1/2}$	F_1	$F_{3/2}$	F_2	$F_{5/2}$	F_3	$F_{7/2}$	F_4	η
1.5	4.398 (−1)	8.176 (−1)	1.2493 (0)	1.7014 (0)	2.1449 (0)	2.5582 (0)	2.9278 (0)	3.2472 (0)	3.5155 (0)	3.7354 (0)	3.9120 (0)	4.0513 (0)	1.5
1.6	4.365 (−1)	8.320 (−1)	1.2931 (0)	1.7839 (0)	2.2720 (0)	2.7324 (0)	3.1486 (0)	3.5116 (0)	3.8192 (0)	4.0732 (0)	4.2786 (0)	4.4415 (0)	1.6
1.7	4.325 (−1)	8.455 (−1)	1.3366 (0)	1.8678 (0)	2.4035 (0)	2.9150 (0)	3.3823 (0)	3.7939 (0)	4.1456 (0)	4.4383 (0)	4.6766 (0)	4.8668 (0)	1.7
1.8	4.281 (−1)	8.582 (−1)	1.3796 (0)	1.9530 (0)	2.5393 (0)	3.1060 (0)	3.6294 (0)	4.0949 (0)	4.4961 (0)	4.8326 (0)	5.1085 (0)	5.3301 (0)	1.8
1.9	4.233 (−1)	8.699 (−1)	1.4222 (0)	2.0394 (0)	2.6794 (0)	3.3056 (0)	3.8903 (0)	4.4154 (0)	4.8719 (0)	5.2580 (0)	5.5767 (0)	5.8344 (0)	1.9
2.0	4.182 (−1)	8.808 (−1)	1.4643 (0)	2.1269 (0)	2.8237 (0)	3.5139 (0)	4.1654 (0)	4.7563 (0)	5.2746 (0)	5.7164 (0)	6.0838 (0)	6.3828 (0)	2.0
2.1	4.126 (−1)	8.909 (−1)	1.5058 (0)	2.2155 (0)	2.9722 (0)	3.7310 (0)	4.4552 (0)	5.1185 (0)	5.7055 (0)	6.2099 (0)	6.6325 (0)	6.9788 (0)	2.1
2.2	4.070 (−1)	9.002 (−1)	1.5468 (0)	2.3051 (0)	3.1249 (0)	3.9571 (0)	4.7600 (0)	5.5028 (0)	6.1662 (0)	6.7408 (0)	7.2258 (0)	7.6261 (0)	2.2
2.3	4.013 (−1)	9.089 (−1)	1.5872 (0)	2.3956 (0)	3.2816 (0)	4.1921 (0)	5.0803 (0)	5.9102 (0)	6.6580 (0)	7.3113 (0)	7.8668 (0)	8.3283 (0)	2.3
2.4	3.954 (−1)	9.168 (−1)	1.6271 (0)	2.4868 (0)	3.4423 (0)	4.4362 (0)	5.4164 (0)	6.3416 (0)	7.1827 (0)	7.9237 (0)	8.5585 (0)	9.0897 (0)	2.4
2.5	3.893 (−1)	9.241 (−1)	1.6663 (0)	2.5789 (0)	3.6070 (0)	4.6895 (0)	5.7689 (0)	6.7978 (0)	7.7419 (0)	8.5804 (0)	9.3044 (0)	9.9145 (0)	2.5
2.6	3.833 (−1)	9.309 (−1)	1.7049 (0)	2.6716 (0)	3.7755 (0)	4.9520 (0)	6.1380 (0)	7.2798 (0)	8.3371 (0)	9.2841 (0)	1.0108 (+1)	1.0807 (+1)	2.6
2.7	3.772 (−1)	9.370 (−1)	1.7430 (0)	2.7650 (0)	3.9480 (0)	5.2238 (0)	6.5241 (0)	7.7885 (0)	8.9700 (0)	1.0037 (+1)	1.0973 (+1)	1.1773 (+1)	2.7
2.8	3.712 (−1)	9.427 (−1)	1.7804 (0)	2.8590 (0)	4.1241 (0)	5.5050 (0)	6.9277 (0)	8.3249 (0)	9.6425 (0)	1.0843 (+1)	1.1903 (+1)	1.2817 (+1)	2.8
2.9	3.654 (−1)	9.478 (−1)	1.8172 (0)	2.9536 (0)	4.3040 (0)	5.7957 (0)	7.3491 (0)	8.8898 (0)	1.0356 (+1)	1.1703 (+1)	1.2903 (+1)	1.3943 (+1)	2.9
3.0	3.595 (−1)	9.526 (−1)	1.8535 (0)	3.0486 (0)	4.4876 (0)	6.0958 (0)	7.7886 (0)	9.4843 (0)	1.1113 (+1)	1.2622 (+1)	1.3976 (+1)	1.5159 (+1)	3.0
3.1	3.537 (−1)	9.569 (−1)	1.8891 (0)	3.1441 (0)	4.6747 (0)	6.4054 (0)	8.2467 (0)	1.0109 (+1)	1.1915 (+1)	1.3601 (+1)	1.5127 (+1)	1.6470 (+1)	3.1
3.2	3.481 (−1)	9 608 (−1)	1.9242 (0)	3.2400 (0)	4.8653 (0)	6.7246 (0)	8.7237 (0)	1.0766 (+1)	1.2763 (+1)	1.4645 (+1)	1.6360 (+1)	1.7882 (+1)	3.2
3.3	3.425 (−1)	9.644 (−1)	1.9588 (0)	3.3363 (0)	5.0595 (0)	7.0534 (0)	9.2199 (0)	1.1455 (+1)	1.3660 (+1)	1.5755 (+1)	1.7681 (+1)	1.9401 (+1)	3.3
3.4	3.370 (−1)	9.677 (−1)	1.9927 (0)	3.4328 (0)	5.2571 (0)	7.3918 (0)	9.7357 (0)	1.2177 (+1)	1.4608 (+1)	1.6937 (+1)	1.9094 (+1)	2.1035 (+1)	3.4
3.5	3.319 (−1)	9.707 (−1)	2.0262 (0)	3.5298 (0)	5.4580 (0)	7.7400 (0)	1.0271 (+1)	1.2933 (+1)	1.5608 (+1)	1.8192 (+1)	2.0605 (+1)	2.2791 (+1)	3.5
3.6	3.267 (−1)	9.734 (−1)	2.0591 (0)	3.6270 (0)	5.6623 (0)	8.0978 (0)	1.0827 (+1)	1.3725 (+1)	1.6662 (+1)	1.9524 (+1)	2.2218 (+1)	2.4676 (+1)	3.6

Appendix H: Periodic Table

PERIODS	GROUP I	GROUP II	Transition elements										GROUP III	GROUP IV	GROUP V	GROUP VI	GROUP VII	GROUP VIII
1	1 H 1.01 $1s^1$																	2 He 4.00 $1s^2$
2	3 Li 6.94 $2s^1$	4 Be 9.01 $2s^2$											5 B 10.81 $2p^1$	6 C 12.01 $2p^2$	7 N 14.01 $2p^3$	8 O 16.00 $2p^4$	9 F 19.00 $2p^5$	10 Ne 20.18 $2p^6$
3	11 Na 22.99 $3s^1$	12 Mg 24.31 $3s^2$											13 Al 26.98 $3p^1$	14 Si 28.09 $3p^2$	15 P 30.97 $3p^3$	16 S 32.07 $3p^4$	17 Cl 35.45 $3p^5$	18 Ar 39.95 $3p^6$
4	19 K 39.10 $4s^1$	20 Ca 40.08 $4s^2$	21 Sc 44.96 $3d^14s^2$	22 Ti 47.88 $3d^24s^2$	23 V 50.94 $3d^34s^2$	24 Cr 52.00 $3d^54s^1$	25 Mn 54.94 $3d^54s^2$	26 Fe 55.85 $3d^64s^2$	27 Co 58.93 $3d^74s^2$	28 Ni 58.69 $3d^84s^2$	29 Cu 63.55 $3d^{10}4s^1$	30 Zn 65.39 $3d^{10}4s^2$	31 Ga 69.72 $4p^1$	32 Ge 72.61 $4p^2$	33 As 74.92 $4p^3$	34 Se 78.96 $4p^4$	35 Br 79.90 $4p^5$	36 Kr 83.80 $4p^6$
5	37 Rb 85.47 $5s^1$	38 Sr 87.62 $5s^2$	39 Y 88.96 $4d^15s^2$	40 Zr 91.22 $4d^25s^2$	41 Nb 92.91 $4d^45s^1$	42 Mo 95.94 $4d^55s^1$	43 Tc (98) $4d^55s^2$	44 Ru 101.07 $4d^75s^1$	45 Rh 102.91 $4d^85s^1$	46 Pd 106.42 $4d^{10}5s^6$	47 Ag 107.87 $4d^{10}5s^1$	48 Cd 112.41 $4d^{10}5s^2$	49 In 114.82 $5p^1$	50 Sn 118.71 $5p^2$	51 Sb 121.76 $5p^3$	52 Te 127.60 $5p^4$	53 I 126.90 $5p^5$	54 Xe 131.29 $5p^6$
6	55 Cs 132.91 $6s^1$	56 Ba 137.33 $6s^2$	57 La 138.91 * $5d^16s^2$	72 Hf 178.49 $5d^26s^2$	73 Ta 180.95 $5d^36s^2$	74 W 183.85 $5d^46s^2$	75 Re 186.21 $5d^56s^2$	76 Os 190.2 $5d^66s^2$	77 Ir 192.22 $5d^76s^2$	78 Pt 195.08 $5d^96s^1$	79 Au 196.97 $5d^{10}6s^1$	80 Hg 200.59 $5d^{10}6s^2$	81 Tl 204.36 $6p^1$	82 Pb 207.2 $6p^2$	83 Bi 208.98 $6p^3$	84 Po (209) $6p^4$	85 At (210) $6p^5$	86 Rn (222) $6p^6$
7	87 Fr (223) $7s^1$	88 Ra 226.03 $7s^2$	89 Ac 227.03 † $6d^17s^2$	104 Rf (261) $6d^27s^2$	105 Db (262) $6d^37s^2$	106 Sg (266) $6d^47s^2$	107 Bh (264) $6d^57s^2$	108 Hs (269) $6d^67s^2$	109 Mt (268) $6d^77s^2$	110 (271)	111 (272)	112 (277)		114 (289)		116 (289)		118 (293)

Key: Atomic number — 26 Fe — Symbol; 58.85 — Atomic mass; Outer electron configuration — $3d^64s^2$

*	58 Ce 140.12 $5d^14f^16s^2$	59 Pr 140.91 $4f^36s^2$	60 Nd 144.24 $4f^46s^2$	61 Pm (145) $4f^56s^2$	62 Sm 150.36 $4f^66s^2$	63 Eu 151.96 $4f^76s^2$	64 Gd 157.25 $5d^14f^76s^2$	65 Tb 158.93 $5d^14f^86s$	66 Dy 162.50 $4f^{10}6s^2$	67 Ho 164.93 $4f^{11}6s^2$	68 Er 167.26 $4f^{12}6s^2$	69 Tm 168.93 $4f^{13}6s^2$	70 Yb 173.04 $4f^{14}6s^2$	71 Lu 174.97 $5d^14f^{14}6s^2$	Lanthanides
†	90 Th 232.04 $6d^27s^2$	91 Pa 231.04 $5f^26d^17s^2$	92 U 238.03 $5f^36d^17s^2$	93 Np 237.05 $5f^46d^17s^2$	94 Pu (244) $5f^66d^07s^2$	95 Am (243) $5f^76d^07s^2$	96 Cm (247) $5f^76d^17s^2$	97 Bk (247) $5f^86d^17s^2$	98 Cf (251) $5f^{10}6d^07s^2$	99 Es (252) $5s^{11}6d^07s^2$	100 Fm (257) $5f^{12}6d^07s^2$	101 Md (258) $5f^{13}6d^07s^2$	102 No (259) $5f^{14}6d^07s^2$	103 Lr (262) $5f^{14}6d^17s^2$	Actinides

Appendix I: General Comments on Project Execution

All projects require definition on the part of the executor so that the problem statement is clear and crisp. This could be done in an introduction section. This definition is the first step toward an oft-stated paradigm: "say what you are going to do, do it, and finally say what you have done." The first component (*say what you are going to do*) makes a good introduction with relevant facts stated and orients the reader to the problem to be solved. The second component (*do it*) can be broken down into subtasks with each subtask given a section, pasting graphics (figures) at appropriate places, and discussing the results with callouts for figures and equations, properly numbered and figures captioned. A good discussion involves the constraints under which output has been acquired, approximations that have been made, and the outcomes with their clear and conspicuous interpretations. The third component (*say what you have done*) is tying things together with outcomes emphasized and perhaps a sentence or two about how the outcomes can be further extended for applications. A good abstract (normally 100–200 words) captures the interest of the readers and compels them to read further. Even though the abstract appears first in any paper, it is written last. An iterative process can be designed to capture the mind in an abstract before writing the paper, complete the paper, and revisits the abstract for identification that it does capture the reader's attention. A *descriptive abstract* orients the reader to what is contained in the paper without stating the outcomes. This is appropriate for a paper in social sciences and humanities. However, in engineering, physical, and biological sciences, it is essential that the abstract is informative, stating not only the methodology followed, but also the major outcomes obtained and their possible extension for future exploration.

The best is to look at some of the sample papers to capture the essence of the organization of a project report. The papers normally do not have a cover sheet, but project reports normally have a cover sheet and table of contents. It is hybrid of the book format and a paper format.

The derivations are avoided in a formal report. Where applicable, the references are given to published sources. However, the meaning and significance of all symbols is always stated. Web of Science or Web of Knowledge do provide opportunities to an End Note environment existing on your computer or on the web. End Note and Math Type equation editor are embedded in MS Word menu bar and are easily called in when needed. The major advantage of End Note is that references are automatically formatted at the end of the paper. Most academic organizations subscribe to the Web of Science.

Google Scholar is another source of information especially for books as Web of Science does not cover books.

Computational and graphical practices can be done by using MATLAB®. It is not the only choice. MAPLE, MATHCAD, and MATHEMATICA are a few other vendors who are pushing their products in the educational market. For the projects stated at the end of each chapter, some of the function subroutines are already available and can be downloaded. For example, MATLAB does provide a function to calculate the Fermi-Dirac (FD) integral or has numerical integration programs to write a function. Nanohub.org site at Purdue University has a number of function subroutines stored that must be checked for their validity with the current version of the software.

The Information Matrix sketched below was suggested by Drs. Clifford B. Felder and James M. Gregory at the Texas Tech University (*The Information Matrix: Taking the Trouble Out of Technical Writing,* Engineering Education, December 1988, p. 163). It is an organization plan to streamline the writing and grading process. The rows of the matrix identify the writer's three main goals: Row 1—Capture the reader's attention; Row 2—present the body of information; and Row 3—credit the sources and highlight the facts that others can use in the future. The columns reveal whether or not the document has met its objectives. Each section of the matrix can be written separately and then combined into an appropriate section(s) of the report. This helps the writer and grader keep the focus on important segments and streamline the process of writing and grading.

The following is a grading form that can be easily duplicated to streamline the process of grading. It can also be used as a check list for the authors to confirm whether he/she said what is to be said in three different ways: Tell what you are going to do; tell what is being done, and tell about the outcomes. Outcomes are major achievements and their broader appeal includes the room for expansion and recommendation for exploration.

These are not the nine sections of the report. Suggested organization can be easily implemented in the Information Matrix. For example, the purpose can be stated in the last paragraph of the introduction section. Similarly, summary, conclusions, and possible extensions can be combined in one section that may contain a single paragraph. Development is the body of the paper that may be broken down into sections or subsections depending on the individual taste and style. Instructors can state their expectations when implementing this grading matrix.

NAME(S)____

Feature ↓	Purpose	Multiplicity	Mix and Match	Achievements	Grade
Say what you are going to say	**Title** (5)	**Abstract** (10)	**Introduction** (5)	Did you capture the reader's attention?	/20
Say it	**Purpose** (5)	**Development** (25)	**Outcomes** (10) Results (5) Discussion (5)	Was the foundation (body) of the paper strong?	/40
Say what you have said	**Conclusions** (5)	**Summary** (5)	**References** (10) Utilized (5) Formatted (5)	Did you cite and recite the other works and effectively used them in the text?	/20
Check points →	Do the conclusions reveal that the purpose was achieved?	Did you tell the story three times and in three different ways?	How well did your results fit or match other references or the overall problem?	Did you check the coherent and logical flow of your thoughts in support of the theme?	/20
				Total	/100

Appendix J: Published Papers by the Author

The book is based on papers published by the author in numerous journals. The author can be reached at vijay.arora@wilkes.edu to get more information on the contents of these papers. Here is the list of these papers for readers to refer to original sources:

1. V. K. Arora and A. Bhattacharyya, A unified bandgap engineering of graphene nanoribbons, *Physica Status Solidi (b)*, 251(11), 2257–2264, 2014.
2. E. Akbari, V. K. Arora, M. T. Ahmadi, A. Enzevaee, M. Saeidmanesh, M. Khaledian, H. Karimi, and R. Yusof, An analytical approach to evaluate the performance of graphene and carbon nanotubes for NH_3 gas sensor applications, *Beilstein Journal of Nanotechnology*, 5, 726–734, 2014.
3. E. Akbari, V. K. Arora, A. Enzevaee, M. T. Ahmadi, M. Khaledian, and R. Yusof, Gas concentration effects on the sensing properties of bilayer graphene, *Plasmonics*, 9, 987–992, 2014.
4. H. C. Chin, A. Bhattacharyya, and V. K. Arora, Extraction of nanoelectronic parameters from quantum conductance in a carbon nanotube, *Carbon*, 76, 451–454, 2014.
5. H. C. Chin, C. S. Lim, W. S. Wong, K. A. Danapalasingam, V. K. Arora, and M. Tan, Enhanced device and circuit-level performance benchmarking of graphene nanoribbon field-effect transistor against a nano-MOSFET with interconnects, *Journal of Nanomaterials*, 2014, Article ID 879813, 14 (http://dx.doi.org/10.1155/2014/879813).
6. Bandgap engineering of carbon allotropes, *Facta Universitatis Series: Electronics and Energetics*, 27(1), 113–127, March 2014. DOI: 10.2298/FUEE1401113A.
7. V. K. Arora and A. Bhattacharyya, Cohesive band structure of carbon nanotubes for applications in quantum transport, *Nanoscale*, 5(22), 10927–10935, 2013.
8. V. K. Arora, M. L. P. Tan, and C. Gupta, High-field transport in a graphene nanolayer, *Journal of Applied Physics*, 112, 114330, 2012.
9. V. K. Arora, OBE and WA: Understanding the paradigm shift on knowledge delivery and management, *Journal of Engineering, Science and Management Education*, 5(II), 430–444, 2012.
10. V. K. Arora, M. S. Z. Abidin, M. L. P. Tan, and M. A. Riyadi, Temperature-dependent ballistic transport in a channel with length below the scattering-limited mean free path, *Journal of Applied Physics*, 111, 054301, 2012.
11. V. K. Arora, M. S. Z. Abidin, S. Tembhurne, and M. A. Riyadi, Concentration dependence of drift and magnetoresistance ballistic mobility in a scaled-down metal–oxide–semiconductor field-effect transistor, *Applied Physics Letters*, 99, 063106, 2011.
12. V. K. Arora, D. Chek, and A. M. bin Hashim, Digital signal propagation delay in a nano-circuit containing reactive and resistive elements, *Solid State Electronics*, 61, 87–92, 2011.

13. M. Riyadi and V. K. Arora, The channel mobility degradation in a nanoscale MOSFET due to injection from the ballistic contacts, *Journal of Applied Physics*, 109, 056103, 2011.
14. R. Qindeel, M. Riyadi, M. T. Ahmadi, and V. K. Arora, Low-dimensional carrier statistics in nanostructures, *Current Nanoscience*, 7(2), 235–239, April 2011.
15. T. Saxena, D. Chek, M. L. P. Tan, and V. K. Arora, Micro-circuit modeling and simulation beyond Ohm's law, *IEEE Transactions on Education*, 54(1), 34–40, 2011.
16. M. A. Riyadi, M. L. P. Tan, A. M. Hashim, and V. K. Arora, Mobility diminution in a nano-MOSFET due to carrier injection from the ohmic contacts, *Enabling Science and Nanotechnology 2010 (AIP Conference Proceedings Journal)*, 1341, 169, 2011.
17. D. C. Y. Chek, M. L. P. Tan, A. M. Hashim, and V. K. Arora, Universal velocity-field characteristics for a nanowire of arbitrary degeneracy, *Enabling Science and Nanotechnology 2010 (AIP Conference Proceedings Journal)*, 1341, 157, 2011.
18. D. C. Y. Chek, M. L. P. Tan, A. M. Hashim, and V. K. Arora, Comparative study of ultimate saturation velocity in zigzag and chiral carbon nanotubes, *Enabling Science and Nanotechnology 2010 (AIP Conference Proceedings Journal)*, 1341, 370, 2011.
19. V. K. Arora, D. C. Y. Chek, M. L. P. Tan, and A. M. Hashim, Transition of equilibrium stochastic to unidirectional velocity vectors in a nanowire subjected to a towering electric field, *Journal of Applied Physics*, 108, 114314-1–114314-8, 2010.
20. R. Vidhi, M. L. P. Tan, T. Saxena, A. M. Hashim, and V. K. Arora, The drift response to high-electric-field in carbon nanotubes, *Current Nanoscience*, 6, 492–495, 2010.
21. V. K. Arora, Educating the future leaders: Integration of engineering, science, and management paradigms, *Journal of Engineering, Science and Management Education*, 1, 4–9, 2010.
22. D. C. Y. Chek, M. L. P. Tan, M. T. Ahmadi, R. Ismail, and V. K. Arora, Analytical modeling of high performance single-walled carbon nanotube field-effect-transistor, *Microelectronics Journal*, 41, 579–584, 2010.
23. M. L. P. Tan, T. Saxena, and V. K. Arora, Resistance blow-up effect in micro-circuit engineering, *Solid State Electronics*, 54, 1617–1624, 2010.
24. I. Saad, M. L. P. Tan, M. T. Ahmadi, R. Ismail, and V. K. Arora, The dependence of saturation velocity on temperature, inversion charge and electric field in a nanoscale MOSFET, *International Journal of Nano Electronics and Materials (IJNeM)*, 3, 17–34, 2010.
25. I. Saad, M. T. Ahmadi, M. A. Riyadi, R. Ismail, and V. K. Arora, Ballistic saturation velocity of quasi-2D low-dimensional nanoscale field effect transistor (FET), *Nanoscience and Nanotechnology, AIP Conference Proceedings Journal*, 1136, 302–306, 2009.
26. I. Saad, M. L. P. Tan, R. Ismail, and V. K. Arora, Nano-physics of transient phenomenon in semiconducting devices and circuits, *Journal Teknologi D (UTM, Malaysia)*, 50.D, 119–125, 2009.
27. I. Saad, M. T. Ahmadi, M. A. Riyadi, R. Ismail, and V. K. Arora, Numerical analysis of carrier statistics in low-dimensional nanostructure device, *Journal Teknologi D (UTM, Malaysia)*, 50.D, 109–117, 2009.
28. V. K. Arora, Engineering the soul of management in the nano era, *Chinese Management Studies (Emerald Journal)*, 3(3), 213–234, 2009.

29. M. L. P. Tan, V. K. Arora, I. Saad, M. T. Ahmadi, and R. Ismail, The drain velocity overshoot in an 80-nm metal–oxide–semiconductor field-effect-transistor, *Journal of Applied Physics*, 105, 074503, 2009.
30. V. K. Arora, Theory of scattering-limited and ballistic mobility and saturation velocity in low-dimensional nanostructures, *Current Nanoscience* (*CNANO*), 5, 227–231, 2009.
31. M. T. Ahmadi, M. L. P. Tan, R. Ismail, and V. K. Arora, The high-field drift velocity in degenerately-doped silicon nanowires, *International Journal of Nanotechnology*, 6, 601–617, 2009.
32. I. Saad, R. Ismail, and V. K. Arora, Investigation on the effects of oblique rotating ion implantation (ORI) method for nanoscale vertical double gate NMOSFET, *Solid State Science and Technology Letters*, 15(2), 69–76, 2008.
33. M. T. Ahmadi, H. H. Lau, R. Ismail, and V. K. Arora, Current–voltage characteristics of a silicon nanowire transistor, *Microelectronics Journal*, 40, 547–549, 2009.
34. I. Saad, M. L. P Tan, A. C. E. Lee, R. Ismail, and V. K. Arora, Scattering-limited and ballistic transport in a nano-CMOS circuit, *Microelectronics Journal*, 40, 581–583, 2009.
35. I. Saad, M. L. P Tan, I. H. Hii, R. Ismail, and V. K. Arora, Ballistic mobility and saturation velocity in low-dimensional nanostructure, *Microelectronics Journal*, 40, 540–542, 2009.
36. M. T. Ahmadi, R. Ismail, M. L. P. Tan, and V. K. Arora, The ultimate ballistic drift velocity in carbon nanotubes, *Journal of Nanomaterials*, Article ID Number 769250, published online, DOI: 10.1155/2008/769250.
37. V. K. Arora, M. L. P. Tan, I. Saad, and R. Ismail, Ballistic quantum transport in a nanoscale metal–oxide–semiconductor field effect transistor, *Applied Physics Letters*, 91, 103510, 2007.
38. M. L. P. Tan, I. Saad, R. Ismail, and V. K. Arora, Enhancement in nano-RC switching delay due to current saturation, *NANO*, 2(4), 233–237, 2007.
39. A. Sharma and V. K. Arora, Velocity-field characteristics in the multivalley model of gallium arsenide, *Journal of Applied Physics*, 97, 093704–093705, April 1, 2005.
40. V. K. Arora and L. Faraone, 21st century engineer–entrepreneur, *IEEE Antennas and Propagation Magazine*, 45(5), 106–114, October 2003.
41. V. K. Arora, Drift-diffusion and Einstein relation for electrons in silicon subjected to a high electric field, *Applied Physics Letters*, 80(20), 3763–3765, May 20, 2002.
42. V. K. Arora, Einstein ratio for electrons in silicon under a high electric field, *TechOnLine Paper* 20220, http://www.techonline.com/community/20220.
43. T. M. Fairus and V. K. Arora, Quantum engineering of nanoelectronic devices: The role of quantum confinement in mobility degradation, *Microelectronics Journal*, 32, 679–686, 2002.
44. V. K. Arora, Quantum engineering of nanoelectronic devices: The role of quantum emission in limiting drift velocity and diffusion coefficient, *Microelectronics Journal*, 31(11–12), 853–859, December 2000.
45. V. K. Arora, Einstein relation in a high electric field, *EEE Search*, Nanyang Technological University, Singapore, January 2000 (Millennium Issue), pp. 1–2.
46. G. Samudra, A. K. F. Yong, T. K. Lee, and V. K. Arora, The role of velocity saturation on switching delay of RC-loaded inverter, *Semiconductors Science and Technology*, 9, 1108–1116, May 1994.

47. V. K. Arora and G. Samudra, Computer-aided engineering in microelectronic processes and design in Singapore, *IEEE Transactions on Education*, 36, (1), 148, 1993.
48. L. S. Tan, S. J. Chua, and V. K. Arora, Velocity-field characteristics in selectively-doped GaAs/$Al_xGa_{1-x}As$ quantum well heterostructures, *Physical Review B* 47(20), 13868–13871, May 1993.
49. V. K. Arora and T. C. Chong, Quantum engineering of modern optoelectronic devices and circuits, *IEEE Search* (Singapore), 170/7/92(MITA(p)), 12–17, August 1992.
50. G. Samudra, S. J. Chua, A. K. Ghatak, and V. K. Arora, High field electron transport for ellipsoidal multivalley band structure of silicon, *Journal of Applied Physics*, 72(10), 4700, 1992.
51. V. K. Arora and S. J. Chua, Quantum engineering of layered semiconductor nanoelectronic structures, *Research News*, Faculty of Engineering, National University of Singapore, 7(3), 6, 1992.
52. V. K. Arora, Velocity response to high electric fields in submicron gate length field effect transistor, *Asia Pacific Engineering Journal* 2(3), 383, 1992.
53. V. K. Arora and W. Pethick, The effect of velocity-field characteristics on the modelling of a sub-micron channel MESFET, *Journal of the Institution of Electronics and Telecommunication Engineers*, 38(2–3), 92–102, 1992.
54. V. K. Arora and V. Choudhry, Integration of liberal learning skills with engineering design skills in microelectronic fabrication, *International Journal of Applied Engineering Education*, 7(3), 231, 1991.
55. V. K. Arora and M. B. Das, Effect of electric-field-induced mobility degradation on the velocity distribution in a submicron-length channel of InGaAs/AlGaAs heterojunction MODFET, *Semiconductor Science and Technology*, 5, 967, 1990.
56. V. K. Arora and M. B. Das, The role of velocity saturation in lifting pinchoff condition in a long-channel MOSFET, *Electronics Letters*, 25, 820, 1989.
57. V. K. Arora, D. S. L. Mui, and H. Morkoc, Mobility degradation and transferred electron effect in gallium arsenide and indium gallium arsenide, *IEEE Transactions on Electron Devices*, ED-34, 1231, 1987.
58. V. K. Arora, D. S. L. Mui, and H. Morkoc, High-field electron drift velocity and temperature in gallium phosphide, *Journal of Applied Physics*, 61, 4704, 1987.
59. V. K. Arora, D. S. L. Mui, and H. Morkoc, Mobility degradation in a quantum-well heterostructure of GaAs/AlGaAs prototype, *Applied Physics Letters*, 50, 1080, 1987.
60. V. K. Arora and V. Choudhry, Field-dependence of the drift velocity and electron population in the satellite valleys in gallium arsenide, *Electronics Letters*, 22, 271, 1986.
61. V. K. Arora and N. Ahmad, Velocity-field profile for *n*-silicon: A theoretical analysis, *IEEE Transactions on Electron Devices*, ED-33, 1075, 1986.
62. V. K. Arora and H. N. Spector, Electrical and optical properties of parabolic semiconducting quantum wells, *Surface Science*, 176, 669, 1986.
63. V. K. Arora and H. N. Spector, Conductivity in intrinsic quantum well wires: Semimetal–semiconductor transition, *Superlattices and Microstructures*, 1, 251, 1985.
64. V. K. Arora, High field distribution and mobility in semiconductors, *Japanese Journal of Applied Physics*, 24, 537, 1985.
65. V. K. Arora, Quantum well wires: Electrical and optical properties, *Journal of Physics C*, 18, 3011, 1985.

66. V. K. Arora and M. A. Al-Massári, Field dependence of mobility and temperature for quasi-hot electrons in semiconductors, *Arabian Journal of Science and Engineering*, 10, 203, 1985.
67. V. K. Arora, Field-broadening-limited hot electron mobility, *Journal of Engineering Sciences*, 11, 319, 1985.
68. V. K. Arora, Quantum mechanical formulation of the Boltzmann transport equation, *Journal of College of Science (KSU)*, 16, 183, 1985.
69. V. K. Arora, Intracollisional field effect in quasi-hot-electron transport, *Physica Status Solidi (b)*, 127, K171, 1985.
70. V. K. Arora and H. N. Spector, Mobility in a quantum well heterostructure limited by scattering from remote impurities, *Surface Science*, 159, 425, 1985.
71. V. K. Arora and A. Naeem, Phonon-scattering-limited mobility in a quantum-well heterostructure, *Physical Review B*, 31, 3887, 1985.
72. V. K. Arora and H. N. Spector, Quantum theory of nonohmic galvanomagnetic effects in semiconductors, *Arabian Journal of Science and Engineering*, 12(4), 465, 1987.
73. V. K. Arora, High-field electron mobility and temperature in bulk semiconductors, *Physical Review B*, 30, 7297, 1984.
74. V. K. Arora and H. N. Spector, Scattering of electrons by a sheet of impurities in a quasi-two-dimensional gas, *Physica Status Solidi (b)*, 126, K171, 1984.
75. V. K. Arora, Transport in quantum-well wires, *Arab Gulf Journal of Scientific Research*, 3, 287, 1985.
76. V. K. Arora and H. N. Spector, Quantum freeze-out of carriers in semimetallic and semiconducting thin wires, *Physica Status Solidi (b)*, 123, 747, 1984.
77. J. Lee, H. N. Spector, and V. K. Arora, Impurity-scattering-limited mobility in a heterojunction, *Journal of Applied Physics*, 54, 6995, 1983.
78. J. Lee, H. N. Spector, and V. K. Arora, Quantum transport in single layered structure for impurity scattering, *Applied Physics Letters*, 42, 363, 1983.
79. V. K. Arora, Nonohmic conductivity for acoustic-phonon scattering, *Physica Status Solidi (b)*, 115, K107, 1983.
80. V. K. Arora, Quantum transport in quasi-one-dimensional systems, *Physica Status Solidi (b)*, 117, 127, 1983.
81. V. K. Arora, Nonlinear quantum transport in semiconductors, *Journal of Applied Physics*, 54, 824, 1983.
82. V. K. Arora, A scientific approach for faculty development in a university system, *Journal of College of Education (KSU)*, 5, 124, 1984.
83. V. K. Arora, Quantum and classical-limit longitudinal magnetoresistance for anisotropic energy surfaces, *Physical Review B*, 26, 7046, 1982.
84. V. K. Arora, Onset of degeneracy in confined systems, *Physical Review B*, 26, 2247, 1982.
85. V. K. Arora and H. N. Spector, Quantum freeze-out in size-limited intrinsic transport, *Journal of Applied Physics*, 54, 831, 1983.
86. V. K. Arora, Computer solution of administrative problems represented by network graphs, *Journal of College of Administrative Sciences (KSU)*, 18, 18, 1982.
87. V. K. Arora, Effect of spin-splitting on the magnetoresistance in parabolic semiconductors, *Physica Status Solidi (b)*, 104, K71, 1981.
88. M. Prasad and V. K. Arora, Quantum transport techniques in two-dimensional systems: Application to cyclotron resonance, *Surface Science*, 113, 333, 1982.

89. V. K. Arora and H. N. Spector, Quantum-limit magnetoresistance in intrinsic semiconductors, *Physical Review B*, 25, 3822, 1982.
90. V. K. Arora and H. N. Spector, Transition in magnetoresistance behavior from classical to quantum regime, *Physical Review B*, 24, 3616, 1981.
91. V. K. Arora, Interference effect in the theory of magnetotransport in the transverse configuration, *Physical Review B*, 24, 1099, 1981.
92. V. K. Arora and M. A. Al-Massári, Quantum transport theory of semiconductors in magnetic fields, *Arabian Journal of Science and Engineering*, 7, 303, 1982.
93. V. K. Arora and H. N. Spector, Quantum-limit magnetoresistance: Piezoelectric phonon scattering, *Physica Status Solidi (b)*, 101, K125, 1980.
94. V. K. Arora, Quantum size effect in thin-wire transport, *Physical Review B*, 23, 5611, 1981.
95. V. K. Arora, Size dependent electric conductivity in semiconducting thin rectangular wires, *Physica Status Solidi (b)*, 105, 707, 1981.
96. V. K. Arora, M. A. Al-Massári, and M. Prasad, Theoretical explanation of observed quantum-limit cyclotron resonance linewidth in InSb, *Physical Review B*, 23, 5619, 1981.
97. V. K. Arora, Quantum size resonance in semiconducting thin films, *Journal of Vacuum Science and Technology*, 20, 94, 1982.
98. V. K. Arora and F. G. Awad, Quantum size effects in semiconductor transport, *Physical Review B*, 23, 5570, 1981.
99. V. K. Arora and M. A. Al-Massári, On transverse magnetoconductivity in strict born approximation, *Solid State Communications*, 36, 191, 1980.
100. V. K. Arora and M. A. Al-Massári, Superoperator theory of magnetoresistance, *Physical Review B*, 21, 876, 1980.
101. V. K. Arora and H. N. Spector, Quantum-limit magnetoresistance for impurity scattering, *Physica Status Solidi (b)*, 94, 323, 1979.
102. V. K. Arora, Quantum-limit cyclotron resonance linewidth in semiconductors, *Physica Status Solidi (b)*, 94, 710, 1979.
103. V. K. Arora and M. A. Al-Massári, Thermoelectric power in high magnetic fields, *International Journal of Magnetism and Magnetic Materials*, 11, 80, 1979.
104. V. K. Arora, M. A. Al-Massári, and M. Prasad, Superoperator theory of cyclotron resonance linewidth in semiconductors, *Physica B*, 106, 311, 1981.
105. V. K. Arora, Quantum theory of electric conductivity, *Acta Ciencia Indica*, 3, 334, 1978.
106. V. K. Arora, D. R. Cassiday, and H. N. Spector, Quantum limit magnetoresistance for acoustic phonon scattering, *Physical Review B*, 15, 5996, 1977.
107. V. K. Arora, C. Munera, and M. Jaafarian, High field magnetoresistance and magnetothermal emf in many-valley semiconductors with ellipsoidal energy surface, *International Journal of Physics and Chemistry of Solids*, 38, 469, 1977.
108. V. K. Arora, Quantum theory of magneto-microwave conductivity, *Physical Review B*, 14, 679, 1976.
109. V. K. Arora and E. Quershi, Quantum theory of microwave conductivity, *Physica Status Solidi (b)*, 77, 77, 1976.
110. V. K. Arora and K. L. Gomber, Quantum transport equation for electric conductivity, *Physica Status Solidi (b)*, 74, K111, 1976.
111. V. K. Arora, Extreme quantum limit magnetotransport in transverse configuration, *Physica Status Solidi (b)*, 75, K65, 1976.

112. V. K. Arora, Ohmic electric conductivity in a magnetic field, *American Journal of Physics*, 44, 643, 1976.
113. V. K. Arora and M. Jaafarian, Effect of non-parabolicity on ohmic magnetoresistance in semiconductors, *Physical Review B*, 13, 4457, 1976.
114. V. K. Arora, Ohmic magnetoresistance for inelastic acoustic phonon scattering in semiconductors, *Physical Review B*, 13, 2532, 1976.
115. V. K. Arora, Effect of inelasticity on magnetoresistance, *Physica Status Solidi (b)*, 73, K93, 1976.
116. V. K. Arora and R. L. Peterson, Quantum theory of ohmic galvano- and thermomagnetic effects in semiconductors, *Physical Review B*, 12, 2285, 1975.
117. V. K. Arora, Linear magnetoresistance in parabolic semiconductors, *Physica Status Solidi (b)*, 71, 293, 1975.
118. V. K. Arora and S. C. Miller, Effect of electron–phonon drag on magnetoconductivity tensor of *n*-germanium, *Physical Review B*, 10, 688, 1974.
119. V. K. Arora and R. L. Peterson, Theory of magnetophonon structure in the longitudinal magnetothermal emf, *Physical Review B*, 9, 4323, 1974.

Appendix K: Final Word

Publication history is replete with biases and conflict of interest that is hard to beat as peer-review process is the only process through which the quality of a publication is judged. The recent case of the Double Nobel winner Linus Pauling is noteworthy for biases that are present. Pauling never accepted the findings of Daniel Shechtman who won the 2011 Nobel Prize in chemistry for his discovery of quasi-crystals, a mosaic-like chemical structure that researchers previously thought was impossible. His keynote paper was rejected, but finally published to uproar in the scientific world. With similar biases present among the so-called leaders with multi-million-dollar grants and self-interests, it is difficult to avoid humiliation to a "small fish in a pond," giving breathing space for them to not only survive, but also thrive in scientific research. The current procedures to assure quality and fairness seem to discourage scientific advancement, especially important innovations, because findings that conflict with the current beliefs are often judged to have defects. The leadership resides with the broad readership of the journals, not with experts who are reluctant to appreciate the alternative paradigms being proposed. This book certainly provides an alternative paradigm that conflicts with the interest of superficial leaders in research.

This book seamlessly covers the physics-based paradigm valid for a wide variety of nanoensembles with regimes from drift diffusion to ballistic transport. The ideas presented take into account quantum confinement, reduced conducting channel length, and asymmetry of the electron distribution as increasingly high fields are encountered. The described voyage from drift diffusion to ballistic transport is consistent with Buttiker's paradigm [1], with each conducting channel a series of micro-resistors each of average length equal to the mfp. The ends of a free-path resistor are virtual thermalizing probes, one end being higher than the other end in the Fermi level (electrochemical potential) by the energy gained in an mfp. Within this scenario, carriers are removed from the device and injected into a virtual reservoir where they are thermalized and reinjected into the ballistic channel with length equal to an mfp. This description is also consistent with Natori's model of ballistic transport in the ballistic domain that is driven by the Fermi level of the contacts [2]. In fact, Mugnaini and Iannaccone [3,4] validate our ballistic model by applying it to a nanoscale MOSFET, both for ND and degenerate statistics. Most undergraduate textbooks use ND statistics in its simplistic form. However, modern nanoensembles are degenerate in their carrier concentration that requires degenerate statistics. With generalized carrier statistics and a broad spectrum from drift diffusion to the ballistic domain, this book will benefit future investigators engaged in the research and development. The ideas presented will encourage creativity and innovation in designing new experiments, developing new models, and engaging new simulations with realistic parameters leading to performance evaluation for possible circuit and system applications. To cite an example, CNT and several of its variation in the form of graphene sheets and ribbons are finding a wide variety of applications in monitoring the environment, sensing chemical elements, developing new indigenous materials, and providing interconnects for a variety of biochips. Novelty of the archetypes presented will trigger dialog with people of all specialties in developing innovative technology and economic uplift through techno-entrepreneurship.

The theory of scattering-limited transport has a rich and interesting history since World War II, starting from the work of Shockley [5] and Ryder and Shockley [6] who observed deviations from Ohm's law leading to current and drift velocity saturation as the electric field (or voltage) across the device is increased. Conwell et al. [7] documented very well the scattering processes that were known until the 1960s. It was the availability of supercomputers in the 1970s that created intense activity in the high-field transport. Monte Carlo simulations were tried and are still in vogue by inputting a number of scattering processes and adjusting parameters until the desired outcome is obtained. The consensus emerging out of Monte Carlo experiments are that the saturation velocity is scattering limited that may be controlled by the optical phonon energy. However, no clear connection of the saturation velocity to the ohmic low-field mobility emerged from these studies. In 1985, a nonequilibrium high-field distribution [8,9] was proposed that for the first time indicated the ballistic nature of the saturation velocity whose magnitude was comparable to the carrier thermal velocity in an ND semiconductor. In this nonequilibrium distribution function, all complex scattering interactions are lumped into a single mfp with a typical value of 100 nm. This inclusion of an mfp in the carrier distribution made clear the anisotropic nature of carrier transport in a high electric field, leading to unidirectional streaming of velocity vectors and velocity saturation in extreme nonequilibrium. Specifically, a critical electric field marks the boundary between scattering-limited ohmic transport and ballistic high-field transport. The corresponding critical voltage for a long-length channel is projected to be several thousand volts in the paradigm presented. If the potential difference across the length of a channel is considerably below this critical voltage, Ohm's law is well obeyed for macroresistors. Therefore, all circuit principles, including voltage and current dividers, power dissipation, and Kirchhoff's laws remain valid in the ohmic regime. This critical voltage may become as low as a few mV for a micro- or nanoscale resistor, making Ohm's law invalid even for voltage as low as 1 V. The formalism of Reference 9 replicated well the observed characteristics when applied to an n-silicon, explained well the negative differential resistivity in gallium arsenide, and most recently explained the *I–V* characteristics of a degenerately doped nanoscale MOSFET, to cite a few prominent applications. The overwhelming mode of thinking in 1985, when Reference 9 was promulgated, was that the carriers become hot and hence follow the ND

Maxwellian statistics. This hot-carrier concept was later disputed [10]. It was specifically shown that the electric field streamlines the otherwise random motion in equilibrium, thereby reducing the randomness in the direction of the applied electric field for holes and opposite for electrons. The field broadening equal to the energy gained in a de Broglie wavelength was shown to overwhelm the collision broadening. This predominance of collision broadening makes the transport ballistic. The critical electric field at which collision effect tends to fade on velocity response is then equal to thermal voltage divided by mfp or equivalently the saturation velocity divided by the mobility. In all these descriptions, the equilibrium FD distribution function transforms into a nonequilibrium one as the Fermi energy (chemical potential) is replaced with the electrochemical potential [9]. The understanding emanating from this book is already disseminated in a series of papers appearing in Appendix J. Readers will have a plenty of opportunities to develop their own mindpower as they go through the prototypes discussed throughout the book. Let the action begin as thinking minds take the future of nanoelectronics in the hands of creative and innovative scholars whose minds are not captured by a few capitalist scientists and editors who use taxpayers money to stop research reaching the independent thinkers. Those who are spiritually charged are directed to read References 11 and 12.

Bell [13] has explained very well the impure practices that prevent others in uncovering the truth. Science of the known and unknown is a discovery process of uncovering the reality. The scientific research enterprise is built on a foundation of trust. More attention must be given by the scientific community to the mechanisms that sustain and transmit the values that are associated with ethical scientific conduct. It is the working scientists themselves who must bear the greatest responsibility for maintaining high standards of conduct. Involvement by the most respected scientists at an institution is necessary for setting standards of conduct, designing educational programs, and responding to alleged violations of ethical norms. However, as the very institution funded by taxpayers' money becomes corrupt, progress is difficult where freedom of the minds is constrained only to ideas of a few pseudoscientists. Mahoney's findings [14] are consistent with these prejudices that prevail. Confirmatory bias is the tendency to emphasize and believe experiences that support one's views and ignore or discredit those that do not conform. In Mahoney's study, 75 journal reviewers were asked to refer manuscripts that described identical experimental procedures but which reported positive, negative, mixed, or no results. In addition to showing poor inter-rater agreement, reviewers were strongly biased against manuscripts that reported results contrary to their theoretical perspective.

Our review process for this book went through similar biases. However, truth prevailed and the author offers his gratitude to all those reviewers who gave a detailed analysis on how to make this book publication ready. In responding to the review process, the integration of mind–body–divine forces as practiced on Silk Route that extended from Mediterranean to China, was discovered. Perhaps, one way of bringing peace, prosperity, and intellectual growth back is to revive the cultural cohesion of the Silk Route model of cosmopolitanism. The Chinese were intrigued by invoking those secular forces with mantra chanting for political empowerment, consolidation of power, and well-being of people. Combining power with virtue was a new concept for them. That is why Buddhism was welcomed in China and Japan. It still continues to be a divine force in Japan as Japanese feel ashamed if the quality of the product or service is not good. Japan has grasped the concept of Total Quality Management. Buddhism provided the cohesive culture and was cosmopolitan. It was indeed gratifying to discover our own "Silk Route" and found so much knowledge through the discovery process. Gratitude means counting our blessings, being thankful, and acknowledging everything that we received on our custom-made Silk Route. It is living our life as if everything were a miracle, being aware nonstop of how we have been blessed by others. Gratitude shifts our focus from what our life lacks to the richness that is already present. As simple as it sounds, gratitude is actually a multifaceted emotion that requires self-reflection, the ability to admit that one is dependent on the help of others, and the humility to realize one's own limitations. Buddha's bounty is ever present on habitually grateful people as all are Children of God, male or female does not matter as God itself is love for all, which cannot be defined through figures or words. Buddha is a symbol of any knowledgeable person. It is the knowledge that removes ignorance and makes us enlightened. In that spirit, we are all Buddhas in discovering the truth. Both science and spirituality help us to discover that truth by a process of enquiry, dialog, teamwork, and open communication without any fear of retribution. It is appropriate to end with the final word as this book goes to press:

> Buddham Sharanam Gachammi (Be in the protection of embodiment of knowledge)
> Dharmam Sharanam Gachammi (Be in the protection of ethical conduct)
> Sangam Sharanam Gachammi (Be in the protection of noble people)

In the spirit of following the path of righteousness, following advice is appropriated:

> Buddham Bhav (Be an enlightened wise person)
> Dharmam Char (Follow the path of righteousness)
> Yogastha Kuru Karmani (Establish yourself in total integration of mind–body–soul, to perform an action)

References

1. M. Büttiker, Role of quantum coherence in series resistors, *Physical Review B*, 33, 3020, 1986.

2. A. Bindal, T. Ogura, N. Ogura, and S. Hamedi-Hagh, Silicon nanowire transistors for implementing a field programmable gate array architecture with scan chain, *Journal of Nanoelectronics and Optoelectronics*, 4, 342–352, 2009.
3. G. Mugnaini and G. Iannaccone, Physics-based compact model of nanoscale MOSFETs—Part I: Transition from drift-diffusion to ballistic transport, *IEEE Transactions on Electron Devices*, 52, 1795–1801, 2005.
4. G. Mugnaini and G. Iannaccone, Physics-based compact model of nanoscale MOSFETs—Part II: Effects of degeneracy on transport, *IEEE Transactions on Electron Devices*, 52, 1802–1806, 2005.
5. W. Shockley, Hot electrons in germanium crystal and Ohm's law, *Bell System Technical Journal*, 30, 990–1034, 1951.
6. E. Ryder and W. Shockley, Mobilities of electrons in high electric fields, *Physical Review*, 81, 139, 1951.
7. E. M. Conwell, F. Seitz, H. Ehrenreich, and D. Turnbull, *High Field Transport in Semiconductors*. New York: Academic Press, 1967.
8. V. K. Arora, High-field electron mobility and temperature in bulk semiconductors, *Physical Review B*, 30, 7297–7298, 1984.
9. V. K. Arora, High-field distribution and mobility in semiconductors, *Japanese Journal of Applied Physics*, 24, 537, 1985.
10. V. K. Arora, Hot electrons: A myth or reality? in *SPIE Proceedings Series*, 563–569, 2002.
11. V. K. Arora and N. Arora, Converging paradigms in behavioral and social engineering, in *Recent Trends in Social and Behavior Sciences*, F. L. Gaol, S. Kadry, M. Taylor, and P. S. Li, Eds. CRC Press/Taylor & Francis Group, 2014, pp. 3–8, 2014. Boca Raton, FL, U.S.A.
12. V. K. Arora, Engineering the soul of management in the nano era, *Chinese Management Studies*, 3, 213–234, 2009.
13. R. Bell, *Impure Science: Fraud, Compromise and Political Influence in Scientific Research*. New York: Wiley, 1992, vol. 1, 1992.
14. M. J. Mahoney, Publication prejudices: An experimental study of confirmatory bias in the peer review system, *Cognitive Therapy and Research*, 1, 161–175, 1977.

Index

D

E